T0250502

Allergic Hypersensitivities Induced by Chemicals

Recommendations for Prevention

Edited by

Joseph G. Vos
National Institute of Public Health and the Environment
Bilthoven, The Netherlands

Maged Younes
International Programme on Chemical Safety
Geneva, Switzerland

Edward Smith
International Programme on Chemical Safety
Geneva, Switzerland

CRC PRESS

Boca Raton London New York Washington, D.C.

Published on behalf of the
World Health Organization Regional Office for Europe
by CRC PRESS LLC

Library of Congress Cataloging-in-Publication Data

Allergic hypersensitivities induced by chemicals : recommendations for
    prevention / edited by J.G. Vos, M. Younes, E. Smith.
        p. cm.
    Includes bibliographical references and index.
    ISBN 0-8493-9226-8 (alk.)
    1. Allergy. I. Vos, J.G. II. Younes, M. (Maged) II. Smith,
    E. (Edward), 1932- . IV. World Health Organization.
RC584.A347 1995
616.2′02—dc20   95-21238
                                                                         CIP

Visit the CRC Press Web site at www.crcpress.com

No claim to original U.S. Government works
International Standard Book Number 0-8493-9226-8
Library of Congress Card Number 95-21238
Printed in the United States of America   2 3 4 5 6 7 8 9 0
Printed on acid-free paper

THE EDITORS

Joseph G. Vos, D.V.M., Ph.D., is head of the Laboratory of Pathology, National Institute of Public Health and the Environment, Bilthoven, The Netherlands, and Professor of Toxicological Pathology at the Faculty of Veterinary Medicine, Utrecht University.

Dr. Vos obtained his training at the Utrecht University receiving the D.V.M. degree in 1967 and the Ph.D. degree in 1972 on a thesis entitled *Toxicology of polychlorinated biphenyls (PCBs) and impurities.* In 1972-73 he spent a year as Visiting Associate at the National Institute of Environmental Health Sciences, Research Triangle Park, NC, studying the immunotoxicity of dioxin. Since 1975 he has been at the National Institute of Public Health and the Environment where he became head of the Laboratory of Pathology in 1979. This position was combined with head of the Laboratory of Toxicology from 1985 to 1987, Director of Immunology from 1987 to 1991, and Director, Basic Health Research Division from 1991 to 1994. In 1994 he became part-time Professor of Toxicological Pathology. He has been joint head of the WHO Collaborating Center for Immunotoxicology and Allergic Hypersensitivity since 1992.

Dr. Vos is a member of a number of professional societies. He served as chairman of the Dutch Society of Toxicology from 1984 to 1988, and has chaired the Animal Pathology Section of the Dutch Society of Pathology since 1990. He chaired the Dutch Health Council Committee "Immunotoxicity of Chemicals," and the WHO Task Group Principles and Methods for Assessing Direct Toxicity Associated with Exposure to Chemicals.

Dr. Vos has authored or co-authored more than 150 original papers, book and article reviews. His major research interests are in immunotoxicity assessment and in the toxicologic pathology of chemicals of environmental concern.

Maged Younes graduated from the University of Tübingen, Germany, where he obtained an M.Sc. degree in 1975, and a Ph.D. degree in 1978, both in biochemistry. He then worked at the Institute of Toxicology, Medical University of Lübeck, Germany, until 1989. During that period, he conducted research in the field of biochemical toxicology, with special emphasis on mechanisms of toxic action, free radical toxicology, and lipid peroxidation, biotransformation reactions, and organ-directed toxicity. In that period, Dr. Younes qualified as an Expert in Toxicology ("Fachtoxikologe") through the German Society of Pharmacology and Toxicology, and was nominated Professor of Toxicology and Biochemical Pharmacology. In summer 1985, he spent a sabbatical at the Commonwealth Institute of Health, University of Sydney.

In 1991, Dr. Younes was nominated head of the Unit on Biochemical Toxicology at the Max von Pettenkofer-Institute of the German Federal Health Office in Berlin. From November 1991 until April 1995, he was the manager responsible for the European Chemical Safety Program of the World Health Organization at the Bilthoven Division of the WHO European Centre for

Environment and Health. Since May 1995, he has been Chief, Assessment of Risk and Methodologies at the International Programme on Chemical Safety.

Dr. Younes is a member of a number of professional societies and has editorial responsibilitites with two scientific journals. He has published some 115 original papers and more than 25 book and review articles.

Dr. Edward Smith, FFOM RCP, is a staff member of the Central Unit of the International Programme on Chemical Safety, located in the World Health Organization, Geneva, Switzerland.

Dr. Smith graduated in medicine and surgery at the University of Bristol, United Kingdom, in 1955. He spent time in the United Kingdom Government Service, and in industry. His specialization is in occupational medicine and he has wide experience in this and in toxicology. He has served on many international groups and committees concerned with chemical toxicity testing, food safety, and chemical risk assessment.

Since 1984, he has been with the International Programme on Chemical Safety, responsible for evaluation of the human health and environmental risks of chemicals and groups of chemicals, for the preparation of monographs on nephrotoxicity and immunotoxicology, and for the development of monographs on chemical risk assessment for human health protection.

CONTRIBUTORS

Wolf-M. Becker
Division of Allergology
Research Institute Borstel
Borstel, Germany

Heidrun Behrendt
Department of Experimental
  Dermatology and Allergology
Universitätskrankenhaus Eppendorf
Hamburg, Germany

Blanche Bellon
INSERM U28
Hôpital Broussais
Paris, France

Philippe Druet
INSERM U28
Hôpital Broussais
Paris, France

Mari-Ann Flyvholm
National Institute of Occupational
  Health
Copenhagen, Denmark

Karl-Heinz Friedrichs
Medical Institute of Environmental
  Hygiene
University of Düsseldorf
Düsseldorf, Germany

J. Garssen
National Institute of Public Health
  and the Environment (RIVM)
Bilthoven, The Netherlands

Janos Gergely
Department of Immunology
Eötvös Lorand University
Göd, Hungary

Bettina Hitzfeld
Department of Experimental
  Dermatology and Allergology
Universitätskrankenhaus Eppendorf
Hamburg, Germany

Meryl H. Karol
Department of Environmental and
  Occupational Health
University of Pittsburgh
Pittsburgh, Pennsylvania, U.S.

Ian Kimber
Research Toxicology Section
Zeneca Central Toxicology
  Laboratory
Alderley Park
Cheshire, United Kingdom

Ursula Krämer
Medical Institute of Environmental
  Hygiene
University of Düsseldorf
Düsseldorf, Germany

Charlotte Madsen
Institute of Toxicology
National Food Agency
Søborg, Denmark

Howard I. Maibach
Department of Dermatology
University of California
San Francisco, California, U.S.

Thomas Maurer
Pharmaceuticals Division
Preclinical Safety
Ciba-Geigy Ltd.
Basel, Switzerland

Torkil Menné
Department of Dermatology
Gentofte Hospital
University of Copenhagen
Hellerup, Denmark

Karin A. Pacheco
Division of Allergy and Immunology
National Jewish Center for
  Immunology and Respiratory
  Medicine
Denver, Colorado, U.S.

Lucette Pelletier
INSERM U28
Hôpital Broussais
Paris, France

Johannes Ring
Department of Experimental Derma-
tology and Allergology
Universitätskrankenhaus Eppendorf
Hamburg, Germany

Lanny J. Rosenwasser
Division of Allergy and Immunology
National Jewish Center for
  Immunology and Respiratory
  Medicine
Denver, Colorado, U.S.

Rik J. Scheper
Department of Pathology
Free University Hospital
Amsterdam, The Netherlands

R. J. Vandebriel
National Institute of Public Health
  and the Environment (RIVM)
Bilthoven, The Netherlands

H. Van Loveren
National Institute of Public Health
  and the Environment (RIVM)
Bilthoven, The Netherlands

Katherine M. Venables
Department of Occupational and
  Environmental Medicine
National Heart and Lung Institute
London, University
London, United Kingdom

B. Mary E. von Blomberg
Department of Pathology
Free University Hospital
Amsterdam, The Netherlands

FOREWORD

Allergic hypersensitization is a reaction produced by an immune response to an antigen that results in tissue inflammation and organ dysfunction. It gives rise to a variety of health disorders. These are common in all countries although the types of allergy and the population groups affected vary from country to country. Allergenic substances, natural and synthetic, are present throughout our environment and the spectrum of allergic disorders is wide. It includes atopic and contact dermatitis, allergic rhinitis ("hayfever"), allergic asthma, occupational asthma, reactions to insect stings, and allergy to drugs and food constituents. Although allergic disorders, except for anaphylactic shock and severe asthma, are seldom fatal, they are an important cause of morbidity and poor health. There is increasing evidence that the prevalence of allergic disorders is rising in many countries, particularly among children where it can limit activity and affect development.

The development of allergic hypersensitization is influenced by genetic as well as environmental factors. Understanding of the mechanisms of this influence is advancing. Predictive testing is being further developed and refined and is applied routinely in the notification of new chemical substances, in the safety testing of formulations, and in the further investigation of the allergenic potential of existing substances. Application of this knowledge will help in the prevention and/or more effective treatment of allergic disorders, contributing significantly to achieving the targets of the health-for-all policy of the World Health Organization.

To make information on allergic hypersensitization widely available, the Regional Office for Europe of the World Health Organization, in cooperation with the International Programme on Chemical Safety and the Commission of the European Communities, assembled a group of international experts to write this book, which is relevant to the work of many groups of health professionals, such as physicians, nurses, health managers, immunologists, and toxicologists. The book covers several aspects of the subject including the basic mechanisms of chemically induced hypersensitization, the mechanistic, clinical, epidemiological as well as diagnostic, curative, and preventive aspects of respiratory and skin allergy, the methodology of predictive testing, and information on gastrointestinal and kidney allergies. The chapters dealing with these issues were presented and reviewed at a consultation held at the WHO European Centre for Environment and Health in Bilthoven, The Netherlands, on 14 17 December 1992 with the generous financial support of the German Federal Ministry for the Environment, Nature Conservation and Nuclear Safety. The consultation also reached conclusions and made recommendations for priority actions by scientists and public health officials to improve our knowledge about allergic hypersensitization and to ensure its prevention as far as possible.

In expressing the World Health Organization's appreciation and gratitude to all those who contributed their time and expertise to this publication, I feel confident that it will be of great value in promoting a healthy environment.

J.E. Asvall
WHO Regional Director for Europe

TABLE OF CONTENTS

SECTION I — INTRODUCTION

SECTION II — RESPIRATORY ALLERGY

SECTION III — SKIN ALLERGY

SECTION IV — OTHER TISSUE ALLERGIES

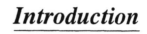

Introduction

Chapter 1

HYPERSENSITIVITIES INDUCED BY CHEMICALS: REPORT ON A WHO CONSULTATION

CONTENTS

I. INTRODUCTION

Exposure to many chemicals results in allergic hypersensitivity, which causes conditions such as contact dermatitis and asthma. Recognizing the growing importance of the problem, in 1982 the WHO Regional Office for Europe and the Commission of the European Communities (CEC), supported by the Government of the Federal Republic of Germany, jointly organized a workshop that evaluated the mechanisms through which chemicals induce allergies and related conditions, related problems in clinical surveillance and epidemiological studies, and possible preventive measures. The participants made specific recommendations on research priorities, the dissemination of information, and preventive measures.

Allergic contact dermatitis resulting from exposure to chemicals is a common occupational health problem. Recognition is increasing that chemicals can

3

also induce allergic sensitization in the respiratory tract. The need to assess the problem and to develop preventive measures became evident with the increase in the amount of chemical substances to which a large number of people are exposed.

During the past 10 years, about 10,000 new substances have been marketed in the European Region, and the prevalence and severity of respiratory hypersensitivity have increased. Asthma mortality worldwide has also increased. In 11 of 14 countries, the death rate from asthma was higher in the 1980s than in the 1970s. In six countries, the increase in asthma mortality was over 20%, although the actual numbers were small (0.13 to 3.63 per 100,000).

Many substances may cause allergic contact sensitization, but only about 100 chemicals frequently cause allergic dermatitis. Diagnostic tests show that about 15 to 20% of the population in certain industrialized countries has contact sensitization to chemicals, although disease is much less common.

Recognizing the extent of the problem, the Regional Office, the International Programme on Chemical Safety (IPCS), and the CEC, organized the Consultation on Recommendations for Preventive Measures Related to Hypersensitivities Induced by Chemicals with the support of the German Federal Ministry for the Environment, Nature Conservation and Nuclear Safety. The Consultation took place from December 14 to 17, 1992, in the division of the WHO European Center for Environment and Health in Bilthoven, The Netherlands. Experts from nine countries, representatives of CEC, the International Life Science Institute (ILSI), and WHO staff attended. Dr. Joseph G. Vos was elected chairperson and Professor Meryl H. Karol, rapporteur. The participants are listed on pages 13 through 15.

The participants were to review progress in scientific knowledge of the prevalence and incidence of clinical disorders related to allergic hypersensitivity, and the causative agents and their mechanisms of action, and to consider diagnostic and predictive testing techniques and other preventive measures. Although the participants focused mainly on allergic hypersensitivity, they also reviewed the disorders that have no immunological basis but are of significant public concern. They gave special emphasis to evaluating the role of man-made environmental chemicals in the development of allergic hypersensitivity, as well as pollutants modulating this response. Finally, a short glossary was developed to reflect the latest developments in the area of allergic hypersensitivities, and to avoid any possible misinterpretation of data. The background papers prepared for the Consultation make up the body of this publication.

II. DISCUSSION

A. APPROACH TO HYPERSENSITIVITIES

The discussion focused on approaches to understanding people's basic defense systems: physical or chemical reactions, nonspecific resistance, and specific immunity, compartmentalized in the skin, respiratory tract, gastrointestinal tract, and other body systems. Allergic reactions are classified according to their mechanisms, which may be summarized in terms of initiation, regulation, and resulting pathology at the sites where the reactions can occur. Mechanisms have most often been described in detail for proteins in animal species. Attention focused on responses in human beings, and responses to compounds with low molecular weight. The discussion mainly emphasized the mechanisms of respiratory and skin allergy, although special attention was paid to the gastrointestinal tract and kidney.

The role of environmental pollution in the development of allergic hypersensitivities is a major cause of concern in the European Region. Although the data show a persistent increase in the prevalence of allergies, causal association with environmental pollution is still under debate. The discussion focused, therefore, on the questions of how to substantiate the necessary epidemiological information related to the population in general and those exposed to pollutants at work.

Further, the participants considered whether pollutants may modulate sensitization, and how they might influence the effector cells in allergic hypersensitivity in sensitized and nonsensitized individuals. Whether pollutants can alter the allergic inflammatory response should receive high priority in future research.

B. RESPIRATORY ALLERGY

1. Mechanisms

Chemical respiratory allergy is an immunologically-mediated reaction in the respiratory tract that is characterized by a variety of symptoms, including rhinitis and asthma. Sensitization is believed to occur via inhalation, but in theory it might result from exposure via other routes, notably dermal contact.

Respiratory reactions in a previously sensitized individual are elicited after exposure via inhalation. By analogy with protein-induced respiratory allergic reactions, allergic responses to chemicals in the respiratory tract are assumed to be associated with a specific IgE antibody. Other immunological effector mechanisms, however, may play a role of equal or greater importance.

The participants focused on the pathology of allergic hypersensitivity related to the respiratory system. In addition to general aspects, rhinitis, asthma, and hypersensitivity pneumonitis received particular attention. Special emphasis was placed on structure-activity relationships.

2. Clinical and Epidemiological Aspects

In addition, the participants reviewed the clinical and epidemiological aspects of respiratory allergic and hypersensitivity responses to chemicals. Rhinitis, asthma, and hypersensitivity pneumonitis/allergic alveolitis (HSP/AA) were described. The respiratory effects of outdoor and indoor pollutants were discussed in some detail, including agents such as ozone (O_3), sulfur dioxide (SO_2), nitrogen dioxide (NO_2), total suspended particles (TSP), cigarette smoke, and formaldehyde. While not directly acting as chemical allergens, these agents may have a profound effect on sensitization and allergic respiratory responses to chemicals. Further, the participants reviewed the respiratory effects of occupational exposure to such agents as isocyanate, epoxy resins, colophony, organic and inorganic dusts, and platinum and other metal salts.

The participants tried to evaluate the scope and prevalence of various respiratory sensitivities, utilizing the available published studies and health statistics. Finally, the problem of multiple chemical sensitivity was discussed. The participants concluded that scientific data are lacking to relate the appearance of hypersensitivity to specific agents in this syndrome.

3. Diagnosis and Treatment

The participants reviewed the standard methods of assessing the severity of rhinitis, asthma, and HSP/AA, and the appropriate medical therapy to relieve symptoms and prevent long-term respiratory damage. They also discussed the contribution of provocation testing and other, more invasive experimental methods of assessing lung inflammation and damage, with an understanding of their value in the studies. Exposure prevention and the use of anti-inflammatory therapies are crucial in both asthma and HSP/AA. Anti-inflammatory therapies are useful in both diseases, although damage in the latter syndrome can rarely be fully reversed. Sentinel surveillance has particular diagnostic and investigative value.

4. Exposure and Dose-Response Relationships

There are two important issues in exposure and dose-response relationships. First, the dose-response relationships of sensitization and of elicitation of respiratory reactions in sensitized individuals may differ widely. Moreover, the concentration required for effective sensitization via different routes may vary. Second, the induction and elicitation of respiratory hypersensitivity and the concentrations of chemical allergens required may vary markedly as a function of a number of factors, including the presence of environmental pollutants (irritant gases) or other response-modifying agents (such as viruses or smoking).

5. Predictive Testing and Preventive Measures

Respiratory allergy to chemicals is a substantial concern in industrial settings, as large numbers of workers can be exposed. Remarkably, the number of chemicals known to have caused problems appears to be limited. Nevertheless, predictive testing is essential for the development of appropriate environmental health protection. The approaches used include: structure-activity studies, and the assessment of *in vitro* binding of reactive chemicals to proteins and of *in vivo* exposures with measurement of serological endpoints, histopathology, and asthmatic episodes (including measurement of hyperresponsiveness of the airways). Some methods rely on endpoints that assume knowledge of the mechanisms of response (such as protein binding and measurement of antibody titer). *In vitro* assessment can provide valuable information, while requiring less use of animals. All methods require validation using positive and negative controls.

In the handling of any chemical, adequate attention should be paid to potential allergenic effects. The participants discussed priorities for the prevention of occupational asthma. They grouped preventive measures into primary, secondary, and tertiary activities. Primary prevention focuses on the control of exposures that may cause asthma. Secondary prevention means detection at a sufficiently early stage to minimize impairment and disability. Tertiary prevention focuses on the provision of appropriate health care for patients with asthma, to avert complications. Specific surveillance programs, as well as other types of epidemiological investigations, are an important guide to appropriate preventive measures.

C. SKIN ALLERGY

1. Mechanisms

Extensive research has recently led to the unravelling of the basic immunological mechanisms of allergic contact hypersensitivity. It identified the major cell types and mediators involved. How T-cells recognize distinct allergens and how these and other inflammatory cells interact to generate inflammation have begun to be understood. This rapid progress contrasts sharply with the slow progress in unravelling the regulatory mechanisms in cell-mediated immunity, including allergic contact hypersensitivity (ACH). The actions of putative suppressor cells are still heavily disputed. So far, no general method of permanent desensitization has been devised.

Molecules capable of interaction with immune-competent cells, as well as mediators, provide promising targets for anti-inflammatory drugs, some of which have already entered clinical trials.

2. Clinical and Epidemiological Aspects

The skin as an organ is involved in a variety of hypersensitivity reactions, whose mechanisms of pathology have not all been clearly established. The most common skin diseases in which allergic or possible allergic reactions play a role comprise: allergic contact dermatitis, including photoallergic reactions, urticaria, angioedema and anaphylaxis, atopic eczema, and certain vasculitic diseases (allergic vasculitis, progressive pigmentary purpura).

Allergic contact dermatitis is a major health problem in many industrialized countries. Epidemiological studies of the general population in some European countries show an estimated prevalence of around 10% for eczema or dermatitis. The figure for contact sensitization is 15 to 20% because not all sensitized individuals display dermatitis. At any time, it is estimated that 2 to 4% of the population may have active allergic contact dermatitis. Contact dermatitis accounts for about one third of all cases of occupational disease. Allergic contact dermatitis may have severe consequences. Allergic diseases often start early in life and become chronic, causing years of ill health.

The disease is caused by a T-cell-mediated allergy to a wide variety of substances (such as metals, rubber chemicals, occupational substances, cosmetics, etc.) and manifests in different ways. In some cases, it can be elicited by oral application of the relevant contact allergen (systemic contact dermatitis).

A subgroup of contact dermatitis is due to photosensitization, in which combined exposure to a chemical and ultraviolet (UV) radiation (mostly UVA) elicits sensitization and disease.

Immediate hypersensitivity reactions of the skin include urticaria, angioedema, and anaphylaxis. They are mostly IgE mediated, but some also depend on nonimmunological mechanisms (pseudoallergic reactions). The most common sensitizing agents are drugs and food additives. Epidemiological studies reveal that pseudoallergic reactions have a prevalence of 1 to 3% in the general population.

Atopic eczema is a skin disease associated with considerable IgE production. This IgE-mediated hypersensitivity has been found in many patients showing allergic reactions, for example, to aeroallergens or foods. Marked exacerbation of eczematous skin lesions has sometimes been observed after oral provocation with food additives. In many countries, the prevalence of atopic eczema seems to have increased up to 10 to 15% during the past decades. The reasons for this are unclear; a putative influence of environmental pollutants, such as tobacco smoke, is a focus of research.

3. Diagnosis and Treatment

The discussion confirmed that the diagnosis of allergic diseases comprises the following steps: history, skin tests, *in vitro* tests, and provocation tests. Some diagnostic procedures need improvement. Therapeutic strategies comprise allergen avoidance, anti-inflammatory therapy (with agents acting at different stages of the cascade of the allergic reaction), and optimal skin care.

Drugs found to be effective in preventing severe T-cell-mediated conditions, such as rejection of a vital organ graft, must be shown to be very safe before use in ACH would seem appropriate. To date, measures to prevent ACH, including taking legal action, prohibiting the use of certain materials, or avoiding contact with allergenic materials, are favored over curative measures. Meanwhile, difficult-to-avoid allergens seem to warrant further studies of the potential value of tolerance induction before possible sensitization.

4. Exposure and Dose–Response Relationships

Many millions of naturally occurring and man-made chemicals are known. About 1000 to 2000 new chemicals are synthesized and introduced into commerce each year. Chemicals may reach the skin either by direct contact or by airborne exposure. Exposure may take place in occupational or other settings. This distinction has important legal implications, but the same allergens often appear both in domestic and occupational environments. Many substances have been described as able to cause allergic contact sensitization, but only about 100 chemicals or naturally occurring substances commonly cause allergic contact dermatitis.

Exposure to allergens depends on age and can be divided into five different types: in occupational settings, domestic work, hobby and leisure activities, topical medicaments, and cosmetic and personal care products.

The pattern of exposure differs from one geographical area to another, depending on degree of industrialization, and the use and availability of medicaments and naturally occurring substances. Tracing exposure to allergens is crucial for both the primary and secondary prevention of allergic contact sensitization.

Whether the sensitization follows exposure to a chemical depends on many factors, of which the exposure concentration of the chemical is the most important. The induction and eliciting hapten concentrations were tabulated for the most important allergens. In general, the concentration for induction of sensitization is much larger than the minimum eliciting concentration. Genetic factors are important in determining whether a person develops allergic contact sensitization after a given exposure. Other factors may modulate the induction and elicitation of sensitization, including the area of exposure, repetition of exposure, diseases of the skin, age, race, and sex.

5. Predictive Testing and Preventive Measures

The participants reviewed *in vivo* predictive testing for contact and photocontact allergens.

As to contact allergy tests, the guidelines of the Organization for Economic Co-operation and Development (OECD) include predictive methods. The main changes in the 1992 version of the guidelines are recommendation of only two protocols (maximization and Bühler test in guinea pigs), a detailed description of the methods, and agreement that a positive result in a mouse screening test (LLNA or MEST) may be definitive evidence of allergenicity, thus reducing any need for additional studies in animals.

Some external factors may influence the induction of sensitization, such as animal husbandry, selection of induction concentrations, and reaction assessment. The relevance for humans of results obtained in guinea pigs was discussed.

Human predictive testing has shown overall agreement with the results of animal testing. Technicians need great expertise in performing the tests to exclude misinterpretation. Even the best predictive tests may be unable to indicate allergic contact sensitization in a small population, although it would become evident if a sufficiently large population were examined.

No guidelines yet exist for photocontact allergy tests.

Preventing allergic contact sensitization is important because it is a common problem, often causing long-standing contact dermatitis, and thus problems in occupational and social life. The total number sensitized in the population depends on the frequency and intensity of exposure. The potential is low for exposure in the general environment to chemicals that can induce allergic contact sensitization. Experience has shown that allergic contact sensitization can be prevented by avoiding contact with a known allergen, regulating exposure concentrations, or taking preventive measures to protect individuals.

Research in dermatology focuses on the prevention of allergic contact sensitization by the main allergens: nickel, chromate, rubber chemicals, preservatives, fragrance materials, and acrylates. Establishing preventive programs for allergic contact sensitization requires:

(a)   information on exposure to chemicals;
(b)   predictive testing;
(c)   diagnostic testing and education;
(d)   epidemiological research and monitoring to set priorities and evaluate the effect of interventions;
(e)   identification of sources of exposure that cause sensitization;
(f)   identification of sensitizing and eliciting concentrations of allergens; and
(g)   research on the induction of tolerance or desensitization.

D. ALLERGIES AFFECTING OTHER ORGAN SYSTEMS
1. Gastrointestinal Allergy

Chemicals with low molecular weight rarely, if ever, cause a true allergic reaction in the gastrointestinal tract, as they seldom, if ever, cause primary sensitization via or in it. Most allergic reactions to substances with high molecular weight (including proteins such as milk, eggs, and fish) induced via the gut occur during childhood. Sensitization to pollen via the respiratory tract (hay fever) can lead to cross sensitivity to nuts, fruits, and vegetables. Contact sensitizers with low molecular weight may induce skin flare reactions in sensitized individuals after ingestion.

Nonallergic reactions to foods exist. Some have known mechanisms, such as a deficiency of the enzyme that digests lactose. Intolerance to food additives has as yet unknown causes. Intolerance reactions have been reported after the ingestion of food additives, such as certain colors, preservatives, and antioxidants. While the mechanisms for such reactions are unknown, they are assumed not to be immunological. Most often, symptoms from the skin or respiratory tract are provoked in atopic subjects. The prevalence of food additive intolerance was found to be 1 to 2% in a group of children in Denmark. In the U.K., a much lower prevalence was found in a group of adults and children who were identified in a different way.

2. Kidney

Some causes and mechanisms of chemically-induced nephropathies in humans were discussed, as chemical sensitivity may occur in the kidney. Few industrial chemicals and environmental pollutants are associated with these nephropathies, but very little research has been done. Knowledge in the renal field would be helpful in understanding the mechanisms of immune-mediated damage in other systems.

III. CONCLUSIONS

1. People are exposed to many chemicals at work, at home, and via the outdoor environment. A limited number of chemicals cause respiratory hypersensitivity, but they are of substantial industrial concern. The prevalence and severity of these problems are increasing. For example, the overall incidence of asthma mortality is increasing worldwide. While asthma mortality caused by chemicals is not a major public health problem, the worldwide trends may be relevant to the morbidity, loss of work, and economic loss resulting from chemical respiratory allergy. Sensitization to chemicals may result from exposure via various routes.

2. Many substances cause allergic contact sensitization, but only about 100 chemicals or naturally occurring substances commonly cause allergic contact dermatitis. In industrialized countries, 15 to 20% of the population has a positive reaction to chemicals with low molecular weight in diagnostic tests. The most important skin sensitizers are metals (nickel, cobalt, and chromate), preservatives (formaldehyde and formaldehyde-releasing substances), isothiazolinones, rubber additives, fragrance materials, plant allergens, and acrylate and epoxy derivatives. The most common elicitors of immediate skin reactions are drugs and food additives.

3. Chemicals with low molecular weight do not induce allergic sensitization via the gastrointestinal tract. Contact sensitizers may induce skin flare reactions after ingestion. Certain food additives may cause pseudoallergic reactions, mainly in the skin and respiratory tract. The mechanism of these reactions is not known.

4. Frequency of sensitization is an exposure-related problem, since the induction concentrations for allergic respiratory and contact hypersensitivity are 10 to 100 times larger than the minimum eliciting concentration.

5. The concentrations of allergens required for induction and elicitation of both respiratory and skin allergy vary markedly as a function of a number of factors that can modulate the response, such as the area exposed, repetition of exposure, and the presence of environmental pollutants.

6. Surveillance programs can give information on the trends and incidence of allergic disease. This is useful in the setting of priorities for control, evaluation, and professional education.

7. Predictive animal testing for allergic assessment has been improved in the past decade. *In vivo* testing indicates only the potential of a chemical

to induce allergy. How genetic and other endogenous factors, such as age, determine susceptibility to sensitization is not fully understood. Although such factors are clearly important for the development of sensitization in individuals, the frequency of sensitization in the population is an exposure-related problem. Hence, the real frequency of allergic sensitization depends on the distribution of the chemical and the exposure of susceptible individuals.

8. During the past 10 years, the understanding of some aspects of chemical allergy has significantly increased. Many uncertainties remain, however, and much remains to be learned. The cellular and molecular mechanisms through which chemicals provoke allergic responses require further clarification. Research in this area will pay dividends in allowing the development and use of improved methods for the clinical diagnosis of chemical allergy, and for the prospective identification and characterization of materials that can induce sensitization. New predictive test methods, with emphasis on *in vitro* assays, can minimize the need to use animals and the trauma to which they may be subjected. Research is needed into the genetic and environmental factors that influence the development of chemical allergy.

IV. RECOMMENDATIONS*

1. Programs for both health professionals and the public should provide information on exposure to chemicals and allergens. Such programs should ensure that up-to-date lists of allergenic compounds are available, that products are labeled to indicate the presence of allergenic chemicals, and that diagnostic testing is made feasible.

2. Sentinel surveillance programs should be used to alert clinicians to potential problems and to prompt early diagnosis, treatment, and intervention.

3. Epidemiological and toxicological research should examine the influence of modulating factors (such as pollutants) on the concentrations of chemicals required to induce and elicit respiratory and/or contact allergy.

4. As part of effective primary prevention, strategies and technologies should be developed to control occupational and nonoccupational exposures to known allergens, and their effectiveness should be evaluated.

5. Surveillance programs, including epidemiological surveys, should estimate the trends and incidence of allergic diseases in different countries and population groups to help identify risk factors and to determine priorities for public health action.

6. Research is needed to develop and validate predictive tests reflecting the mechanisms of respiratory and dermal allergic hypersensitivity (includ-

---

* The participants considered recommendations 1 through 6 to have the highest priority.

ing photoallergy). Great attention should be paid to making accurate estimates of dose-response relationships in experimental models. Comparison with data on humans should further validate current predictive methods using experimental animals.

7. As part of secondary prevention, diagnostic reagents and tests should be standardized, and objective, noninvasive methods for evaluating patients with suspected allergy to environmental chemicals should be further developed.

8. Prospective employees and groups with occupational exposure to sensitizing chemicals at work should not be tested, as it will do nothing to prevent sensitization.

9. The nature of the immunobiological mechanisms that result in the induction and regulation of chemical allergy (through sensitization or tolerance) should be described in more detail. Of particular importance is investigation of how chemical allergens are processed by and presented to the immune system, and the influence of different routes of exposure on the development of sensitization.

10. The true personal and financial costs of human disease due to environmental chemical allergens should be analyzed.

11. The use of animals in testing should be minimized, and emphasis placed on development of *in vitro* test procedures.

12. New problems allegedly linked to hypersensitivity should receive rigorous scientific investigation to confirm the existence and nature of syndromes such as multiple chemical sensitivities and pseudoallergic reactions. Research should also facilitate the early detection of new and unknown allergens and chemicals in the environment and diet.

V. PARTICIPANTS

Temporary Advisers

Professor Vladimir Bencko
Head, Institute of Hygiene, the First Medical Faculty, Charles University, Prague, Czechoslovakia
Professor Heidrun Behrendt
Medical Institute of Environmental Hygiene, Düsseldorf, Federal Republic of Germany
Professor A. D. Dayan
D.H. Department of Toxicology, St. Bartholomew's Hospital Medical College, St. Bartholomew's Centre for Research, London, United Kingdom
Professor Leonardo Fabbri
Pulmonary Diseases, University of Ferrara, Ferrara, Italy

Dr. Mari-Ann Flyvholm
Danish National Institute of Occupational Health, Copenhagen, Denmark
Dr. Janos Gergely
Department of Immunology, University of L. Eötvös, Göd, Hungary
Dr. Meryl H. Karol
Department of Environmental and Occupational Health, University of
Pittsburgh, PA, USA
Dr. Ian Kimber*
Section Head, Cell and Molecular Biology, ICI Central Toxicology
Laboratory, Macclesfield, Cheshire, United Kingdom
Dr. H. Van Loveren
National Institute of Public Health and the Environment, Bilthoven, The
Netherlands
Dr. Charlotte Madsen
Institute of Toxicology, National Food Agency, Søborg, Denmark
Dr. Thomas Maurer
Ciba-Geigy Ltd., Preclinical Safety, Basel, Switzerland
Professor Torkil Menné
Department of Dermatology, Gentofte Hospital, Hellerup, Denmark
Dr. Lucette Pelletier
INSERM U 28, Hôpital Broussais, Paris, France
Professor Dr. Johannes Ring
Universität Hamburg, Universitäts-Krankenhaus Eppendorf, Hamburg,
Federal Republic of Germany
Dr. R. J. Scheper
VU/University Hospital, Pathological Institute, Amsterdam,
The Netherlands
Dr. Katherine M. Venables
Senior Lecturer and Honorary Consultant Physician, National Heart and
Lung Institute, Department of Occupational and Environmental
Medicine, London, United Kingdom
Dr. Joseph G. Vos
National Institute of Public Health and the Environment (RIVM),
Bilthoven, The Netherlands

Representatives of Other Organizations

Commission of the European Communities (CEC)

Dr. G. Aresini
Health and Safety Directorate, Commission of the European
Communities, Luxembourg

---

* Also representing European Chemical Industry, Ecology and Toxicology Centre (ECETOC).

International Life Science Institute (ILSI)

Dr. Lanny J. Rosenwasser
  National Jewish Center for Immunology and Respiratory Medicine,
  Denver, CO, USA

World Health Organization

Regional Office for Europe

Ms. Yvonne Hoogland
  Secretary, WHO European Centre for Environment and Health, Bilthoven
  Division, The Netherlands

Dr. Dinko Kello
  Regional Adviser, Toxicology and Food Safety

Ms. Lena Klos
  Secretary, Toxicology and Food Safety

Dr. Maged Younes
  Toxicologist, WHO European Centre for Environment and Health,
  Bilthoven Division, The Netherlands

Headquarters

Dr. Edward Smith
  Medical Officer, International Programme on Chemical Safety, Geneva,
  Switzerland

Respiratory Allergy

Chapter 2

HYPERSENSITIVITY REACTIONS: DEFINITIONS, BASIC MECHANISMS, AND LOCALIZATIONS

J. Garssen, R. J. Vandebriel, I. Kimber, and H. Van Loveren

CONTENTS

I. INTRODUCTION

The essential function of the immune system is defense against neoplastic cells and infectious agents such as parasites, viruses, fungi, and bacteria. The immune system can be divided into two parts. Nonspecific defense is facilitated by a physical/chemical barrier found in the skin and mucosal tissues of the respiratory and gastrointestinal tracts. In addition to the cells of the physical/chemical barrier, such as epithelial and mucosal cells, natural killer cells, macrophages, and polymorphonuclear cells also play important roles in nonspecific immune responses. The specific or adaptive immune system has evolved in vertebrates and comprises organs, tissues, cells, and molecules devoted to host defense. The cardinal characteristics of the adaptive immune system are memory, specificity, and the ability to distinguish self from nonself.

Specific immunity can be subdivided further into a humoral and a cellular immune system. In the humoral system, B-lymphocytes, and in the cellular immune system, T-lymphocytes, play the predominant role. B-lymphocytes originate from the bone marrow, and after maturing into plasma cells, produce immunoglobulins that can bind antigens via the combining sites of their variable regions. Each mature B-lymphocyte (plasma cell) can produce antibodies of only a single specificity. Diversity in antibody specificity is created by somatic recombination of genes, encoding for different regions of the immunoglobulin molecules. In general, antibody binding serves to identify the antigen. Antigen-antibody complexes are cleared with the help of complement and phagocytic cells. Complement-immunoglobulin complexes are responsible for lysis of pathogenic cells. Complement-mediated cytotoxicity can be induced via two routes, the classical complement and the alternative pathways. Both routes will lead to the cleavage of C3, the central event in the complement system. Another set of plasma proteins becomes assembled into the structures (membrane attack complexes) responsible for the lytic lesions in the membranes of foreign pathogens such as bacteria. The first step leading to C3 cleavage by the classical route is complexing of antigen to its specific immunoglobulin (IgG and IgM). C3 cleavage induced by the alternative route does not require complexed antibody. In this case, probably carbohydrate components of the cell membrane of the microorganism initiates the complement activation cascade. In contrast, immunoglobulins may also directly neutralize the action of bacterial toxins without complement activation.

T-lymphocytes recognize antigen only as short stretches of amino acids, presented by gene products of the major histocompatibility complex (MHC).

MHC molecules are generally polymorphic between individuals. Within one individual or within one inbred strain of animals, the MHC molecules are identical. A T-lymphocyte specific for ovalbumin, for example, will recognize this protein only in the context of a certain MHC molecule. This phenomenon is called MHC restricted recognition. MHC and antigen are co-recognized by the T-cell receptor (TCR). The TCR is the antigen recognition molecule of T-lymphocytes. MHC products can be subdivided into MHC class I and MHC class II molecules. MHC class I molecules are present on all nucleated cells of an organism, whereas MHC class II molecules are present constitutively only on B-lymphocytes, macrophages, dendritic cells, and endothelial cells. CD4$^+$ lymphocytes (helper T-lymphocytes) see antigen only when it is presented by MHC class II molecules. CD8$^+$ lymphocytes (cytotoxic T-lymphocytes) see antigen only when it is presented by MHC class I molecules. CD4 and CD8 molecules are involved in antigen recognition by T-lymphocytes as they bind to a nonpolymorphic part of the MHC class II and MHC class I molecule, respectively. MHC class I molecules present antigens predominantly of intracellular origin (e.g., viral peptides that are synthesized inside an infected cell), whereas MHC class II molecules present antigens from extracellular origin only (e.g., from extracellular bacteria). Diversity in TCR specificity is created by somatic recombination of genes, encoding for different regions of the TCR. Although T-lymphocytes play the predominant role in cellular immunity, polymorphonuclear cells, macrophages, and natural killer cells also play roles in the cellular immune response. However, these latter three cell types act nonspecifically in contrast to the very specific action of T-lymphocytes. The directed activity of these nonspecific cells is often brought about by T-lymphocytes that recognize foreign structures and attract or stimulate nonspecific cells such as macrophages to it. Antigen presentation takes place very quickly upon the entry of antigen into lymphoid tissue. Langerhans dendritic cells are responsible for the early recruitment and activation of naive CD4$^+$ T-lymphocytes. Also, B-lymphocytes with immunoglobulin molecules on their surface with specificity for the antigen can also participate by binding the antigen (protein), processing it, and presenting it to memory CD4$^+$ T-lymphocytes. The interaction of the antigen-presenting cell (APC) with CD4$^+$ T-lymphocytes is facilitated by cell adhesion molecules, such as lymphocyte function antigen-1 (LFA-1), on the T-lymphocyte with its ligands ICAM-1 and ICAM-2 on the antigen-presenting cell, and the CD2 molecule with its counterpart LFA-3. This process is reciprocal: both T-lymphocytes and antigen-presenting cells will be stimulated.

The precise mechanisms responsible for the presentation of low molecular weight compounds such as nonpeptides are unknown at present. It is suggested that some of these small molecular weight compounds can act as a hapten and bind to protein carriers leading to an immune response against this carrier/ hapten complex, e.g., against a hapten bound to self-albumin or a hapten bound to the MHC itself.

In addition to cellular immunity, T-lymphocytes also play an important role in humoral immune responses. Most humoral immune responses depend on helper T-cell activity. Isotype switch, an important feature of type I hypersensitivity, depends also on activity of specific helper T-cells.

Quantitatively, the most important physiologic routes for the entry of antigens into the body is directly via the skin, through the respiratory tract, or the gastrointestinal tract. For the defense against pathological agents infiltrating the body, an extensive local immune system is available in the skin, the respiratory tract, and the gastrointestinal tract.

A. THE SKIN IMMUNE SYSTEM

The skin can be seen as one of the major organs of defense. The physical/chemical barrier serves as the first line of defense. Its principal physical function is that of a barrier to water and water-soluble compounds, to the photons of sunlight, as well as to trauma caused mechanically and by potentially pathogenic microorganisms. Its resistance to exogenous harmful influences is due mainly to the physicochemical properties of its outermost layer, the corneal (horny) substance of the epidermis. With respect to defense against water and water-soluble compounds, the barrier function is ascribed to the production of sphingolipids between the outer layers of the stratum corneum in the epidermis. If an antigen or allergen is able to penetrate the outermost layer of the epidermis, an additional defense mechanism is necessary. This second line of defense is ascribed to the immune system. Several lines of evidence support the fact that in the skin the immune system is compartmentalized in the so-called skin-associated lymphoid tissue (SALT)[1] or skin immune system (SIS).[2] Since, in contrast to rodents, in humans dendritic epidermal T-cells (DEC) are not present, the latter denomination is perhaps most adequate. Lines of evidence for a compartmentalized skin immune system include:

(a) Identification of functional T-cell subsets that exhibit epidermotropism
(b) Presence of epidermal antigen-presenting cells, named Langerhans cells (LC)
(c) Demonstration of antigen-presenting function of keratinocytes following immune or inflammatory reactions
(d) Presence of Thy 1[+] DEC cells with antigen-presenting functions (absent in men)
(e) Demonstration that some topically applied agents, like contact allergens, induce immune responses
(f) Demonstration that responsive T-cells recognize antigens within the microenvironment provided by the skin
(g) Characterization of keratinocyte and LC-derived cytokines that influence immune T-cell differentiation and hematopoiesis
(h) Identification of unique immunoregulatory circuits within the skin

Changes in the SIS may affect the immune system at other sites because there is a constant recirculation of lymphoid cells from the skin to draining lymph nodes and other lymphoid organs via blood and lymph vessels. Immune surveillance by recirculating lymphocytes enhances protection against pathogens, toxins, and neoplasms. Compartmentalized immune circuits depend on lymphocyte recirculation, to increase the random chance of an antigen-specific T-cell interacting with an antigen-bearing APC, and to amplify immunity by directing effector cells to the site of antigen deposition.

There is a role for SIS in the immunologic surveillance against skin tumors and cutaneous infections. Morphological data have emphasized that cellular immunity in the skin plays a role in the rejection of skin tumors in humans. Examination of the infiltrate surrounding basal cell carcinomas (BCCs) and squamous cell carcinomas (SCCs) shows predominantly T-lymphocytes and only a few B-lymphocytes.[3-5] Patients with SCCs, but not BCCs, have increased levels of circulating Th-lymphocytes.[6] The role of Langerhans cells in immunity against human skin tumors is unknown, but increased levels of LCs are found near and in SCCs and BCCs.[7] LCs presumably play a role as antigen-presenting cells to initiate immune responses directed to tumor cells, and there are indications that LCs may inhibit squamous cell growth.[8]

As mentioned above, cellular immunity in SIS is important in skin infections. HIV-1 infected individuals who have reduced CD4/CD8 ratios suffer from various bacterial,[9-12] viral,[13-15] and fungal[16-19] infections. HIV is also able to reduce the frequency of LCs in the skin[20,21] resulting in a decreased possibility of interaction between CD4+ T-lymphocytes and LCs, which is thought to be important for the resistance against skin infections.

B. THE RESPIRATORY IMMUNE SYSTEM

The nonspecific immune function of the respiratory tract is performed by the epithelium, cilia (ciliary movement), nonspecific phagocytes, natural killer cells, mucus, lysozyme, lactoferrin, and acute phase proteins. The specific part is characterized by an extensive lymphoid apparatus including, for example, lymphocytes of B- and T-cell subpopulations and antigen-presenting cells, essential to generate an immune response.

Lymphoid tissue is detectable throughout the whole respiratory tract from the nasopharynx to bronchioli and alveolar ducts. Typical lymph nodes with afferent and efferent channels, a capsule, follicles, and germinal centers are found only in the region of the trachea and major bronchi (tracheobronchial and hilar lymph nodes). These serve as the regional lymph nodes for the entire lung, receiving lymphatic drainage from most of the respiratory tract. The hilar lymph nodes contain greater concentrations of IgA- and IgE-producing cells than other lymph nodes,[22,23] probably because of their proximity to mucosal surfaces. In other respects, they appear to be identical to other lymph nodes.[24]

Lymphoid nodules contain follicle-like structures but lack capsules. These structures occur in the walls of large- and medium-sized bronchi and extend from subepithelium through submucosa to the peribronchial connective tissue. On the air-side these structures are mostly covered by a single layer of flattened, nonciliated epithelium that is infiltrated by lymphocytes (lymphoepithelium). In the gastrointestinal tract, similar structures can be found. These lymphoid structures are called mucosal associated lymphoid tissue (MALT), GALT (gut associated lymphoid tissue) for the gastrointestinal tract, and BALT (bronchus associated lymphoid tissue) for the respiratory tract. BALT was described extensively for the first time by Bienenstock et al.[25,26] Recently Sminia et al.[27] reviewed the data on MALT. MALT in the lower part of the respiratory tract (BALT) occurs at well-defined sites between arteries and bronchus epithelium. In addition, MALT is not restricted to the lower part of the respiratory tract. Tonsils in man also have similarities with MALT, and recently also nasal-associated lymphoid tissue (NALT) has been described in the rat.[28] However, NALT has not been described in humans.[28]

Lymphocytes are also present throughout peripheral lung tissue. Diffuse collections of lymphocytes and phagocytes are distributed along the entire submucosa and lamina propria of the airways, and are located close to both lymphatic and blood vessels. Particularly prominent in bronchioles are discrete aggregates of lymphatic tissue adjacent to morphologically specialized lymphoepithelium. These structures bring respiratory epithelium, lymphatic tissue, and blood vessels into close proximity. It is suggested that these structures serve as filters for the removal of antigenic material to the lymphoid apparatus. These structures form a potential mechanism for the initial interaction of inhaled antigens with lymphoid tissue. If pathogenic material passes the nonspecific defense system (first-line defense: mucociliar clearance, phagocytes), it encounters the epithelial cell layer. This epithelial layer forms the second line of aspecific (mechanical) defense. Recently, it has been found that epithelial cells can present antigen to immunocompetent memory cells. It was suggested that epithelial cells can express Ia (class II MHC)[29] and that there is a correlation between Ia expression on epithelial cells and the number of intraepithelial lymphocytes present.[30,31] A special type of cell is found in the epithelium that covers the lymphoid follicles (nodules). These cells (M cells, membraneous epithelial cells) are found also in the epithelium covering BALT.[29] M cells take up pathogenic material from the airway lumen and discharge it into the extracellular space where it can be picked up by infiltrating cells (dendritic cells or macrophages) that will present the antigen to local intraepithelial lymphocytes (IEL). Intraepithelial cells are mainly T-lymphocytes. Among these T-cells the suppressor/cytotoxic subtype $CD8^+$ usually predominates in humans.[32] In general, mature circulating T-cells have receptors that are composed of α and β chains, in contrast to the IEL T-cells in mice which are Thy-1 and express primarily receptors composed of γ and δ chains.[33] During development of mature T-cells in the thymus, $\gamma\delta$ T-cells precede $\alpha\beta$ T-cells,[34] and IEL lymphocytes may therefore be regarded as immature T-lymphocytes. Recently, Abraham et al.[35] reported that nearly all the intraparenchymal lung T-cells bear αβ T-cell receptors, of which the major part stained with either CD4 or CD8. This contrasts with Augustin et al.,[36] who described the presence of γδ T-cells that are mostly CD4 and CD8 negative. In this latter study, lymphocytes were obtained by EDTA treatment of lung slices, and therefore probably represent an intraepithelial cell population. The biological role of cells bearing the γδ T-cell antigen receptor is as yet unclear. The strategic locations of these "immature-like" cells have led to the suggestion that γδ cells could constitute a primary defense mechanism in the vicinity of large surfaces of contact with the environment.

Additionally, many lymphoid cells can be found in lung lavage fluid. The cellular component of bronchoalveolar cells varies among animal species, but generally consists primarily of alveolar macrophages, moderate proportions of lymphocytes and plasma cells, and a minor fraction of granulocytes and mast cells. In certain pathological circumstances like asthma and in smokers, the numbers (and percentages) of certain cell subtypes can change.[37,38]

C. THE GASTROINTESTINAL IMMUNE SYSTEM

In the gastrointestinal tract, the defense system has many similarities with that of the respiratory tract. In addition to a nonspecific first line of defense, there is a second line of defense, in the form of an adaptive immune function. The first line of defense is characterized by a physical/chemical barrier. This barrier is ascribed to an epithelial cell layer, bacteriostatic enzymes, phagocytic cells, and mucus secretion. The specific defense mechanism in the gastrointestinal tract comprises a compartmentalized immune system which has many characteristics in common with the immune system found in the airways. GALT consists of the Peyer's patches, solitary lymphoid nodules, and lymphocytes in the lamina propria and epithelium of the intestinal villi. GALT comprises one of the major parts of the mucosal associated immune system[25,26] and can be divided into afferent and efferent limbs.[39] The afferent limb plays a pivotal role in antigen-specific responses against pathogens from the intestinal lumen and is located mainly in the Peyer's patches. Immune responses here will result in the production of specific IgA and IgM antibody,[39-43] and the appearance of antigen-specific T-lymphocytes.[44,45] The efferent limb of GALT consists of lymphocytes in the intestinal epithelium and lamina propria. IgA and IgM producing B-cells and certain T-cells mature in the efferent limb and home specifically to their intestinal locations. The intraepithelial cells are, in the main, similar to the intraepithelial lymphoid cells found in the respiratory epithelia (γδ T-cells).

D. IMMUNOLOGICAL HYPERSENSITIVITY REACTIONS

Specific immune function has evolved to provide host defense mechanisms and protection from pathogenic microorganisms and malignant disease. The functional integrity of the adaptive immune system can be subverted by certain chemicals. The adverse effects which result from exposure to chemicals may

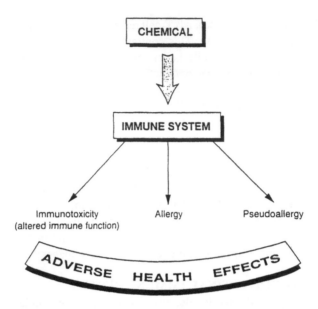

FIGURE 2-1. Diagram of adverse reactions to the immune system due to chemical exposure.

be subtle and result in changes in the functional activity of the immune system and alterations in host resistance. Alternatively, chemicals may provoke allergic responses. Allergy may be defined as the adverse health effects which result from specific immune responses, directed usually against exogenous antigens (Figure 2-1).

In pseudoallergy, specific immunity does not play a pivotal role. Aspecific stimuli can induce mast cell degranulation, for example, leading to the symptoms of allergic disease such as asthma rhinitis and skin reactions. These triggers may be responsible for pseudoallergic reactions such as some neuropeptides including substance P, somatostatin, neurotensin, and nerve growth factors.

Adverse effects due to allergy are mediated by different "specific" immune responses and are classified according to the immune mechanism that is primarily responsible for these adverse reactions. These hypersensitivity reactions can be subdivided and classified into four or five subtypes.[46] The majority of all these different subtypes of hypersensitivity reactions can occur, for example, in the airways, gut, and skin (Table 2-1).

Allergic diseases can be induced by high molecular weight compounds, usually proteins. However, small molecular weight compounds can also induce allergic reactions. Although less is known about the precise mechanisms of allergy induced by chemicals, it is clear that they can induce specific IgE production, causing T-cell-mediated immune responses.

Type I, immediate hypersensitivity, is mediated by IgE and is pertinent to the pathogenesis of hay fever, atopic (allergic/extrinsic) asthma, and urticaria, for example.

TABLE 2-1

Examples of Allergic Diseases Occurring in the Skin, Lungs, and Gut

	Type I	Type II	Type III	Type IV
Skin:	Urticaria, Contact-urticaria	Skin Purpura, Emphigus, Bullous pemphigoid	Allergic vasculitis Erythema nodosum Erythema multiforme	Contact hypersensitivity Eczema?
Lungs:	Atopic asthma, Atopic rhinitis	Good Pasture's syndrome*	Allergic alveolitis	Hard metal asthma, Sarcoidosis
Gut:	Food allergy?	Hemolytic anemia**	(Dermatitis) herpetiformis***	Food allergy?, Crohn's disease?

*   Also in the kidney.
** Ingested penicillin will bind to erythrocytes leading to autoantibodies against erythrocytes.
*** Occurring in both skin and gut.

Type II hypersensitivity refers to cytotoxic antibodies produced by the host against his own tissues (autoantibodies). This mechanism appears to be responsible, for example, for Goodpasture's syndrome. In this disease, autoantibodies (in majority IgG) against a basement membrane antigen are formed.

Type III hypersensitivity involves tissue damage secondary to the formation of immune complexes. This reaction (Arthus reaction) is responsible, for example, for hypersensitivity pneumonitis or allergic alveolitis and possibly for late-phase reactions in asthma.

Type IV hypersensitivity is exemplified by the well-known delayed hypersensitivity cutaneous reaction to tuberculin. This type of hypersensitivity may possibly be responsible for granulomatous lung diseases, such as tuberculosis and sarcoidosis, and for inflammatory reactions found in some forms of chronic obstructive pulmonary diseases (COPD). Contact hypersensitivity, for example nickel allergy induced by earrings, is an example of a skin-associated Type IV hypersensitivity reaction. Type IV hypersensitivity is T-cell dependent and is characterized by erythema and induration which appears 1 or 2 days after antigen contact in sensitized individuals.

In addition, another subtype of autoimmune-mediated hypersensitivity reactions is recognized. In this subtype noncomplement binding antibodies directed against certain cell surface components may actually stimulate rather than destroy the cell. This subtype can therefore be recognized as stimulatory hypersensitivity. An example is thyrotoxicosis, a disease in which autoantibodies overstimulate thyroid cell receptors. In healthy persons only the thyroid-stimulating hormone (TSH) is active in stimulating the thyroid receptor.

In summary, Types I, II, and III depend on the interaction of antigen with antibodies and tend to be called immediate- or intermediate-type reactions. Type

IV involves receptors bound to the lymphocyte surface, and because of the longer time course, this has been referred to as delayed-type hypersensitivity.

E. GENETIC FACTORS IN ALLERGIC DISEASE

For several reasons, previous genetic analyses of allergic asthma, for example, which have depended largely on the expression of a clinically recognizable trait, have been contradictory and difficult to interpret.[47-49] Asthma is generally considered to represent a complicated disease in which the etiology of reversible airway obstruction is not necessarily the same for each patient.[47,50-53] It is likely that the genes responsible may not be the same in various forms of asthma. Also, evidence suggests that at least several genes interact to produce the asthmatic phenotype (polygenic).[47,50,51]

Townley et al.[54] demonstrated that a certain allele (S allele) is associated with bronchial hyperresponsiveness, which is one of the characteristics of bronchial asthma. Asthmatic families (AFs) and normal families (NFs) were studied to determine the relationship between bronchial hyperresponsiveness and α-1-antitrypsin protease inhibitor phenotype (Pi phenotype). It is demonstrated that α-1-antitrypsin variants are determined by a set of fully penetrant codominant alleles (MM, MS, MZ, and ZZ). The study of Townley et al. demonstrated a relationship between the MS phenotype and bronchial hyperresponsiveness.[54] The other phenotypes (MM, MZ, ZZ) were less related to bronchial hyperresponsiveness. In addition, atopy, defined as an increased tendency to form IgE antibodies to common allergens, can be inherited, and there may be an increased incidence of hyperresponsiveness in atopic individuals.[55-58]

Additionally, it is hypothesized that contact dermatitis may also have a genetic base. Nickel is a major example of a simple chemical contactant implicated in allergic contact dermatitis, particularly in industrialized countries and especially among women.[59] There have been several reports on heritability of contact dermatitis in humans.[60] In fact about 10% of the general population in Scandinavia suffers from nickel allergy.[61,62] A significant difference has been found in the concordance rate between monozygotic and dizygotic twins among female Danish twins with a history of nickel allergy.[63] Olerup and Emtestam[60] reported that allergic contact dermatitis in response to nickel is associated with Taq 1 HLA-DQA allelic restriction fragment. Furthermore, DQ, and presumably also DR, may restrict the proliferative response to nickel.[64]

II. REGULATION OF HYPERSENSITIVITY

In 1986, the existence of two CD4+ Th cell subsets was discovered in mice, and designated Th1 and Th2.[65] Their identification has greatly improved the understanding of the regulation of immune effector functions, not in the least on Type I and Type IV hypersensitivity responses. These Th subsets cannot be differentiated on the basis of a cell surface marker. They produce, however, defined patterns of cytokines that lead to strikingly different T-cell functions.[66] Roughly speaking, Th2 cells are more efficient B-cell helpers, especially in the

TABLE 2-2
Cytokine Production Phenotypes of Mouse T Cells

	TH1	TH2	TH0	THP
IL-2	++	—	++	++
IFN-γ	++	—	++	—
TNF-β[1]	++	—		
GM-CSF	++	+	++	—
TNF-α	++	+		
IL-3	++	++	++	—
IL-4	—	++	++	—
IL-5	—	++	++	—
IL-6	—	++		
IL-10	—	++	++	—

[1]   Lymphotoxin.

Adapted from Mosmann, T. R. and Moore, K. W., *Immunol. Today*, 12, A49, 1991.

production of IgE, whereas Th1 mediate delayed-type hypersensitivity (DTH). In addition, they cross-regulate by producing mutually antagonistic cytokines. Recently, it has become clear that they also exist in humans.[67]

We will discuss first the regulation of hypersensitivity in mice, since in this species it has been investigated most thoroughly. In addition to Th1 and Th2 cells, additional cytokine production phenotypes of CD4+ cells exist. They are, however, characterized less well. Most resting T-cells mainly produce IL-2 on first contact with antigen, and differentiate within a few days into cells producing multiple cytokines,[68-70] such as IL-4 and IFN-γ. Mosmann and Moore suggested, in addition to Th1 and Th2, the existence of ThP and Th0 cells.[71] ThP cells (IL-2 producing) are the virgin Th cells, producing only or predominantly IL-2, and Th0 cells are in the process of differentiation, producing cytokines of both Th1 type, such as IL-2 and IFN-γ, and Th2 type, such as IL-4, IL-5, and IL-10. Experiments *in vitro* with mouse[72] and human[73] cells and *in vivo*,[70] especially when using strong immunogens, support the hypothesis that the differentiation from ThP occurs *in vivo*. The pathways of differentiation from ThP are, however, unclear. In addition, it is unknown whether there is one common ThP or that ThP is already committed to a particular cytokine pattern before exposure to antigen. The cytokine production profiles of the four different Th subsets are shown in Table 2-2.[65,71,74-76] In conclusion, it can be suggested that Th1 and Th2 cells represent the most differentiated populations of the CD4+ phenotype, which develop following prolonged exposure to antigen or following stimulation by potent immunogens.

At least two mechanisms can influence the selective differentiation of the T-cell subsets. First, the cytokines that are present during differentiation, in particular IFN-γ and IL-4, may greatly influence the type of Th that will be generated. IFN-γ augments development of Th1-type responses and IL-4 promotes differentiation of Th2 cells.[77,78] Second, the type of APC is thought to influence the type of immune response.[79-82] Upon activation, Th cells transiently express p39 on their surface. In the case of Th2 cells, p39 interacts with CD40 that is present constitutively on the surface of B-cells. The interaction of p39 with CD40, and TCR with antigen/MHC class II, together lead to production of IL-4, IL-5, and IL-6 by Th2 cells, stimulating B-cells to antibody production. Th1 cells, on the other hand, may interact with macrophages. A pair of cell surface molecules analogous to p39/CD40 have as yet not been identified; however, the interaction of Th1 cells with macrophages leads to IFN-γ production by Th1 cells, stimulating macrophages to produce monokines. It has been suggested[83] that the production of IFN-γ and lymphotoxin by Th1 cells lyse B-cells, further suggesting that B-cells cannot present antigen to Th1 cells.

The difference in APC, macrophage vs. B-cell, that preferentially activates Th1 or Th2, suggests differences in antigen requirements for activation, for example, large, particulate antigens requiring phagocytosis for Th1, and low antigen concentration for Th2. Whereas moderate concentrations of antigen preferentially activate Th1, extremely high concentrations are suggested to inhibit Th1 and select for Th2.[84-86]

If Th1 and Th2 clones are stimulated by immobilized anti-CD3 (in the absence of APC), both types produce their respective cytokine pattern. The proliferative responses are, however, very different between the two types of clones. Whereas Th2 clones show good proliferative responses, Th1 clones not only fail to do so, but are even rendered incapable of proliferating in response to exogenously added IL-2.[84,85,87] These Th1 clones are in a state of anergy or tolerance.[88]

For the interaction of APC and Th cells, IL-1 is a co-stimulator for Th2, but not Th1 cells. The nature of the co-stimulator for Th1 is as yet unknown.[88]

Th1 and Th2 differ in signal transduction, suggesting a pharmacological target for regulation of these subsets *in vivo*.

IFN-γ inhibits the proliferation of Th2 responding to either IL-2 or IL-4, but does not inhibit Th1.[89,90] IL-10 inhibits the synthesis of cytokines by Th1 cells, and although growth factor requirement is not affected, the reduction in IL-2 synthesis can lead to decreased proliferation.[91] It has been shown recently in *in vitro* human systems[92] that IL-10 can suppress the antigen-presenting capacity of monocytes by downregulation of MHC class II. IL-10 had no effect on the antigen-presenting capacity of B-cells or downregulation of their MHC class II. These results suggest a mechanism for the general observation that macrophages/dendritic cells preferentially stimulate Th1, whereas B-cells preferentially stimulate Th2.[84]

Both Th1 and Th2 can mediate B-cell help, but Th2 cells are much more efficient,[93-97] especially in the production of IgE[93] and IgG,[98] whereas Th1 cells mediate a DTH reaction.[99] Especially when strong immune responses are considered, responses involving Th1 and Th2 appear to be mutually inhibitory.[66]

IL-2 is a T-cell growth factor (TCGF) that mediates autocrine proliferation of Th1,[100] whereas the TCGF IL-4 mediates autocrine proliferation of Th2. Interestingly, it has been shown that IL-4 is the major TCGF produced by T-cells from lymphoid organs that drain mucosal tissues, whereas IL-2 is the major TCGF produced by T-cells from other lymphoid organs.[101] Involvement for dehydroepiandosterone in this site/tissue-specific control on lymphokine production was suggested.[101] Dihydrotestosterone and 1,25-dihydroxyvitamin D3 also change the cytokine production pattern of T-cells.[102]

Until recently, the existence of Th1 and Th2 in humans was questioned. A predominant fraction of CD4+ T-cell clones was found to produce IL-2, IL-4, and IFN-γ, although the quantities varied considerably.[103,104] Keeping in mind the findings in mice (see above), it was thought that unrestricted profiles are mainly a property of T-cells that are not yet committed to a certain differentiation pathway. Consequently, functional heterogeneity of CD4+ cells should most likely be found in chronically stimulated responders. Therefore, two categories of patients were extensively studied,[105] namely with nickel hypersensitivity, exemplary of type IV hypersensitivity, and with housedust mite, *Dermatophagoides pteronysinnus*, (Dp) hypersensitivity, exemplary of type I hypersensitivity.

Most Dp-specific T-cell clones from peripheral blood[106] and lesional skin biopsies[107] of Dp-allergic patients showed a Th2-like production profile. In contrast to nickel hypersensitivity, a clear difference in cytokine production profile (i.e., an elevated IL-4/IFN-γ ratio) was found between allergic patients and control individuals. Dp-specific clones from atopic patients induce IgE production (also see below). It was shown that this production is dependent on a high IL-4/IFN-γ ratio, and not dependent on the origin of B-cells. Only Dp-specific IgE (and not, for example, tetanus toxoid or Candida albicans-specific IgE) was elevated in atopic Dp-allergic patients.

The majority of allergen-specific human T-cell clones produce IL-4 and IL-5, but not IFN-γ. Virtually all T-cell clones specific for bacterial components, which were derived from the same patients, produced large amounts of IL-2 and IFN-γ, and few produced IL-4 and/or IL-5.[106,108] In a subsequent study,[109] antigen specific T-cell clones were derived for the bacterial antigen purified protein derivate (PPD) from *Mycobacterium tuberculosis* and for the helminth antigen *Toxocara canis* excretory-secretory (TES). Most PPD-specific clones produce IL-2 and IFN-γ, but not IL-4 and IL-5, whereas most TES-specific clones produce IL-4 and IL-5, but not IL-2 and IFN-γ. This study shows nicely that in the course of natural immunization, certain infectious agents preferentially expand T-cell subsets. PPD expands Th1, paralleling the (Th1-mediated) tuberculin DTH, whereas TES expands Th2, paralleling the (Th2-mediated) parasite infection.

In a large series of human T-cell clones, all Th1 clones lysed EBV-transformed autologous B-cells pulsed with the specific antigen, and the decrease of Ig production correlated with the lytic activity of Th1 clones against autologous antigen-presenting B-cell targets.[67] This suggests an important mechanism for downregulation of antibody responses *in vivo*.

Almost all nickel-specific T-cell clones produce TNF-α, granulocyte/macrophage colony-stimulating factor (GM-CSF), IL-2, and high levels of IFN-γ, but low or undetectable levels of IL-4 and IL-5,[110] thus resembling Th1 cells. Nickel induces DTH in the skin of allergic patients. Since IFN-γ is an important mediator for DTH, IFN-γ may be essential to DTH. However, no clear difference in the cytokine production profiles between allergic patients and control individuals was found.

Chemicals differ with respect to the type or types of allergic reactions they will elicit. Dearman et al. have shown recently that chemicals known to cause respiratory hypersensitivity induce in mice preferential Th2-type responses and IgE antibody. In contrast, chemical contact allergens which are known or suspected not to cause respiratory hypersensitivity, induce instead immune responses consistent with a selective activation of Th1 cells.[111-115]

Table 2-3 summarizes cytokine activity.

A. REGULATION OF IgE SYNTHESIS BY IL-4 AND IFN-γ

Atopy is associated with enhanced serum titers of allergen-specific IgE. The production of IgE is heightened and sustained by B-cells in atopic patients. IL-2 secreted by Th cells is necessary for the production of all isotypes of immunoglobulins.[105] Activated B-cells are induced by IL-4 to undergo immunoglobulin heavy-chain rearrangements to the ϵ-constant region, resulting in synthesis of IgE.[116,117] So far, IL-4 appears to be the only cytokine capable of mediating this isotype switch, which is blocked very efficiently by IFN-γ.[118] IFN-γ induces switching to γ2a.[97,119,120] Enhancement of IgG4, elevated in parallel with IgE, has not yet been shown to result from an increase in isotype switching to γ4.[121,123] Because IL-4 and IFN-γ are produced by Th2 and Th1 cells, respectively,[65] a response that involves mainly Th2 cells should produce a large amount of IgE, whereas responses involving mainly Th1 cells such as DTH reactions should be nonpermissive for IgE production. *In vivo* experiments have confirmed these predictions.[123] IL-4 deficient mice lack IgE and IgG1 responses,[124] whereas transgenic mice constitutively producing IL-4 show elevated serum IgE levels.[125,126] Injection of mice with anti-IgD antibodies results in a strong stimulation of both B- and T-cell populations, leading to polyclonal antibody production and very high IgE levels. Anti-IL-4 antibodies dramatically reduce IgE levels after anti-IgD immunization, whereas anti-IFN-γ antibodies elevate IgE levels even further. Similarly, administration of IFN-γ results in considerable inhibition of the IgE response. Because the anti-IgD immunization leads to a response that involves high levels of Th2 cytokines, all of these results are consistent with the effects of IL-4 and IFN-γ on IgE synthesis as defined by *in vitro* model systems. Similar correlations between Th2-like responses and high IgE production are seen during several parasite infections.

B. EOSINOPHILIA AND IL-5

Anti-IgD immunizations, as well as many parasite infections, induce high levels of IgE and very high levels of circulating eosinophils. Because IL-5 has been implicated as a major growth and differentiation factor for eosinophils by *in vitro* experiments, the association of IgE and eosinophilia may be explained by the association of IL-4 and IL-5 as products of Th2 cells.[127] Supporting evidence has been provided by experiments *in vivo*, in which administration of anti-IL-5 during a strong anti-parasitic immune response completely abrogated eosinophilia,[128] and from studies of transgenic mice that express high levels of IL-5. The major abnormality in these animals is the presence of extremely high levels of eosinophils in the blood and various lymphoid organs.[129] Patients with filaria-induced eosinophilia exhibit a significantly greater frequency of IL-5-producing T-cells than uninfected individuals.

C. THE RELATIONSHIP BETWEEN Th2 CELLS AND TYPE I HYPERSENSITIVITY

In mice, in addition to enhancing IgE production via IL-4, Th2 cells also influence other features of allergic reactions. First, IL-3, IL-4, and IL-10 are mast cell growth factors that act in synergy, at least *in vitro*,[130,131] and second, IL-5 induces the proliferation and differentiation of eosinophils *in vitro* and *in vivo*.[128,132] In addition, IL-3 and IL-4 have been shown to enhance the secretory function of murine mast cells.[133] So, Th2 cell activation not only increases the level of IgE synthesized, but also potentially increases the number of cells that can bind IgE, which will degranulate in response to allergen challenge.

Mast cells and basophils produce IL-4.[134-138] It is hypothesized that IL-4 produced by these cells induces the development of Th2 cells, and that these cells in turn produce IL-4. In addition, mast cells are an important source of IL-5.[134,136]

D. IL-12 DRIVES THE IMMUNE RESPONSE TOWARD Th1

During the past two years, the pivotal role of the cytokine IL-12 in the differentiation of Th cells toward Th1 has become evident from both *in vitro* and *in vivo* studies.[139] IL-12 is produced by T-cells and macrophages and stimulates the production of IFN-γ from T-cells and NK-cells.[140,221] IL-12 enhances Th1 cell expansion in cell lines from atopic patients.[141] The presence of IL-12 during primary stimulation of naive CD4+ cells skews the response in the direction of Th1 differentiation.[142] These results suggest that IL-12 may be the IL-4 equivalent for the differentiation of Th1 cells. IL-10 has been shown to inhibit lymphocyte IFN-γ production by suppressing IL-12 synthesis in accessory cells.[143] A variety of pathogens that are associated with Th1 development have been shown to induce IL-12 production.[144]

IL-12 may have therapeutic applications in a wide range of diseases, such as the treatment of allergies in which a Th2 response mediates immunopathol-

TABLE 2-3
List of Cytokines

Cytokine	Produced by	Acts on	Ref.
IL-1[a]	Macrophages	Thymocytes	201
	T	T	
	B	B	
	Fibroblasts	Hemopoietic cells	
	Keratinocytes	Endothelial cells	
	Brain astrocytes	Hepatocytes	
	Microglial cells	Fibroblasts	
		Bone marrow	
		stromal cells	
IL-2	T (mainly T_h)	T	201
		Thymocytes	
		Macrophages	
		B	
		NK	
IL-3	T	Mast cells	201, 202
		Granulocytes	
		Macrophages	
		Eosinophils	
		Megakaryocytes	
		Erythrocytes	
IL-4	T (T_h2)	T	201
		Thymocytes	
		B	
		Macrophages	
		Mast cells	
		Hemopoietic cells	
	Mast cells		134–136
	Basophils		137, 138
IL-5	T	B	201
		Eosinophils	
	Mast cells		134, 136
IL-6	T	T	201
	Macrophages	Thymocytes	
	Fibroblasts	B (some transformed)	
	Astrocytes	Hemopoietic cells	
	Stromal cells from thymus	Granulocytes	
		Macrophages	
IL-7	Stromal cells from	T	201
	thymus and bone marrow	Thymocytes	
		Pro- and pre-B	
	Spleen		
		Hepatocytes	
IL-8	T	T	203
	Monocytes	Neutrophils	
	Fibroblasts	Basophils	
	Endothelial cells		
	Keratinocytes		

TABLE 2-3 (continued)
List of Cytokines

Cytokine	Produced by	Acts on	Ref.
IL-9	T_h	T_h Mast cells Thymocytes Erythrocytes	204–210
IL-10	T_h2 T_h Thymocytes B Monocytes Mast cells	T CTL Thymocytes B Monocytes/ macrophages Mast cells	92, 211, 212
IL-11	Stromal cells	Hemopoietic Progenitors (e.g., megakaryocyte) B	213–215
IL-12[b]	PBL Macrophages B-cells	T NK	216, 221 139, 140, 144, 221
IL-13	Activated T-cells	Monocytes B-cells	146, 222–224
GM-CSF	T Macrophages Fibroblasts Stromal cells Endothelial cells	Neutrophils Macrophages Eosinophils Megakaryocytes Erythrocytes	201, 202
G-CSF	Lymphocytes Fibroblasts Endothelial cells	Neutrophils Erythrocytes	201, 202
M-CSF	Lymphocytes Fibroblasts Endothelial cells Stromal cells	Macrophages	202
IFN-γ	T	Epithelial cells Endothelial cells Connective tissue Macrophages Monocytes Tumor cells	201
TNF-α[c]	Mononuclear Phagocytes	Tumor cells	201

[a]   IL-1 consists of two distinct but related molecules: IL-1α and IL-1ß.

[b]   IL-12 is a heterodimer.

[c]   Lymphotoxin is homologous to TNF-α. It binds to the same receptor and has common biological activities.

ogy, but also in infections and malignancies. IL-12 has been shown to function as an adjuvant.[145]

E. IL-13, AN IL-4-LIKE CYTOKINE

Information on the newly described cytokine IL-13 is only based on limited information about its activities *in vitro*. As it shares biological activities with IL-4, these activities will, however, be briefly discussed. IL-13 is produced by activated T-cells. The activities *in vitro* of IL-13 are similar to IL-4 with two major exceptions. First, IL-13 does not act on T-cells. Second, IL-13 does not act on murine B-cells.[146] Similar to IL-4 and IL-10, IL-13 inhibits the production by LPS-stimulated monocytes of proinflammatory cytokines, chemokines, and hematopoietic growth factors. In contrast to IL-10, however, IL-13 upregulates the antigen-presenting capacity of monocytes. Similar to IL-4, IL-13 inhibits transcription of IFN-α and both the α- and β-chain of IL-12. Thus, IL-13 may (like IL-4) suppress the development of Th1 cells through downregulation of IFN-α and IL-12 production by monocytes, favoring the generation of Th2 cells. Also, in the mouse IL-13 inhibits production of proinflammatory cytokines and expression of IL-12 α- and β-chain mRNA. Murine IL-13 does not affect macrophage antigen-presenting capacity. Similar to IL-4, IL-13 acts on human B-cells in inducing class switch to production of IgG4 and IgE and inducing CD23 surface expression.[147,148] Following activation of T-cells, IL-13 is produced earlier and for much longer periods than IL-4.[149] Thus, IL-13 may play an important role in the regulation of enhanced IgE synthesis in allergic patients. In contrast to IL-4, murine and human IL-13 do not induce IgE synthesis in murine B-cells. Importantly, this may restrict the use of mice as an animal model for allergy.

In summary, IL-13 may favor development of Th2 cells, consistent with the induction of IgG4 and IgE synthesis. Determination of the actual role of IL-13 requires more information on the biological effects *in vivo*.

F. MAST CELLS/MEDIATORS IN
HYPERSENSITIVITY REACTIONS

One of the most important cell types in allergic disease is the mast cell. Mast cells are widely distributed in organ systems such as the integumentary system, respiratory, gastrointestinal and urogenital tracts, and various other organs.[150-152] Upon activation, mast cells release a variety of mediators. The most well-studied example of mast cell degranulation is induced by protein allergens in IgE-sensitized mast cells participating in hypersensitivity reactions of the immediate, or anaphylactic type. Upon sensitization, protein allergens induce specific antibodies of the IgE class. Mast cells possess a variety of surface membrane receptors, among those a high number that specifically bind certain domains[153] of the Fc region of the IgE antibody with high affinity.[154-157] Bridging of these Fc-ε receptors by cross-linking of the bound IgE by the bi-

or multivalent allergen, or anti-IgE, or otherwise,[158] results in a transmembrane signal that causes activation of membrane-associated enzymes, leading to degranulation and mediator release.[159] Apart from this classical IgE-mediated release of mast cell products, activation and mediator release can be initiated by a number of other potent secretagogues, such as products generated during inflammatory reactions like anaphylatoxins C3a and C5a (complement breakdown products) and factors released by neutrophils, eosinophils, macrophages, and lymphocytes.[160,161] Mast cells contain specific complement component receptors.[162] Additionally, there is substantial evidence that antigen-specific T-cell factors are capable of stimulating mast cell mediator release during delayed-type hypersensitivity reactions.[163-165] It is conceivable that in these cases, too, activation of the mast cell is achieved through specific receptors for the T-cell factors.[166]

Mediators released by mast cells after stimulation by one or other of the mechanisms as described above can be divided into at least five major groups of factors.

1. Vasoactive mediators. The most common vasoactive amine is histamine in humans. Histamine acts by binding to specific histamine receptors leading to smooth muscle contraction and an increase in vascular permeability, leading to leakage of serum and egress of leukocytes from the blood stream. Activation of a special subtype of histamine receptors can lead to inhibition of cellular immunity, histamine release by basophils, T-cell-mediated cytotoxicity, and granular leukocyte chemotaxis and stimulation of T suppressor activity.[167,168] In rodents, the major vasoactive amine is serotonin. It can cause smooth muscle contraction and increase vascular permeability. Other important vasoactive mediators released by mast cells are some prostaglandins (PGD2, PGE2, PGI2) and leukotrienes.[167-169] Platelet activating factor (PAF) is another mast cell-derived mediator that can induce increased vasopermeability and also neutrophil and platelet activation.[168,170]

2. Chemotactic mediators. Mast cells can release chemotactic factors specific for neutrophils and eosinophils (NCF and ECF). These factors are particularly relevant during IgE-mediated allergic responses. Histamine can also modulate leucocyte migration.

3. Enzymes. Several enzymes released by mast cells such as chymase, tryptase, acid hydrolases, and arylsulfatase play a major role in tissue damage and repair processes.

4. Proteoglycans. Heparin proteoglycan is a potent anticoagulant,[169,171] inhibits activation of the alternative complement pathway,[169] and stimulates angiogenesis.[168,172]

5. Cytokines, such as IL-4.[114]

Figure 2-2 summarizes these relationships.

G. TYPE I HYPERSENSITIVITY

Type I hypersensitivity, or anaphylactic hypersensitivity, is important in allergic reactions. Nearly 10% of the western population suffers from some form of type I hypersensitivity reaction. The majority of these reactions are strongly localized anaphylactic react2ions to extrinsic allergens such as grass pollen, birch pollen, cat dander, house dust mite (*Dermatophagoides pteronysinnus*), ragweed, wasp and bee stings, penicillin, etc.

After first contact with the specific antigen, such as grass pollen, antibody-producing cells will produce specific IgE and/or IgG4, both of which can bind to mast cells and basophils. Subsequent contact with the allergen will trigger the whole cascade of some allergic reactions. Contact of the allergen with IgE-armed inflammatory cells, such as mast cells and to some extent basophils, will result in release of inflammatory mediators. In the bronchial tree, nasal mucosa, and conjunctiva this will induce rhinitis and/or asthma. Awareness of the importance of sensitization to food allergens in the gut has increased. Contact of the food with IgE-armed mast cells in the gastrointestinal track may induce local type I-like reactions, resulting in diarrhea and vomiting or may allow the allergen to enter the body by causing alterations in gut permeability after mediator release. Allergen/antibody complexes may cause lesions at other places, e.g., in the lung, skin, and eye. Thus, eating strawberries may induce anaphylactic reactions at the skin or in the airways.

Mast cells play a pivotal role in type I hypersensitivity reactions. Mast cells display high affinity receptors for the Fc part of IgE molecules. In addition to IgE, IgG4 also can bind to specific receptors for this immunoglobulin on mast cell surfaces. However, the majority of type I hypersensitivity is based upon the interaction between IgE and mast cells. In addition to mast cells, basophils also

FIGURE 2-2. (Opposite page) Development of Th1 and Th2 cell subsets in response to infectious diseases. Invading pathogens (bacteria, protozoa, viruses, or helminths) interact with cells of the innate immune system, thereby initiating cytokine production. The balance of cytokines present in the microenvironment during early stages of the immune response determines the direction of the immune response, with macrophages and NK cells a major source of IFN-γ, and mast cells and basophils a major source of IL-4. $\gamma\delta$ T-cells are capable of producing both IL-4 and IFN-γ. IFN-γ drives the immune response towards Th1. This response is characterized by production of IFN-γ, IL-2, and IL-12. These cytokines enhance cell-mediated immunity. IL-4 drives the immune response towards Th2. This response is characterized by production of IL-4, IL-5, IL-6, and IL-10. These cytokines enhance humoral immunity. At least four cross-regulatory cytokines exist: IFN-γ, IL-12, IL-4 , and IL-10. They enhance the differentiation of the T-helper cell subset from which they were derived (+) and antagonize the other subset (–). (Adapted from Haak-Frendscho, M. and Balwit, J. M., *Promega Notes*, 45, 21, 1994.) Several bacteria and protozoa (and likely some viruses) stimulate the production of IL-12 by macrophages or other cells, thereby driving the immune response towards Th1. Other pathogens, particularly helminths, stimulate the production of IL-4, thereby driving the immune response towards Th2. The characteristics of the pathogen that determine the predominant direction of the immune response are unknown.[144] *Abbreviations:* APC, antigen presenting cell; baso, basophil; MC, mast cell; Mϕ, macrophage; NK, natural killer cell.

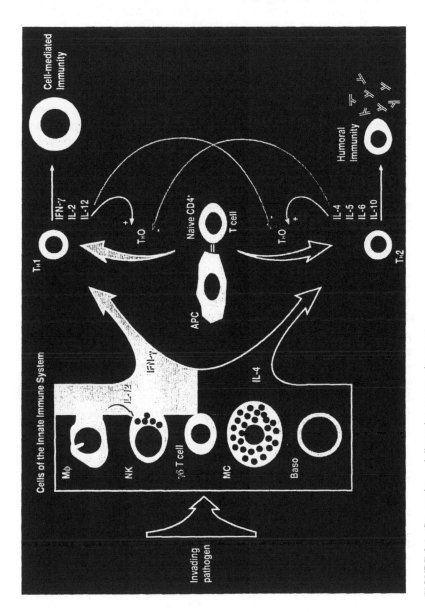

FIGURE 2-2. (See caption and discussion on opposite page.)

have IgE receptors. However, the affinity of the IgE receptors on mast cells is much higher.

The essential role of IgE in the pathogenesis of allergic diseases like atopic asthma has stimulated considerable study into the mechanisms by which IgE synthesis can be regulated. Using several animal models, it has been demonstrated that cooperation between T- and B-cells is essential for the induction of IgE synthesis by B-cells.[173] T-helper cells stimulate and T-suppressor cells inhibit IgE production by B-cells.[174,175]

Although it is clear that the balance between Th1- and Th2-type responses, and in particular the balance between IFN-γ and IL-4, will play a major role in the regulation of IgE antibody production, other mechanisms may be important also. For instance, soluble IgE binding factors have been described which either directly or indirectly influence the production of IgE.[176-178]

Cross-linking of IgE, for example, by divalent hapten will trigger mediator release. Under normal circumstances release of mast cell mediators helps to orchestrate the development of a defensive acute inflammatory reaction, for example, during parasitic infections. But under certain conditions such as in atopic diseases, bronchoconstrictive and vasodilatory effects predominate and may become distinctly threatening.

H. EXAMPLES OF TYPE I-LIKE
HYPERSENSITIVITY REACTIONS

1. Skin

Type I-like hypersensitivity reactions occur primarily at sites were the allergen is applied. The most common example of skin-associated type I hypersensitivity is urticaria. Urticaria is common, occurring in almost 10 to 15% of the population at some time. It is due mainly to increased vascular permeability resulting in an erythematous or edematous swelling reaction of the dermis. Several mast cell mediators play a pivotal role in this disease. In the skin, mast cells are available at high densities of approximately 5000 to 8000 mast cells per mm^3. The appearance of urticaria-like reactions can occur after the ingestion of food allergens, especially in children. Swelling of the lips and tongue can occur immediately after eating the allergen, and vomiting may occur somewhat later if it is ingested. Contact urticaria may also occur if the allergen is applied directly to the skin.

2. Pulmonary Tract

In the pulmonary tract, type I hypersensitivity is responsible, for example, for allergic/atopic asthma and allergic rhinitis. Asthma affects approximately 4 to 10% of the western population. In the U.K., this disease is responsible for about 2000 deaths per year. More than half of the number of asthma cases are induced by type I-like hypersensitivity responses.

3. Gastrointestinal Tract

In the gastrointestinal tract, type I-like reactions can occur, resulting in food allergy. Within one hour or even within some minutes after eating the food allergen, symptoms such as perioral erythema, oral irritation, swelling of the lips, tongue and pharynx, and nausea and vomiting may occur. In some cases even anaphylaxis may occur. Later, after eating the food allergen, allergic reactions at other sites than the intestinal tract can occur, such as eczema-like responses in the skin. The responsible subtype for these delayed (food) allergy responses is doubtful (type III or IV ??).

Diagnosis of type I hypersensitivity is confirmed using the skin prick test or Prausnitz-Kustner test (passive cutaneous anaphylaxis). The allergen in the prick test is applied into the skin leading eventually to a skin weal and flare response. Radio-allergosorbent tests of patient sera for specific IgE can confirm the diagnosis of type I hypersensitivity.

I. TYPE II HYPERSENSITIVITY

Type II hypersensitivity is induced mainly by autoantibodies and is therefore denoted as (auto)antibody-dependent cytotoxic hypersensitivity. Antibodies directed, for example, against membranes or cell surface antigens will bind to effector cells with Fc receptors. Activation of these "antigen"-antibody complexes will activate complement component C3, leading to cell lysis or membrane damage. Antibodies directed against cell surface antigens cause cell death not only by complement-dependent lysis but also by adherence reactions, leading to phagocytosis or through nonphagocytic extracellular killing by certain lymphoreticular cells (antibody-dependent cell-mediated cytotoxicity). Type II-like hypersensitivity reactions may be responsible, for example, for autoimmune hemolytic anemia, Goodpasture's syndrome, hemolytic disease of newborns (HDNB), and Hashimoto's thyroiditis. Mismatched organ transplantation or blood transfusion may also induce type II-like hypersensitivity responses, leading to the killing of cells via immunoglobulin/complement-dependent mechanisms.

J. EXAMPLES OF TYPE II HYPERSENSITIVITY REACTIONS
1. Skin

An example of a type II cytotoxic reaction in the skin is skin purpura. This disease is characterized by a hypersensitivity reaction against circulating platelets, leading to thrombocytopenia. Another type II skin-associated disease is pemphigus that can be life threatening. This autoimmune disease is characterized by fragile blisters (acantholysis) which break to leave denuded skin. This disease is due to IgG autoantibodies. The autoantibodies can be detected in serum and within the epidermis. Another example is bullous pemphigoid. In these patients autoantibodies against an antigen in the lamina lucida of the basement membrane at the dermo-epidermal junction are present. The antigen is possibly produced by keratinocytes.

2. Pulmonary Tract

The most common example of a type II-like hypersensitivity response in the lungs is Goodpasture's syndrome. Actually this autoimmune disease is predominantly a kidney associated disease. But due to the resemblance between the glomerular basement membranes and alveolar basement membranes, this subtype of hypersensitivity can occur in both the airways and the kidneys. In this disease, autoantibodies (IgG1 and IgG4) are produced against antigens expressed on various basement membranes. The production of these autoantibodies is probably triggered by viral infections, exposure to hydrocarbon fumes, and/or smoking.

3. Gastrointestinal Tract

The gastrointestinal tract may serve as a port of entry of antigens, resulting in type II hypersensitivity-like reactions. Penicillin ingested may bind, for example, to erythrocytes and induce the formation of IgG antibodies against erythrocyte-membrane coupled penicillin. This may lead to a form of hemolytic anemia. A similar mechanism may be responsible for the induction of other type II-like diseases.

K. TYPE III HYPERSENSITIVITY

Type III or immune complex-mediated hypersensitivity results from the deposition of immune complexes in blood vessel walls and tissues. Antigen-antibody binding can ultimately lead to the formation of insoluble complexes. These complexes may be directly responsible for local inflammatory reactions and platelet aggregation. If the antibody-antigen complex activates complement components, this will lead to the attraction of polymorphonuclear cells. These inflammatory cells can trigger a cascade, leading to an intense inflammatory reaction. On the other hand, activated complement components can stimulate mast cell degranulation and can activate IL-1 production by macrophages. The localized inflammatory reactions induced by type III-like reactions are known as Arthus reactions. This reaction, an edematous and inflammatory response, peaks at 3 to 8 hr after antigen provocation. The cells that are dominant in the inflammatory reaction are polymorphonuclear cells. Sometimes this is followed by infiltration of mononuclear cells, 24 to 48 hr after antigen injection. It is hypothesized that type IV hypersensitivity mechanisms are involved in some forms of type III hypersensitivity, such as Pigeon Fancier's disease (a form of extrinsic allergic alveolitis).

Diagnosis of type III hypersensitivity can be performed using the hemagglutination and precipitin tests (*in vitro*) or the Arthus reaction (skin) or bronchoprovocation (lungs) tests (*in vivo*).

L. EXAMPLES OF TYPE III HYPERSENSITIVITY REACTIONS
1. Skin

Most examples of skin-associated type III reactions result from immune complex deposition in blood vessel walls of the dermis. However in some skin-associated type III induced diseases, immune-complexes are formed extravascularly. Immune complexes will localize especially in vessels because of the action of vasoactive amines such as histamine and endothelial cell surface receptors. In some cases, clearing of these complexes by macrophages is inhibited. In the blood vessel walls, complexes can induce an intense inflammation, resulting in damage of the vessels and thus vasculitis. Also other organs such as the gut and kidney may be involved in type III hypersensitivity. In addition to allergic vasculitis, other examples of skin-associated type III reactions are erythema nodosum and erythema multiforme.

2. Pulmonary Tract

Intrapulmonary type III reactions are responsible for a great number of occupational allergic diseases. The Farmer's Lung is one important example. Inhalation of dust from moldy hay induces severe respiratory problems within 6 to 8 hr after inhalation. These patients are sensitized mainly to thermophilic actinomycetes. Another example is the Pigeon Fancier's Lung. It is demonstrated that the antigen is a serum protein present in bird feces. It is suggested that in type III hypersensitivity reactions type I-like responses are involved in the initiating phase of the whole process, leading to extrinsic allergic alveolitis. In allergic alveolitis such as the Farmer's Lung, Pigeon Fancier's Lung, and Mushroom Worker's Lung, specific antibodies against the inducing antigen are detectable (mostly IgG). It is suggested that T-cells play a pivotal role in allergic alveolitis because there are many granulomatous inflammatory reactions detectable and the CD4:CD8 ratio in the broncho-alveolar lavage is increased. The number of mast cells is also increased. The exact role of mast cells in this disease is not exactly known.

3. Gastrointestinal Tract

An example of type III hypersensitivity in the gut can be found in dermatitis herpetiformis patients. In fact, it is an example of type III occurring both in the skin and gut. The gut lesions as well as the skin lesions (respectively, blisters and villous atrophy) will decline if a gluten-free diet is used. This indicates that the antigen may be a gluten compound. In these patients, IgA immune complexes appear to be of particular importance.

M. TYPE IV, DELAYED-TYPE HYPERSENSITIVITY
1. Subtypes

Delayed-type hypersensitivity is a T-cell-dependent immune phenomenon characterized by an inflammatory reaction, at the site of antigen provocation (mostly the skin), that is maximal 1 to 2 days following antigen exposure. The

DTH reaction was observed initially by Koch in the dermis of tuberculous individuals toward tuberculin.[179] This is referred to as classic DTH. DTH reactions could in turn be subdivided into four subtypes.[180] Tuberculin-type hypersensitivity is denoted as the classic DTH type. This classic DTH reaction is induced by injection of protein antigens from, for example, the tubercle bacillus. The DTH response can be elicited months after priming (sensitization) and the reaction (erythema, induration) peaks 1 to 2 days after subsequent local antigen deposition. The local infiltrate consists of mononuclear cells. In mice there is also a large polymorphonuclear component.[181] A second subtype of DTH is contact hypersensitivity. This subtype resembles the classic tuberculin subtype reaction, but initially is predominantly an epidermal reaction, caused by sensitizing chemicals.[182,183] Jones-Mote hypersensitivity is characterized by a cutaneous infiltration of basophils.[184] This response differs in several aspects from classic DTH responses. In contrast to the classic DTH reaction (tuberculin subtype) and the contact hypersensitivity subtype reaction, serum can transfer the Jones-Mote reaction. It peaks somewhat earlier than the other subtypes, i.e., 24 hr after local antigen treatment. The last subtype of DTH is granulomatous hypersensitivity. In this form of DTH, localized inflammatory responses are composed predominantly of mononuclear cells (especially macrophages). The infiltrating cells arise after chronic stimulation with persistent foreign materials or living agents.[185] In contrast to the other subtypes the reaction time is at least a few weeks.

N. THE CELLULAR CASCADE OF DELAYED-TYPE HYPERSENSITIVITY

The basis of all subtypes of DTH reactions is the same. When mice are sub- or epicutaneously sensitized, a maximal immune response occurs 5 to 7 days after immunization. In contrast, if animals are immunized (sensitized) with certain antigens in complete Freund's adjuvant, there is a long-lasting (weeks or months) capacity of these animals to mount T-dth cell-mediated responses. DTH reactivity is long-lasting in man without Freund's complete adjuvant (FCA). A concomitant feature of DTH is a blastogenic response of cells in the T-cell-dependent area (paracortex) of lymph nodes draining the sensitization site. Much later also, B-cell-dependent areas have an activated appearance and antibody will be produced.[181] The sensitized T-dth cells recirculate through the whole body and can recognize antigen in the context of class II (MHC) on antigen-presenting cells if the animal is locally exposed to the same antigen for a second time (challenge). Antigen/class II recognition triggers the specific T-dth cells to produce lymphokines (lymphocyte-derived regulator proteins) such as IL-2. IL-2 induces the secretion of other lymphokines and stimulates proliferation of the activated lymphoid cells that express the IL-2 receptor (T-help cells, T-dth cells). Some of these lymphokines are capable of attracting other cell types to the vicinity of antigen deposition and activating them. In summary, the lymphokines, released during the DTH reaction, enhance vascular permeability, recruit inflammatory cells, activate macrophages (MAF, macrophage activating factor), have a direct anti-viral activity (IFN-γ), act cytotoxically (lymphotoxin, TNF-β), induce expression of MHC (IFNγ), and prevent migration of cells from the challenge site (macrophage migration inhibition factor).[183,186] The lymphokine-attracted cells include cells from the monocyte-macrophage lineage and to some extent, other mononuclear cells as well as neutrophils. The polymorphonuclear component in mice is more pronounced than in humans and guinea pigs.[181] With regard to restriction and the interaction of T-cells with antigen-presenting cells, there is a basic resemblance of most T-dth and T-help cells in mice. They are mainly Thy-1, Lyt-1, CD3+, CD4+, CD8+, and IL-2 receptor positive T-cells, restricted to the recognition of antigen in the context of class II (I-A) region gene products for both priming and activation of the immune cell.[187] As described previously, it is likely that Th1 cells are of greatest significance during the induction and elicitation phases of delayed-type sensitivity.

In murine contact sensitivity models (picryl chloride, oxazolone) it was found that DTH responses are actually due to the sequential activities of two different antigen-specific T-cells (Thy-1+, Lyt-1+, Lyt-2−).[188,189] Similar mechanisms have been suggested to occur in man also.[190] The division in these two different T-cell subtypes was not based upon the differences found between Th1 and Th2 cells. The two different T-cell subsets in DTH mediate separate early and late components of DTH that are accompanied by skin swelling responses at 2 and 24 to 48 hr after challenge.[165,188,189] The T-cell responsible for the early phase produces an antigen-binding, specific T-cell factor called picryl chloride factor (PCl-F) in PCl contact sensitivity, or oxazolone factor (Ox-F) in oxazolone contact sensitivity.[191-193] Within 1 day after sensitization, T-cells in the draining lymph nodes produce this antigen-specific factor.[192,193] The antigen-specific factor is detectable in the circulation and sensitizes mast cells in the extravascular tissue, such as the skin.[194,195] The T-cell factors producing T-cells are Thy-1-, and Lyt-1-positive, but are negative for CD4, CD8, and CD3. This means that DTH-initiating T-cells do not belong to the conventional T-helper cell subpopulations.[196] Interestingly, in humans also antigen-specific, CD3 negative T-cells have been described.[197] In contrast to the non-MHC restricted DTH-initiating "T-cell," the classic DTH cell is effected by the Th-1 cell.[99] Characteristic surface markers of the late-acting classic T-dth cell (in contact sensitivity) are Thy-1, Lyt-1, and CD3. In addition, the late-acting DTH cells bear IL-2 receptors.[196]

O. EXAMPLES OF TYPE IV HYPERSENSITIVITY REACTIONS
1. Skin
The major example of skin type IV hypersensitivity is contact hypersensitivity. Antigens that can induce type IV hypersensitivity are, for example, nickel, chromium, beryllium, rubber, and many more. Whether atopic eczema belongs partly to type IV is still not resolved. It is probably that this form of skin hypersensitivity is due to a combination of type I and IV mechanisms.

2. Pulmonary Tract

There are several indications that certain forms of chronic pulmonary diseases may be induced by type IV-like immune reactions. Examples are hard metal asthma, pulmonary disorders due to exposure to some small molecular weight compounds such TDI, sarcoidosis, and many more. Recently Garssen et al. demonstrated that T-cell-mediated immunity, using a mouse model for pulmonary delayed-type hypersensitivity reactions, can play a role in the induction of asthmatic disorders such as bronchial hyperresponsiveness, increment in pulmonary resistance, pulmonary inflammation, and increased pulmonary vascular permeability.[198,199]

3. Gastrointestinal Tract

Whether in the gut type IV-like responses play an important role is not known. In Crohn's disease, granulomatous inflammatory reactions are frequently observed indicating that type IV-like responses may play an important role. In addition to type I, type IV may play a partial role in food allergy. But this is still highly controversial.

III. CONCLUSIONS AND RECOMMENDATIONS

Allergic diseases are, together with cardiovascular and neoplastic diseases, responsible for many serious illnesses. Allergic diseases may affect more than 10% of the population and yet remain underdiagnosed. The prevalence of allergic diseases is increasing, and in some countries mortality due to these diseases has risen. Most studies are restricted to the induction of allergy by high molecular weight compounds, such as pollen and housedust mites, etc. Mechanisms responsible for the induction of allergy to small molecular weight compounds are less well established. Increasing knowledge of the mechanisms leading to allergic diseases induced by high and small molecular weight compounds will provide an insight into the pathogenesis of allergic diseases and improve opportunities for predictive testing, prevention, and therapy.

REFERENCES

1. **Streilein, J. W.**, The skin associated lymphoid tissue (SALT): The next generation, in *Skin Immune System (SIS)*, Bos, J. D., Ed., CRC Press, Boca Raton, FL, 1990, 25.
2. **Bos, J. D., Das, P. K., and Kapsenberg, M. L.**, The skin immune system (SIS), in *Skin Immune System (SIS)*, Bos, J. D., Ed., CRC Press, Boca Raton, FL, 1990, chap. 1.
3. **Bustmante, R., Schmitt, D., and Pillet, C.**, Immunoglobulin producing cells in the inflammatory infiltrates of cutaneous tumours, Immunologic identification *in situ, J. Invest. Dermatol.*, 68, 346, 1977.
4. **Vial, J., Bustmante, R., and Thivolet, J.**, Characterization of mononuclear cells in the inflammatory infiltrates of cutaneous tumours, *Br. J. Dermatol.*, 87, 515, 1986.

5. **Kohchiyama, A., Oka, D., and Ueki, H.,** Immunohistologic studies of squamous cell carcinomas. Possible participation of Leu-7+ (Natural Killer) cells as antitumour effector cells, *J. Invest. Dermatol.*, 87, 515, 1986.

6. **Vena, G. A., Angelini, G., D'Ovidio, R., Pastore, A., and Meneghini, C. L.,** Monocyte Fc-Ig receptors expression and soluble suppressor factor in skin squamous cell carcinoma, *Acta Dermatol.Ven.*, 63, 507, 1983.

7. **McArdle, J. P., Knight, B. A., Halliday, G. M., Muller, H. K., and Rowden, G.,** Quantative assessment of Langerhans cells in actinic keratoses, Bowen diseases, keratoacanthoma, squamous cell carcinoma and basal cell carcinoma, *Pathology*, 18, 212, 1986.

8. **Muller, H. K. and Halliday, G. M.,** The skin immune system and immunosurveillance, in *Skin Immune System (SIS)*, Bos, J. D., Ed., CRC Press, Boca Raton, FL, 1990, 447.

9. **Kaplan, M. H., Sadick, N., McNutt, N. S., Meltzer, M., Sarngadharan, M. G., and Pahwa, S.,** Dermatologic findings and manifestations of acquired immunodefiency syndrome (AIDS), *J. Am. Acad. Dermatol.*, 16, 485, 1987.

10. **Weismann, J., Petersen, C. S., Sondergaard, J., and Wantzin, G. L.,** *Skin Signs in AIDS*. Munksgaard, Copenhagen, 1988.

11. **Penneys, N. S. and Hicks, B.,** Unusual cutaneous lesions associated with acquired immunodefiency syndrome, *J. Am. Acad. Dermatol.*, 13, 845, 1985.

12. **Muhlemann, M. F., Anderson, M. G., Paradinas, F. J., Key, P. R., Dawson, S. G., Evans, B. A., Murray-Lyon, I. M., and Cream, J. J.,** Early warnings signs in AIDS and persistent generalized lymphadenopathy, *Br. J. Dermatol.*, 114, 419, 1986.

13. **Siegal, F. P., Lopez, C., Hammer, G. S., Brown, A. E., Kornfeld, S. J., Gold, J., Hassett, J., Hirschmann, S. Z., Cunningham-Rundles, C., Adelsberg, B. R., Parham, D. M., Siegal, M., Cunningham-Rundles, S., and Armstrong, D.,** Severe acquired immunodefiency in male homosexuals, manifested by chronic perianal ulcerative herpes simplex lesions, *N. Engl. J. Med.*, 305, 1439, 1981.

14. **Resnick, L. and Herbst, J. S.,** Dermatological (non-Kaposi sarcoma) manifestations associated with HLTV-III/LAV infection, in *AIDS: Modern Concepts and Therapeutic Challenges*, Broder, S., Ed., Marcel Dekker, New York, 1987, 285.

15. **Silverman, S., Migliorati, C. A., Lozada-Nur, F., Greenspan, D., and Conant, M. A.,** Oral findings in people with or at high risk for AIDS: a study of 375 homosexual males, *J. Am. Dent. Assoc.*, 112, 187, 1986.

16. **Klein, R. S., Haris, C. A., Small, C. B., Moll, B., Lesser, M., and Friedland, G. H.,** Oral candiadasis in high-risk patients as the initial manifestations of acquired immunodefiency syndrome, *N. Engl. J. Med.*, 9, 354, 1984.

17. **Borton, L. K. and Wintroub, B. U.,** Disseminated cryptococcosis presenting as herpetiform lesions in a homosexual man with acquired immunodefiency syndrome, *J. Am. Acad. Dermatol.*, 10, 387, 1984.

18. **Hazelhurst, J. A. and Vismer, H. F.,** Histoplasmosis presenting with unusual skin lesions in acquired immunodefiency syndrome (AIDS), *Br. J. Dermatol.*, 113, 345, 1985.

19. **Lipstein-Kresch, E., Isenberg, H. D., Singer, C., Cooke, O., and Greenwald, R. A.,** Disseminated Sporothrix schenkii infection with arthritis in a patient with acquired immunodeficiency syndrome, *J. Rheumatol.*, 12, 805, 1985.

20. **Belsito, D. V., Sanchez, M. R., Baer, R. L., Valentine, F., and Thorbecke, G. J.,** Reduced Langerhans cell Ia antigen and ATPase activity in patients with acquired immunodefiency syndrome, *N. Engl. J. Med.*, 310, 1279, 1985.

21. **Daniels, T. E., Greenspan, D., Greenspan, J. S., Lennette, E., Schiodt, M., Petersen, V., and Souza, Y.,** Absence of Langerhans cells in oral hairy leukoplakia, an AIDS-associated lesion, *J. Invest. Dermatol.*, 89, 178, 1987.

22. **Tada, T. and Ishizaka, K.,** Distribution of IgE-forming cells in lymphoid tissues of the human and monkey, *J. Immunol.*, 104, 377, 1970.

23. **Tomasi, T. B. and Grey, H. M.,** Structure and function of IgA, *Progr. Allergy*, 16, 81, 1972.

24. **Kaltreider, H. B., Turner, F. N., and Salmon, S. E.**, A canine model for comparative study of respiratory and systemic immunologic reactions, *Am. Rev. Resp. Dis.*, 111, 257, 1975.
25. **Bienenstock, J., Johnston, N., and Perey, D. Y. E.**, Bronchial lymphoid tissue. I. Morphologic characteristics, *Lab. Invest.*, 28, 686, 1973.
26. **Bienenstock, J., Johnston, N., and Perey, D. Y. E.**, Bronchial lymphoid tissue. II. Functional characteristics, *Lab. Invest.*, 28, 693, 1973.
27. **Sminia, T., Van der Brugge-Gamelkoorn, G. J., and Jeurissen, S. H. M.**, Structure and function of bronchus-associated lymphoid tissue, *Crit. Rev. Immunol.*, 9, 119, 1989.
28. **Kuper, C. F., Koornstra, P. J., Hameleers, D. M. H., Biewenga, J., Spit, B. J., and Duijvestijn, A. M., Breda Vriesman, P. J. C., and Sminia, T.**, The role of nasopharyngeal lymphoid tissue, *Immunol. Today*, 13, 219, 1992.
29. **Van der Brugge-Gamelkoorn, G. J., Van de Ende, M., and Sminia, T.**, Changes occurring in the epithelium covering rat bronchus associated lymphoid tissue after intratracheal challenge with horse radish peroxidase, *Cell. Tissue Res.*, 245, 439, 1986.
30. **Bland, P. W.**, MHC class II expression by the gut epithelium, *Immunol. Today*, 9, 174, 1988.
31. **Cerf-Benussan, N., Quaroni, A., Kurnick, J. T., and Bhan, A. K.**, Intraepithelial lymphocytes modulate the expression by intestinal epithelial cells, *J. Immunol.*, 132, 2244, 1984.
32. **Fournier, M. and Lebargy, F., le Roy Ladurie, F., Lenormand, E., and Pariente, R.**, Intraepithelial T lymphocyte subsets in the airways of normal subjects and of patients with chronic bronchitis, *Am. Rev. Resp. Dis.*, 140, 737, 1989.
33. **Goodman, T. and Lefrancois, L.**, Expression of γδ T cell receptor in intestinal CD8 intraepithelial lymphocytes, *Nature*, 33, 855, 1988.
34. **Pardoll, D. M., Kruisbeek, A. M., Fowlkes, B. J., Coligan, J. E., and Schwartz, R. H.**, The unfolding story of T cell receptor γ, *FASEB J.*, 1, 103, 1987.
35. **Abraham, E., Freitas, A. A., and Coutinho, A. A.**, Purification and characterization of intraparenchymal lung lymphocytes, *J. Immunol.*, 144, 2117, 1990.
36. **Augustin, A., Kubo, R. T., and Sim, G. K.**, Resistant pulmonary lymphocytes expressing the γδ T cell receptor, *Nature*, 340, 239, 1989.
37. **Cherniack, R. M.**, Bronchoalveolar lavage constituents in healthy, idiopathic pulmonary fibrosis, and selected comparison groups, *Am. Rev. Resp. Dis.*, 141, 169, 1990.
38. **Metzger, J., Zavala, D., Richerson, H. B., Moseley, P., Iwamota, P., Monick, M., Sjoerdsma, K., and Hunninghake, G. W.**, Local allergen challenge and bronchoalveolar lavage of allergic asthmatic lungs, *Am. Rev. Resp. Dis.*, 135, 433, 1987.
39. **Kagnoff, M. F.**, Immunology of the digestive system, in *Physiology of the Gastrointestinal Tract*, Johnson L. R., Ed., Raven Press, New York, 1987, 1699.
40. **Rudzik, O., Clancy, R. L., Perey, D. Y. E., Day, R. P., and Bienenstock, J.**, Repopulation with IgA-containing cells of bronchial and intestinal lamina propria after transfer of homologous Peyer's patch and bronchial lymphocytes, *J. Immunol.*, 114, 1599, 1975.
41. **Mcdermott, M. R. and Bienenstock, J.**, Evidence for a common mucosal immunologic system. I. Migration of B immunoblasts into intestinal, respiratory and genital tissues, *J. Immunol.*, 122, 1892, 1979.
42. **Kawanishi, H., Saltzman, L. E., and Strober, W.**, Mechanisms regulating IgA class-specific immunoglobulin production in murine gut-associated lymphoid tissues. I. T cell derived from Peyer's patches that switch sIgM B cells to sIgA B cells in vitro, *J. Exp. Med.*, 157, 433, 1983.
43. **Kawanishi, H., Saltzman, L. E., and Strober, W.**, Mechanisms regulating IgA class-specific immunoglobulin production in murine gut-associated lymphoid tissues. II. Terminal differentiation of postswitch sIgA bearing Peyer's patch B cells, *J. Exp. Med.*, 158, 649, 1983.

44. **Guy-Grand, D., Griscelli, C., and Vassalli, P.,** The mouse gut T lymphocyte, a novel type of T cell. Nature, origin, and traffic in mice in normal and graft-versus-host conditions, *J. Exp. Med.,* 158, 649, 1978.

45. **Dunkley, M. L. and Husband, A. J.,** Distribution and functional characteristics of antigen-specific helper T cells arising after Peyer's patch immunization, *Immunology,* 61, 475, 1987.

46. **Coombs, R. R. A. and Gell, P. G. H.,** The classification of allergic reactions underlying disease, in *Clinical Aspects of Immunology,* Gell, P. G. H. and Coombs, R. R. A., Eds., Davis, Philadelphia, 1963, 317.

47. **Bias, W. B.,** The genetic basis of asthma, in *Asthma Physiology, Immunopharmacology and Treatments,* Gell, P. G. H. and Coombs, R. R. A., Eds., Academic Press, New York, 1973, 317.

48. **Bray, G. W.,** The hereditary factor in asthma and other allergies, *Br. Med. J.,* 1, 384, 1930.

49. **Tips, R. L.,** A study of the inheritance of atopic hypersensitivity in man, *Am. J. Hum. Genet.,* 6, 328, 1954.

50. **Bias, W. B. and Marsh, D. G.,** The genetic basis of asthma: current studies of the genetics of IgE-mediated immune response in man, in *Asthma,* 2nd Ed., Academic Press, New York, 1977, chap. 2.

51. **Black, P. L. and Marsh, D. G.,** *Bronchial Asthma: Mechanisms and Therapeutics,* Little Brown, Boston, 1973, 53.

52. **Cifford, R. D., Pugsley, A., Radfor, M., and Holgate, S. T.,** Symptoms, atopy, and bronchial response to metacholine in parents with asthma and their children, *Arch. Dis. Child,* 62, 66, 1987.

53. **Ronchetti, R., Lucarini, N., Lucarelli, P., Martinez, F., Macri, F., Carapella, E., and Bottini, A.,** A genetic basis for heterogenenity of asthma syndrome in pediatric ages: adenosine deaminase phenotypes, *J. Allergy Clin. Immunol.,* 74, 81, 1984.

54. **Townley, R. G., Southard, J. G., Radford, P., Hopp, R. J., Bewtra, A. K., and Ford, L.,** Association of MS Pi Phenotype with airway hyperresponsiveness, *Chest,* 98, 594, 1990.

55. **Marsh, D. G., Meyers, D. A., and Bias, W. B.,** The epidemiology and genetics of atopic allergy, *N. Engl. J. Med.,* 305, 1551, 1981.

56. **Sibbald, B., Horn, M. E. C., Brain, E. A., and Gregg, I.,** Genetic factors in childhood asthma, *Thorax,* 35, 672, 1980.

57. **Cockcroft, D. W., Murdock, K. Y., and Berscheid, B. A.,** Relationship between atopy and bronchial responsiveness to histamine in a random population, *Ann. Allergy,* 53, 26, 1984.

58. **Turner, K. J., Dowse, G. K., Stewart, G. A., and Alpers, M. P.,** Studies on bronchial hyperreactivity, allergic responsiveness, and asthma in rural and urban children, of the highlands of Papua New Guinea, *J. Allergy Clin. Immunol.,* 77, 558, 1986.

59. **Menne, T. and Holm, N. V.,** Hand eczema in nickel-sensitive female twins. Genetic predisposition and environmental factors, *Contact Dermatitis,* 9, 289, 1983.

60. **Olerup, O. and Emtestam, L.,** Allergic contact dermatitis to nickel is associated with a Taq I HLA-DQA allelic restriction fragment, *Immunogenetics,* 28, 310, 1988.

61. **Kieffer, M.,** Nickel sensitivity: relationship between history and patch test reaction, *Contact Dermatitis,* 5, 398, 1979.

62. **Peltonen, L.,** Nickel sensitivity in the general population, *Contact Dermatitis,* 5, 27, 1979.

63. **Geczy, A. F. and Weck, A. L. D. E.,** Genetic control of sensitization to structurally unrelated antigens and its relationship to histocompatibility antigens in guinea pigs, *Immunology,* 28, 331, 1975.

64. **Emtestam, L., Marcusson, J. A., and Moller, E.,** HLA class II restriction specificity for nickel-reactive T lymphocytes, *Acta Derm. Venereol.,* 68, 395, 1988.

65. **Mosmann, T. R., Cherwinski, H., Bond, M. W., Giedlin, M. A., and Coffman, R. L.,** Two types of murine helper T cell clone. I. Definition according to profiles of lymphokine activities and secreted proteins, *J. Immunol.,* 136, 2348, 1986.

66. **Mosmann, T. R. and Coffman, R. L.,** Heterogeneity of cytokine secretion patterns and functions of helper T cells, *Adv. Immunol.,* 46, 111, 1989.

67. **Romagnani, S.,** Human T_H1 and T_H2 subsets: doubt no more, *Immunol. Today,* 12, 256, 1991.

68. **Swain, S. L., Weinberg, A. D., and English, M.,** CD4+ T cell subsets. Lymphokine secretion of memory cells and of effector cells that develop from precursors in vitro, *J. Immunol.,* 144, 1788, 1990.

69. **Weinberg, A. D., English, M., and Swain, S. L.,** Distinct regulation of lymphokine production is found in fresh versus in vitro primed murine helper T cells, *J. Immunol.,* 144, 1800, 1990.

70. **Street, N. E., Schumacher, J. H., Fong, T. A. T., Bass, H., Fiorentino, D. F., Leverah, J. A., and Mosmann, T. R.,** Heterogeneity of mouse helper T cells. Evidence from bulk cultures and limiting dilution cloning for precursors of Th1 and Th2 cells, *J. Immunol.,* 144, 1629, 1990.

71. **Mosmann, T. R. and Moore, K. W.,** The role of IL-10 in crossregulation of T_H1 and T_H2 responses, *Immunol. Today,* 12, A49, 1991.

72. **Swain, S. L., McKenzie, D. T., Weinberg, A. D., and Hancock, W.,** Characterization of T helper 1 and 2 cell subsets in normal mice. Helper T cells responsible for IL-4 and IL-5 production are present as precursors that require priming before they develop into lymphokine-secreting cells, *J. Immunol.,* 141, 3445, 1988.

73. **Salmon, M., Kitas, G. D., and Bacon, P. A.,** Production of lymphokine mRNA by CD45R+ and CD45R- helper T cells from human peripheral blood and by human CD4+ T cell clones, *J. Immunol.,* 143, 907, 1989.

74. **Cherwinski, H. M., Schumacher, J. H., Brown, K. D., and Mosmann, T. R.,** Two types of mouse helper T cell clone. III. Further differences in lymphokine synthesis between Th1 and Th2 clones revealed by RNA hybridization, functionally monospecific bioassays, and monoclonal antibodies, *J. Exp. Med.,* 166, 1229, 1987.

75. **Fiorentino, D. F., Bond, M. W., and Mosmann, T. R.,** Two types of mouse T helper cell. IV. Th2 clones secrete a factor that inhibits cytokine production by Th1 clones, *J. Exp. Med.,* 170, 2081, 1989.

76. **Brown, K. D., Zurawski, S. M., Mosmann, T. R., and Zurawski, G.,** A family of small inducible proteins secreted by leukocytes are members of a new superfamily that includes leukocyte and fibroblast-derived inflammatory agents, growth factors, and indicators of various activation processes, *J. Immunol.,* 142, 679, 1989.

77. **Swain, S. L.,** Regulation of the development of distinct subsets of CD4+ T cells, *Res. Immunol.,* 142, 14, 1991.

78. **Romagnani, S.,** Induction of Th1 and Th2 responses: a key role for the "natural" immune response, *Immunol. Today,* 13, 379, 1992.

79. **Janeway, C. A., Jr. and Golstein, P.,** Lymphocyte activation and effector functions, *Curr. Opin. Immunol.,* 4, 241, 1992.

80. **Weaver, C. T., Hawrylowicz, C. M., and Uranue, E. R.,** T helper cell subsets require the expression of distinct costimulatory signals by antigen-presenting cells, *Proc. Natl. Acad. Sci. U.S.A.,* 85, 8181, 1988.

81. **Gajewski, T. F., Pinnas, M., Wong, T., and Fitch, F. W.,** Murine Th1 and Th2 clones proliferate optimally in response to distinct antigen-presenting cell populations, *J. Immunol.,* 146, 1750, 1991.

82. **Chang, T.-H., Shea, C. M., Vrioste, S., Thompson, R. C., Boom, W. H., and Abbas, A. K.,** Heterogenicity of helper/inducer T lymphocytes. III. Responses of IL-2 and IL-4-producing (Th1 and Th2) clones to antigens presented by different accessory cells, *J. Immunol.,* 145, 2803, 1990.

83. **Tite, J. P. and Janeway, C. A., Jr.,** Cloned helper T cells can kill B lymphoma cells in the presence of specific antigen: Ia restriction and cognate *vs.* noncognate interactions in cytolysis, *Eur. J. Immunol.,* 14, 878, 1984.

84. **Gajewski, T. F., Schell, S. R., Nau, G., and Fitch, F. W.,** Regulation of T-cell activation: differences among T-cell subsets, *Immunol. Rev.,* 111, 79, 1989.

85. **Williams, M. E., Lichtman, A. H., and Abbas, A. K.,** Anti-CD3 antibody induces unresponsiveness to IL-2 in Th1 clones but not in Th2 clones, *J. Immunol.,* 144, 1208, 1990.

86. **Pfeiffer, C., Murray, I., Madri, J., and Bottomly, K.,** Selective activations of Th1- and Th2-like cells *in vivo* response to human collagen IV, *Immunol. Rev.,* 123, 65, 1991.

87. **Williams, I. R. and Unanue, E. R.,** Costimulatory requirements of murine Th1 clones. The role of accessory cell-derived signals in responses to anti-CD3 antibody, *J. Immunol.,* 145, 85, 1990.

88. **Schwartz, R. H.,** A cell culture model for T lymphocyte clonal energy, *Science,* 248, 1349, 1990.

89. **Gajewski, T. F. and Fitch, F. W.,** Anti-proliferative effect of IFN-γ in immune regulation. I. IFN-γ inhibits the proliferation of Th2 but not Th1 murine helper T lymphocyte clones, *J. Immunol.,* 140, 4245, 1988.

90. **Fernandez-Botran, R., Sanders, V. M., Mosmann, T. R., and Vitetta, E. S.,** Lymphokine-mediated regulation of the proliferative response of clones of T helper 1 and T helper 2 cells, *J. Exp. Med.,* 168, 543, 1988.

91. **Magilavy, D. B., Fitch, F. W., and Gajewski, T. F.,** Murine hepatic accessory cells support the proliferation of Th1 but not Th2 helper T lymphocyte clones, *J. Exp. Med.,* 170, 985, 1989.

92. **De Waal Malefyt, R., Haanen, J., Spits, H., Roncarolo, M.-G., Te Velde, A., Figdor, C., Johnson, K., Kastelein, R., Yssel, H., and De Vries, J. E.,** Interleukin 10 (IL-10) and viral IL-10 strongly reduce antigen-specific human T cell proliferation by diminishing the antigen-presenting capacity of monocytes via downregulation of class II major histocompatibility complex expression, *J. Exp. Med.,* 174, 915, 1991.

93. **Coffman, R. L., Seymour, B. W. P., Lebman, D. A., Hiraki, D. D., Christiansen, J. A., Shrader, B., Cherwinsky, H. M., Savelkoul, H. F. J., Finkelman, F. D., Bond, M. W., and Mosmann, T. R.,** The role of helper T cell products in mouse B cell differentiation and isotype regulation, *Immunol. Rev.,* 102, 5, 1988.

94. **Kim, J., Woods, A., Becker-Dunn, E., and Bottomly, K.,** Distinct functional phenotypes of cloned Ia-restricted helper T cells, *J. Exp. Med.,* 162, 188, 1985.

95. **Boom, W. H., Liano, D., and Abbas, A. K.,** Heterogeneity of helper/inducer T lymphocytes. II. Effects of interleukin 4- and interleukin 2-producing T cell clones on resting B lymphocytes, *J. Exp. Med.,* 167, 1350, 1988.

96. **Killar, L., MacDonald, G., West, J., Woods, A., and Bottomly, K.,** Cloned, Ia-restricted T cells that do not produce interleukin 4 (IL 4)/B cell stimulatory factor 1 (BSF-1) fail to help antigen-specific B cells, *J. Immunol.,* 138, 1674, 1987.

97. **Coffman, R. L. and Carty, J.,** A T cell activity that enhances polyclonal IgE production and its inhibition by interferon-γ, *J. Immunol.,* 136, 949, 1986.

98. **Mosmann, T. R. and Coffman, R. L.,** T_H1 and T_H2 cells: different patterns of lymphokine secretion lead to different functional properties, *Annu. Rev. Immunol.,* 7, 145, 1989.

99. **Cher, D. J. and Mosmann, T. R.,** Two types of murine helper T cell clone. II. Delayed-type hypersensitivity is mediated by T_H1 clones, *J. Immunol.,* 138, 3688, 1987.

100. **Meuer, S. C., Hussey, R. E., Cantrell, D. A., Hodgdon, J. C., Schlossman, S. F., Smith, K. A., and Reinherz, E. L.,** Triggering of the T3-Ti antigen-receptor complex results in clonal T-cell proliferation through an interleukin 2-dependent autocrine pathway, *Proc. Natl. Acad. Sci. U.S.A.,* 81, 1509, 1984.

101. **Daynes, R. A., Araneo, B. A., Dowell, T. A., Huang, K., and Dudley, D.,** Regulation of murine lymphokine production in vivo. III. The lymphoid tissue microenvironment exerts regulatory influences over T helper cell function, *J. Exp. Med.,* 171, 979, 1990.

102. **Daynes, R. A., Meikle, A. W., and Araneo, B. A.,** Locally active steroid hormones may facilitate compartmentalization of immunity by regulating the types of lymphokines produced by helper T cells, *Res. Immunol.*, 142, 40, 1991.
103. **Maggi, E., Del Prete, G., Macchia, D., Parronchi, P., Tiri, A., Chrétien, I., Ricci, M., and Romagnani, S.,** Profiles of lymphokine activities and helper function for IgE in human T cell clones, *Eur. J. Immunol.*, 18, 1045, 1988.
104. **Paliard, X., De Waal Malefijt, R., Yssel, H., Blanchard, D., Chrétien, I., Abrams, J., De Vries, J., and Spits, H.,** Simultaneous production of IL-2, IL-4, and IFN-γ by activated human CD4+ and CD8+ cell clones, *J. Immunol.*, 141, 849, 1988.
105. **Kapsenberg, M. L., Wierenga, E. A., Bos, J. D., and Jansen, H. M.,** Functional subsets of allergen-reactive human CD4+ T cells, *Immunol. Today*, 12, 392, 1991.
106. **Wierenga, E. A. and Snoek, M., De Groot, C., Chrétien, I., Bos, J. D., Jansen, H. M., and Kapsenberg, M. L.,** Evidence for compartmentalization of functional subsets of CD4+ T lymphocytes in atopic patients, *J. Immunol.*, 144, 4651, 1990.
107. **Van Der Heijden, F. L., Wierenga, E. A., Bos, J. D., and Kapsenberg, M. L.,** High frequency of IL-4-producing CD4+ allergen-specific T lymphocytes in atopic dermatitis patients, *J. Invest. Dermatol.*, 97, 389, 1991.
108. **Parronchi, P., Macchia, D., Piccinni, M.-P., Biswas, P., Simonelli, C., Maggi, E., Ricci, M., Ansari, A. A., and Romagnani, S.,** Allergen- and bacterial antigen-specific T-cell clones established from atopic donors show a different profile of cytokine production, *Proc. Natl. Acad. Sci. U.S.A.*, 88, 4538, 1991.
109. **Del Prete, G. F., De Carli, M., Mastromauro, C., Biagotti, R., Macchia, D., Falagiani, P., Ricci, M., and Romagnani, S.,** Purified protein derivative of *Mycobacterium tuberculosis* and excretory-secretory antigen(s) of *Toxocara canis* expand in vitro human T cells with stable and opposite (type 1 T helper or type 2 T helper) profile of cytokine production, *J. Clin. Invest.*, 88, 346, 1991.
110. **Kapsenberg, M. L., Wierenga, E. A., Stiekema, F. E. M., Tiggelman, A. M. B. C., and Bos, J. D.,** Th1 lymphokine production profiles of nickel-specific CD4+ T-lymphocyte clones from nickel contact allergic and non-allergic individuals, *J. Invest. Dermatol.*, 98, 59, 1992.
111. **Dearman, R. J. and Kimber, I.,** Differential stimulation of immune function by respiratory and contact chemical allergens, *Immunology*, 72, 562, 1991.
112. **Dearman, R. J. and Kimber, I.,** Divergent immune responses to respiratory and contact chemical allergens: antibody elicited by phthalic anhydride and oxazolone, *Clin. Exp. Allergy*, 22, 241, 1992.
113. **Dearman, R. J., Spence, L., and Kimber, I.,** Characterization of murine immune responses to allergenic diisocyanates, *Toxicol. Appl. Pharmacol.*, 112, 190, 1992.
114. **Dearman, R. J., Basketter, D. A., Coleman, J. W., and Kimber, I.,** The cellular and molecular basis for divergent allergic responses to chemicals, *Chem-Biol. Interact.*, 84, 1, 1992.
115. **Kimber, I. and Dearman, R. J.,** The mechanisms and evaluation of chemically induced allergy, *Toxicol. Lett.*, 64\65, 79, 1992.
116. **Coffman, R. L., Ohara, J., Bond, M. W., Carty, J., Zlotnik, A., and Paul, W. E.,** B cell stimulatory factor-1 enhances the IgE response of lipopolysaccharide-activated B cells, *J. Immunol.*, 136, 4538, 1986.
117. **Rothman, P., Lutzker, S., Cook, W., Coffman, R., and Alt, F. W.,** Mitogen plus interleukin 4 induction of Cε transcripts in B lymphoid cells, *J. Exp. Med.*, 168, 2385, 1988.
118. **Romagnani, S.,** Regulation and deregulation of human IgE synthesis, *Immunol. Today*, 11, 316, 1990.
119. **Lebman, D. A. and Coffman, R. L.,** Interleukin 4 causes isotype switching to IgE in T cell-stimulated clonal B cell cultures, *J. Exp. Med.*, 168, 853, 1988.

120. **Stavnezer, J., Radcliffe, G., Lin, Y.-C., Nietupski, J., Berggren, L., Sitia, R., and Severinson, E.,** Immunoglobulin heavy-chain switching may be directed by prior induction of transcripts from constant-region genes, *Proc. Natl. Acad. Sci. U.S.A.,* 85, 7704, 1988.

121. **Ishizaka, A., Sakiyama, Y., Nakanishi, M., Tomizawa, K., Oshika, E., Kojima, K., Taguchi, Y., Kandil, E., and Matsumoto, S.,** The inductive effect of interleukin-4 on IgG4 and IgE synthesis in human peripheral blood lymphocytes, *Clin. Exp. Immunol.,* 79, 392, 1990.

122. **Lundgren, M., Persson, U., Larsson, P., Magnusson, C., Smith, C. I. E., Hammarström, L., and Severinson, E.,** Interleukin 4 induces synthesis of IgE and IgG4 in human B cells, *Eur. J. Immunol.,* 19, 1311, 1989.

123. **Finkelman, F. D., Holmes, J., Katona, I. M., Urban, J. F., Jr., Beckmann, M. P., Park, L. S., Schooley, K. A., Coffman, R. L., Mosmann, T. R., and Paul, W. E.,** Lymphokine control of in vivo immunoglobulin isotype selection, *Annu. Rev. Immunol.,* 8, 303, 1990.

124. **Kühn, R., Rajewsky, K., and Müller, W.,** Generation and analysis of interleukin-4 deficient mice, *Science,* 254, 707, 1991.

125. **Tepper, R. I., Levinson, D. A., Stanger, B. Z., Campos-Torres, J., Abbas, A. K., and Leder, P.,** IL-4 induces allergic-like inflammatory disease and alters T cell development in transgenic mice, *Cell,* 62, 457, 1990.

126. **Burstein, H. J., Tepper, R. I., Leder, P., and Abbas, A. K.,** Humoral immune functions in IL-4 transgenic mice, *J. Immunol.,* 147, 2950, 1991.

127. **Gulbenkian, A. R., Egan, R. W., Fernandes, X., Jones, H., Kreutner, W., Kung, T., Payvandi, F., Sullivan, L., Zurcher, J. A., and Watnik, A. S.,** Interleukin-5 modulates eosinophil accumulation in allergic guinea pig lung, *Am. Rev. Respir. Dis.,* 146, 163, 1992.

128. **Coffman, R. L., Seymour, B. W. P., Hudak, S., Jackson, J., and Rennick, D.,** Antibody to interleukin-5 inhibits helminth-induced eosinophilia in mice, *Science,* 245, 308, 1989.

129. **Dent, L. A., Strath, M., Mellor, A. L., and Sanderson, C. J.,** Eosinophilia in transgenic mice expressing interleukin 5, *J. Exp. Med.,* 172, 1425, 1990.

130. **Mosmann, T. R., Bond, M. W., Coffman, R. L., Ohara, J., and Paul, W. E.,** T-cell and mast cell lines respond to B-cell simulatory factor 1, *Proc. Natl. Acad. Sci. U.S.A.,* 83, 5654, 1986.

131. **Thompson-Snipes, L., Dhar, V., Bond, M. W., Mosmann, T. R., Moore, K. W., and Rennick, D. M.,** Interleukin 10: a novel stimulatory factor for mast cells and their progenitors, *J. Exp. Med.,* 173, 507, 1991.

132. **Sanderson, C. J., O'Garra, A., Warren, D. J., and Klaus, G. G. B.,** Eosinophil differentiation factor also has B-cell growth factor activity: proposed name interleukin 4, *Proc. Natl. Acad. Sci. U.S.A.,* 83, 437, 1986.

133. **Coleman, J. W., Holliday, M. R., Kimber, I., Zsebo, K. M., and Galli, S. J.,** Regulation of mouse peritoneal mast cell secretory function by stress cell factor, IL-3 or IL-4, *J. Immunol.,* 150, 556, 1993.

134. **Plaut, M., Pierce, J. H., Watson, C. J., Hanley-Hyde, J., Nordan, R. P., and Paul, W. E.,** Mast cell lines produce lymphokines in response to cross-linkage of FcɛRI or to calcium ionophores, *Nature,* 339, 64, 1989.

135. **Bradding, P., Feather, I. H., Howarth, P. H., Mueller, R., Roberts, J. A., Britten, K., Bews, J. P. A., Hunt, T. C., Okayama, Y., Heusser, C. H., Bullock, G. R., Church, M. K., and Holgate, S. T.,** Interleukin 4 is localized to and released by human mast cells, *J. Exp. Med.,* 176, 1381, 1992.

136. **Bradding, P., Feather, I. H., Wilson, S., Bardin, P. G., Heusser, C. H., Holgate, S. T., and Howarth, P. H.,** Immunolocalization of cytokines in the nasal mucosa of normal and perennial rhinitic subjects. The mast cell as a source of IL-4, IL-5, and IL-6 in human allergic mucosal inflammation, *J. Immunol.,* 151, 3853, 1993.

137. **Brunner, T., Heusser, C. H., and Dahinden, C. A.,** Human peripheral blood basophils primed by interleukin-3 (IL-3) produce IL-4 in response to immunoglobulin E receptor stimulation, *J. Exp. Med.*, 177, 605, 1993.

138. **Seder, R. A., Paul, W. E., Dvorak, A. M., Sharkis, S. J., Kagey-Sobotka, A., Niv, Y., Finkelman, F. D., Barbieri, S. A., Galli, S. J., and Plaut, M.,** Mouse splenic and bone marrow cell populations that express high affinity Fc_ϵ receptors and produce interleukin 4 are highly enriched in basophils, *Proc. Natl. Acad. Sci. U.S.A.*, 88, 2835, 1991.

139. **Scott, P.,** Selective differentiation of CD4$^+$ T helper cell subsets, *Curr. Op. Immunol.*, 5, 391, 1993.

140. **D'Andrea, A., Rengaraju, M., Valiante, N. M., Chehimi, J., Kubin, M., Aste, M., Chan, S. H., Kobayashi, M., Young, D., Nickbarg, E., Chizzonite, R., Wolf, S. F., and Trinchieri, G.,** Production of natural killer cell stimulatory factor (interleukin 12) by peripheral blood mononuclear cells, *J. Exp. Med.*, 176, 1387, 1992.

141. **Manetti, R., Parronchi, P., Giudizi, M. G., Piccinni, M.-P., Maggi, E., Trinchieri, G., and Romagnani, S.,** Natural killer cell stimulatory factor (Interleukin 12 [IL-12]) induces T helper type 1 (Th1)-specific immune responses and inhibits the development of IL-4 producing Th cells, *J. Exp. Med.*, 177, 1199, 1993.

142. **Hsieh, C.-S., Macatonia, S. E., Tripp, C. S., Wolf, S. F., O'Garra, A., and Murphy, K. M.,** Development of T_H1 CD4$^+$ T cells through IL-12 produced by *Listeria*-induced macrophages, *Science*, 260, 547, 1993.

143. **D'Andrea, A., Aste-Amezaga, M., Valiante, N. M., Ma, X., Kubin, M., and Trinchieri, G.,** Interleukin 10 (IL-10) inhibits human lymphocyte interferon γ-production by suppressing natural killer cell stimulatory factor/IL-12 synthesis by accessory cells, *J. Exp. Med.*, 178, 1041, 1993.

144. **Scott, P.,** IL-12: Initiation cytokine for cell-mediated immunity, *Science*, 260, 496, 1993.

145. **Afonso, L. C. C., Scharton, T. M., Vieira, L. Q., Wysocka, M., Trinchieri, G., and Scott, P.,** The adjuvant effect of interleukin-12 in a vaccine against *Leishmania major*, *Science*, 263, 235, 1994.

146. **Zurawski, G. and De Vries, J. E.,** Interleukin 13, an interleukin 4-like cytokine that acts on monocytes and B cells, but not on T cells, *Immunol. Today*, 15, 19, 1994.

147. **Punnonen, J., Aversa, G., Cocks, B. G., McKenzie, A. N. J., Menon, S., Zurawski, G., de Waal Malefyt, R., and de Vries, J. E.,** Interleukin 13 induces interleukin 4-independent IgG4 and IgE synthesis and CD23 expression by human B cells, *Proc. Natl. Acad. Sci. U.S.A.*, 90, 3730, 1993.

148. **Cocks, B. G., de Waal Malefyt, R., Galizzi, J.-P., de Vries, J.E., and Aversa, G.,** IL-13 induces proliferation and differentiation of human B cells activated by the CD40 ligand, *Int. Immunol.*, 5, 657, 1993.

149. **Yssel, H., de Waal Malefyt, R., Roncarolo, M.-G., Abrams, J. S., Lahesmaa, R., Spits, H., and de Vries, J. E.,** IL-10 is produced by subsets of human CD4$^+$ T cell clones and peripheral blood T cells, *J. Immunol.*, 149, 2378, 1992.

150. **Metcalf, D. D., Donlon, M. A., and Kaliner, M.,** The mast cell, *CRC Crit. Rev. Immunol.*, 3, 23, 1981.

151. **Ehrlich, P.,** Beitrage zur Kenntniss der Anilinfarbungen und ihre Verwendung in der mikroskopischen Technik, *Arch. Mikrosk. Anat. Entwicklungsmech.*, 13, 263, 1977.

152. **Lee, T. D. G., Swieter, M., Bienenstock, J., and Befus, A. D.,** Heterogeneity in mast cell populations, *Clin. Immunol. Rev.*, 4, 143, 1985.

153. **Helm, B., Marsch, P., Vercelli, D., Padlan, E., Gould, H., and Geha, R.,** The mast cell binding site on human immunoglobulin E, *Nature*, 331, 180, 1988.

154. **Sterk, A. and Ishizaka, T.,** Binding properties of IgE receptors on normal mouse mast cells, *J. Immunol.*, 128, 838, 1982.

155. **Metzger, H., Rivnay, B., Henkart, M., Kanner, B., Kinet, J. P., and Perez-Montfort, R.,** Analysis of the structure and function of the receptor for immuglobulin E, *Mol. Immunol.*, 21, 1167, 1984.

156. **Froese, A.,** Receptors for IgE on mast cells and basophils. *Progr. Allergy,* 34, 142, 1984.
157. **Lee, T. D. G., Sterk, A., Ishizaka, T., Bienenstock, J., and Befus, A. D.,** Number and affinity of receptors for IgE on enriched populations of isolated rat intestinal mast cells, *Immunology,* 55, 363, 1985.
158. **Stanworth, D. R.,** The role of non-antigen receptors in mast cell signaling processes, *Mol. Immunol.,* 21, 1183, 1984.
159. **Ishizaka, T. and Ishizaka, K.,** Activation of mast cells for mediator release through IgE receptors, *Progr. Allergy,* 34, 188, 1984.
160. **Liu, M. C., Proud, D., Lichjtenstein, L. M., MacGlashan, D. W., Schleimer, R. P., Adkinson, N. F., Kagey-Sobotka, A., Schulman, E. S., and Plaut, M.,** Human lung macrophage-derived histamine releasing activity is due to an IgE-dependent factor(s), *J. Immunol.,* 136, 2588, 1986.
161. **MacDonald, S. M., Lichtenstein, L. M., Proud, D., Plaut, M., Naclerio, R. M., MacGlashan, D. W., and Kagey-Sobotka, A.,** Studies of IgE-dependent histamine releasing factors: heterogeneity og IgE, *J. Immunol.,* 139, 506, 1987.
162. **Sher, A. and McIntyre, S. L.,** Receptors for C3 on mast cells, *J. Immunol.,* 119, 722, 1977.
163. **Askenase, P. W., Bursztajn, S., Gershon, M. D., and Gershon, R. K.,** T cell dependent mast cell degranulation and release of serotonin in murine delayed type hypersensitivity, *J. Exp. Med.,* 152, 1358, 1980.
164. **Kops, S. K., Van Loveren, H., Rosenstein, R. W., Ptak, W., and Askenase, P. W.,** Mast cell activation and vascular alterations in immediate hypersensitivity-like reactions induced by a T cell-derived antigen binding factor, *Lab. Invest.,* 50, 421, 1984.
165. **Van Loveren, H., Meade, R., and Askenase, P. W.,** An early component of delayed type hypersensitivity mediated by T cells and mast cells, *J. Exp. Med,* 157, 1604, 1983.
166. **Kops, S. K., Ratzlaff, R. E., Meade, R., Iverson, G. M., and Askenase, P. W.,** Interaction of antigen-specific T-cell factors with unique "receptors" on the surface of mast cells: demonstration in vitro by an indirect rosetting technique, *J. Immunol.,* 136, 4515, 1986.
167. **Wasserman, S. I.,** The human lung mast cell, *Environ. Health Perspect.,* 55, 259, 1984.
168. **Schwartz, L. B. and Austen, K. F.,** Structure and function of the chemical mediators of mast cells, *Progr. Allergy,* 34, 2721, 1984.
169. **Lewis, R. A. and Austen, K. F.,** Mediation of local homeostasis and inflammation by leukotrienes and other mast cell-dependent compounds, *Nature,* 293, 103, 1981.
170. **Pinckard, R. N.,** Platelet activating factor, *Hosp. Pract.,* 18, 67, 1983.
171. **Scully, M. F., Ellis, V., and Kakkar, V. V.,** Localization of heparin in mast cells, *Lancet,* 2, 718, 1986.
172. **Bienenstock, J., Tomioka, M., Stead, R., Ernst, P., Jordana, M., Gauldie, J., Dolovich, J., and Denburg, J.,** Mast cell involvement in various inflammatory processess, *Am. Rev. Resp. Dis.,* 135, s5, 1987.
173. **Ito, Y., Ogita, T., and Suko, M.,** IgE levels in nude mice, *Int. Arch. Allergy Appl. Immunol.,* 58, 474, 1979.
174. **Tada, T., Taniguchi, M., and Okumura, K.,** Regulation of homocytotropic antibody formatin in the rat. II. Effects of X irradiation, *J. Immunol.,* 106, 1012, 1971.
175. **Taniguchi, M. and Tada, T.,** Regulation of homocytotropic antibody formation in the rat. IV. Effect of various immunosuppressive drugs, *J. Immunol.,* 107, 579, 1971.
176. **Ishizaka, K. and Sandberg, K.,** Formation of IgE binding factors by human T lymphocytes, *J. Immunol.,* 125, 1692, 1981.
177. **Leung, D. Y. M. and Geha, R. S.,** Regulation of the human IgE antibody response, *Int. Rev. Immunol.,* 2, 75, 1987.
178. **Bonnefoy, J.-Y., Pochon, S., Aubry, J.-P., Graber, P., Gauchat, J.-F., Jansen, K., and Flores-Romo, L.,** A new pair of surface molecules involved in human IgE regulation, *Immunol. Today,* 14, 1, 1993.
179. **Waksman, B.,** Delayed (cellular) hypersensitivity, in *Immunological Diseases,* Sainten, M., Ed., Little, Brown, Boston, 1971, 220.

180. **Greene, M. I., Schatten, A., and Bromberg, J. S.,** Delayed hypersensitivity, in *Fundamental Immunology*, Paul, W. E., Ed. Raven Press, New York, 1984, 685.

181. **Crowle, A. J.,** Delayed hypersensitivity in the mouse, *Adv. Immunol.*, 20, 197, 1975.

182. **Asherson, G. L. and Zembala, M.,** Contact sensitivity in the mouse. IV. The role of lymphocytes and macrophages in passive transfer and the mechanism of their interaction, *J. Exp. Med.*, 132, 1, 1971.

183. **Van Blomberg, B. M. E. and Scheper, R. J.,** in *Dermatoxicology*, 4th ed., Marzielli, F. N. and Maibach, H. I., Eds., Hemisphere Publishing, New York, 1993.

184. **Askenase, P. W.,** Cutaneous basophil hypersensitivity in contact-sensitized guinea pigs. I. Transfer with immune serum, *J. Exp. Med.*, 138, 1144, 1973.

185. **Epstein, W. L.,** Granulomatous hypersensitivity, *Progr. Allergy*, 11, 36, 1984.

186. **Geczy, C. L.,** The role of lymphokines in delayed type hypersensitivity reactions, *Springer Semin. Immunopath.*, 7, 321, 1984.

187. **Askenase, P. W.,** Effector and regulatory mechanisms in delayed-type hypersensitivity, in *Allergy — Principles and Practice*, 2nd ed., Middleton, E. Jr., Reed, C. E., Ellis, E. F., and Atkinson, N. F., Eds., C. V. Mosby, St. Louis, 1983, 147.

188. **Van Loveren, H. and Askenase, P. W.,** DTH is mediated by a sequence of two different T cell activities, *J. Immunol.*, 133, 2397, 1984.

189. **Van Loveren, H., Kato, K., Meade, R., Green, D. R., Horowitz, M., Ptak, W., and Askenase, P. W.,** Characterization of two different Lyt-1 T cell populations that mediate delayed-type hypersensitivity, *J. Immunol.*, 133, 2402, 1984.

190. **Trial, J.,** Adoptive transfer of early and late delayed-type hypersensitivity reactions mediated by human T cells, *Regional Immunol.*, 2, 14, 1989.

191. **Askenase, P. W., Rosenstein, R. W., and Ptak, W.,** T cells produce an antigen binding factor with in vivo activity analogous to IgE antibody, *J. Exp. Med.*, 157, 862, 1983.

192. **Askenase, P. W., Van Loveren, H., Rosenstein, R. W., and Ptak, W.,** Immunologic specificity of antigen-binding T cell derived factors that transfer mast cell dependent immediate hypersensitivity-like reactions, *Monogr. Allergy*, 18, 249, 1983.

193. **Ptak, W., Askenase, P. W., Rosenstein, R. W., and Gershon, R. K.,** Transfer of an antigen specific immediate hypersensitivity-like reaction with an antigen binding factor produced by T cells, *Proc. Natl. Acad. Sci. U.S.A.*, 79, 1969, 1973.

194. **Van Loveren, H., Krauter-Kops, S., and Askenase, P. W.,** Different mechanisms of release of vasoactive amines by mast cells occur in T cell dependent compared to IgE-dependent cutaneous hypersensitivity responses, *Eur. J. Immunol.*, 14, 40, 1984.

195. **Van Loveren, H., Ratzlaff, R. E., Kato, K., Meade, R., Ferguson, T., Iverson, M., Janeway, C. A., and Askenase, P. W.,** Immune serum from mice contact sensitized with picrylchloride contains an antigen specific T cell factor that transfers immediate cutaneous reactivity, *Eur. J. Immunol.*, 16, 1203, 1986.

196. **Herzog, W. R., Ferreri, N. R., Ptak, W., and Askenase, P. W.,** The DTH-initiating Thy 1 cell is double negative (CD4 , CD8) and CD3 and express IL-3 receptors, but not IL-2 receptors, *J. Immunol.*, 143, 3125, 1989.

197. **Ciccone, E., Viale, O., Pende, D., Malnati, M., Biassoni, R., Melioli, G., Moretta, A., Long, E. O., and Moretta, L.,** Specific lysis of allogeneic cells after activation of CD3 lymphocytes in mixed lymphocyte culture, *J. Exp. Med.*, 168, 2403, 1988.

198. **Garssen, J., Nijkamp, F. P., Wagenaar, Sj. Sc., Zwart, A., Askenase, P.W., and Van Loveren, H.,** Regulation of delayed type hypersensitivity like responses in the mouse lung, determined with histological procedures: serotonin, T cell suppressor inducer factor and high dose tolerance regulate the magnitude of T cell dependent inflammatory reactions, *Immunology*, 68, 51, 1989.

199. **Garssen, J., Nijkamp, F. P., Van De Vliet, H., and Van Loveren, H.,** T-cell mediated induction of airway hyperreactivity in mice, *Am. Rev. Resp. Disease*, 144, 931, 1991.

200. **Haak-Frendscho, M. and Balwit, J. M.,** Revolutions in the $T_{H}1/T_{H}2$ paradigm of T helper cell subsets: implications for the future, *Promega Notes*, 45, 21, 1994.
201. **Arai, K., Lee, F., Miyajima, A., Miyatake, S., Arai, N., and Yokota, T.,** Cytokines: coordinators of immune and inflammatory responses, *Annu. Rev. Biochem.*, 59, 783, 1990.
202. **Metcalf, D.,** Control of granulocytes and macrophages: molecular, cellular, and clinical aspects, *Science*, 254, 529, 1991.
203. **Matsushima, K. and Oppenheim, J. J.,** Interleukin 8 and MCAF: novel inflammatory cytokines inducible by IL 1 and TNF, *Cytokine*, 1, 2, 1989.
204. **Uyttenhove, C. and Simpson, R. J., Van Snick, J.,** Functional and structural characterization of P40, a mouse glycoprotein with T-cell growth factor activity, *Proc. Natl. Acad. Sci. U.S.A.*, 85, 6934, 1988.
205. **Van Snick, J., Goethals, A., Renauld, J.-C., Van Roost, E., Uyttenhove, C., Rubira, M. R., Moritz, R. L., and Simpson, R. J.,** Cloning and characterization of a cDNA for a new mouse T cell growth factor (P40), *J. Exp. Med.*, 169, 363, 1989.
206. **Hültner, L., Druez, C., Moeller, J., Uyttenhove, C., Schmitt, E., Rüde, E., Dörmer, P., and Van Snick, J.,** Mast cell growth-enhancing activity (MEA) is structurally related and functionally identical to the novel mouse T cell growth factor P40/TCGFIII (interleukin 9), *Eur. J. Immunol.*, 20, 1413, 1990.
207. **Suda, T., Murray, R., Fischer, M., Yokota, T., and Zlotnik, A.,** Tumor necrosis factor-α and P40 induce day 15 murine fetal thymocyte proliferation in combination with IL-2, *J. Immunol.*, 144, 1783, 1990.
208. **Renauld, J.-C., Goethals, A., and Houssiau, F., Van Roost, E., and Van Snick, J.,** Cloning and expression of a cDNA for the human homologue of mouse T cell and mast cell growth factor P40, *Cytokine*, 2, 9, 1990.
209. **Donahue, R. E., Yang, Y.-C., and Clark, S. C.,** Human P40 T-cell growth factor (interleukin-9) supports erythroid colony formation, *Blood*, 75, 2271, 1990.
210. **Williams, D. E., Morrissey, P. J., Mochizuki, D. Y., De Vries, P., Anderson, D., Cosman, D., Boswell, H. S., Cooper, S., Grabstein, K. H., and Broxmeyer, H. E.,** T-cell growth factor P40 promotes the proliferation of myeloid cell lines and enhances erythroid burst formation by normal murine bone marrow cells in vitro, *Blood*, 76, 906, 1990.
211. **Zlotnik, A. and Moore, K. W.,** Interleukin 10, *Cytokine* 3, 366, 1991.
212. **De Waal Malefyt, R., Abrams, J., Bennett, B., Figdor, C. G., and De Vries, J. E.,** Interleukin 10 (IL-10) inhibits cytokine synthesis by human monocytes: an autoregulatory role of IL-10 produced by monocytes, *J. Exp. Med.*, 174, 1209, 1991.
213. **Paul, S. R., Bennett, F., Calvetti, J. A., Kelleher, K., Wood, C. R., O'Hara, R. M., Jr., Leary, A. C., Sibley, B., Clark, S. C., Williams, D. A., and Yang, Y.-C.,** Molecular cloning of a cDNA encoding interleukin 11, a stromal cell-derived lymphopoietic and hematopoietic cytokine, *Proc. Natl. Acad. Sci. U.S.A.*, 87, 7512, 1990.
214. **Musashi, M., Yang, Y.-C., Paul, S. R., Clark, S. C., Sudo, T., and Ogawa, M.,** Direct and synergistic effects of interleukin 11 on murine hemopoiesis in culture, *Proc. Natl. Acad. Sci. U.S.A.*, 88, 765, 1991.
215. **Yin, T., Schendel, P., and Yang, Y.-C.,** Enhancement of in vitro and in vivo antigen-specific antibody responses by interleukin 11, *J. Exp. Med.*, 175, 211, 1992.
216. **Stern, A. S., Podlaski, F. J., Hulmes, J. D., Pan, Y.-C. E., Quinn, P. M., Wolitzky, A. G., Familletti, P. C., Stremlo, D. L., Truitt, T., Chizzonite, R., and Gately, M. K.,** Purification to homogeneity and partial characterization of cytotoxic lymphocyte maturation factor from human B-lymphoblastoid cells, *Proc. Natl. Acad. Sci. U.S.A.*, 87, 6808, 1990.
217. **Gately, M. K., Desai, B. B., Wolitzky, A. G., Quinn, P. M., Dwyer, C. M., Podlaski, F. J., Familletti, P. C., Sinigaglia, F., Chizonnite, R., Gubler, U., and Stern, A. S.,** Regulation of human lymphocyte proliferation by a heterodimeric cytokine, IL-12 (cytotoxic lymphocyte maturation factor), *J. Immunol.*, 147, 874, 1991.

218. **Gubler, U., Chua, A. O., Schoenhaut, D. S., Dwyer, C. M., McComas, W., Motyka, R., Nabavi, N., Wolitzky, A. G., Quinn, P. M., Familletti, P. C., and Gately, M. K.,** Coexpression of two distinct genes is required to generate secreted bioactive cytotoxic lymphocyte maturation factor, *Proc. Natl. Acad. Sci. U.S.A.*, 88, 4143, 1991.

219. **Kobayashi, M., Fitz, L., Ryan, M., Hewick, R. M., Clark, S. C., Chan, S., Loudon, R., Sherman, F., Perussia, B., and Trinchieri, G.,** Identification and purification of natural killer cell stimulatory factor (NKSF), a cytokine with multiple biologic effects on human lymphocytes, *J. Exp. Med.*, 170, 827, 1989.

220. **Wolf, S. F., Temple, P. A., Kobayashi, M., Young, D., Dicig, M., Lowe, L., Dzialo, R., Fitz, L., Ferenz, C., Hewick, R. M., Kelleher, K., Herrmann, S. H., Clark, S. C., Azzoni, L., Chan, S. H., Trinchieri, G., and Perussia, B.,** Cloning of cDNA for natural killer cell stimulatory factor, a heterodimeric cytokine with multiple biologic effects on T and natural killer cells, *J. Immunol.*, 146, 3074, 1991.

221. **Chan, S. H., Perussia, B., Gupta, J. W., Kobayashi, M., Pospisil, M., Young, H. A., Wolf, S. F., Young, D., Clark, S. C., and Trinchieri, G.,** Induction of interferon γ production by natural killer cell stimulatory factor: characterization of the responder cells and synergy with other inducers, *J. Exp. Med.*, 173, 869, 1991.

222. **Brown, K. D., Zurawski, S. M., Mosmann, T. R., and Zurawski, G.,** A family of small inducible proteins secreted by leukocytes are members of a new superfamily that includes leukocyte and fibroblast-derived inflammatory agents, growth factors, and indicators of various activation processes, *J. Immunol.*, 142, 679, 1989.

223. **McKenzie, A. N. J. and Culpepper, J. A., de Waal Malefyt, R., Brière, F., Punnonen, J., Aversa, G., Sato, A., Dang, W., Cocks, B. G., Menon, S., de Vries, J. E., Banchereau, J., and Zurawski, G.,** Interleukin 13, a T-cell-derived cytokine that regulates human monocyte and B-cell function, *Proc. Natl. Acad. Sci. U.S.A.*, 90, 3735, 1993.

224. **Minty, A., Chalon, P., Derocq, J.-M., Dumont, X., Guillemot, J.-C., Kaghad, M., Labit, C., Leplatois, P., Liauzun, P., Miloux, B., Minty, C., Casellas, P., Loison, G., Lupker, J., Shire, D., Ferrara, P., and Caput, D.,** Interleukin-13 is a new human lymphokine regulating inflammatory and immune responses, *Nature*, 362, 248, 1993.

Chapter 3

MECHANISMS OF CHEMICAL RESPIRATORY ALLERGY

Ian Kimber

CONTENTS

I. INTRODUCTION

Allergy may be defined operationally as the adverse health effects which result from the stimulation by antigenic materials of specific immune responses in susceptible individuals. This chapter focuses on allergic reactions provoked in the respiratory tract by chemical allergens.

Allergy is an important cause of occupational respiratory illness. A scheme for the Surveillance of Work Related and Occupational Respiratory Disease (SWORD) was established in the U.K. at the beginning of 1989.[1] During the period of the survey the most frequently diagnosed cause of work-related respiratory illness was found to be asthma, which accounted for 26% of all new notifications. In these cases, the agents suspected of causing occupational asthma ranged from organic dusts and animal proteins to formulated products, industrial chemicals, and metallic salts.[1] Acute allergic respiratory disease resulting in rhinitis and/or asthma is associated classically with type 1 (immediate-type) hypersensitivity reactions. However, as will be discussed later, other immunological mechanisms may also be of importance. It is upon asthma and related symptoms induced by exposure to chemicals that this article will concentrate.

It should be noted that other forms of respiratory allergic disease are recognized and are of importance in the context of occupational health. The SWORD project group found that 6% of new cases of occupational respiratory illness were diagnosed as allergic alveolitis and were usually associated with farming.[1] Extrinsic allergic alveolitis or hypersensitivity pneumonitis is well known and usually classified in terms of either the relevant occupation or causative allergenic material. For instance, Farmer's Lung is caused by *Micropolyspora faeni* and Cheese Worker's Lung by antigens from

Penicillium roqueforti. Nearly all occupational cases are due to the inhalation of organic dusts of animal, vegetable, or bacterial origin. The immunological responses which characterize such diseases are representative of type III hypersensitivity reactions and result from inflammation caused by immune complexes formed locally between antibody and the inducing antigen. In most cases, respiratory symptoms are displayed by the sensitized individual within hours of inhalation exposure. The main effector molecule is IgG antibody and in most instances, sensitization is associated with the presence of allergen-specific antibodies of this class in the plasma.

One conclusion drawn from the SWORD project was that the true incidence of occupational respiratory illness may be some three times greater than recognized previously. The overall rate for work-related asthma was some 22 per million, but in certain occupational groups, higher incidences were found; 159 per million among welders, solderers, and electronic assemblers, 409 per million among workers involved in the manufacture and processing of plastics, and 639 per million for automobile and spray painters.[1]

The increased identification of occupational respiratory allergic disease and asthma is attributable partly to a changing industrial environment and the introduction of new chemicals and new manufacturing processes.[2-5] A variety of chemicals has been found to cause occupational respiratory allergy.[5,6] Among these are acid anhydrides such as tetrachlorophthalic anhydride, trimellitic anhydride, phthalic anhydride, and hexahydrophthalic anhydride,[7-10] certain isocyanates including toluene diisocyanate, diphenylmethane diisocyanate, and hexamethylene diisocyanate,[11-15] some reactive dyes,[16-18] and platinum salts.[19,20]

The characteristics of respiratory allergic hypersensitivity to chemicals are of specific pulmonary reactions which occur usually in only a proportion, and sometimes only a small proportion, of the exposed population. Respiratory reactions can be provoked in sensitized individuals by atmospheric concentrations of the chemical allergen which were tolerated previously and which fail to cause similar symptoms in those who are nonsensitized.[21] Almost invariably there is a latent period between the onset of exposure and the development of respiratory symptoms.[22,23]

Allergy, by definition, results from the stimulation of a specific immune response by the causative agent. It is frequently assumed that sensitization of the respiratory tract to chemicals occurs largely, or exclusively, by inhalation exposure. This is not necessarily the case, however. There is no reason to suppose that induction of the quality of immune response required for effective respiratory sensitization cannot occur via other routes of exposure. There is evidence that occupational respiratory sensitization may be induced by dermal exposure to chemical allergen resulting from spillage or splashing.[24] Moreover, it has been found that respiratory allergic reactions can be provoked in guinea pigs sensitized previously by either topical or intradermal exposure to the same chemical allergen.[25-27] The foregoing is not intended to imply, however, that inhalation is not an important route of exposure for chemical respiratory sensitization.

There is no doubt that the acute onset of respiratory symptoms associated with allergic hypersensitivity to protein aeroallergens such as pollen, housedust mite, and animal dander is due to homocytotropic, primarily IgE, antibody. In the case of chemical respiratory allergens, however, a universal association between pulmonary hypersensitivity and the presence of specific IgE antibody has failed to emerge.[28] IgE antibody specific for all recognized chemical respiratory allergens has been found, but sometimes only in a proportion of those who display the clinical symptoms of hypersensitivity. One explanation is, of course, that immunological effector mechanisms, other than those mediated by IgE, may elicit respiratory hypersensitivity reactions. The role of cell-mediated immune processes in respiratory chemical allergy, which may act independently of, or in concert with, IgE antibody, will be addressed later. An alternative explanation is that in some cases the failure to establish an association between respiratory hypersensitivity and the presence of allergen-specific IgE antibody results from the use of inappropriate or insufficiently sensitive serological detection methods. In this context, it is of interest that the use of skin prick tests may, as an adjunct or alternative to the radioallergosorbent test (RAST), provide information of value in the immunologic investigation of chemical respiratory allergy. It has been reported recently that evidence for the presence of specific IgE antibody was found by skin prick testing in a group of individuals who displayed respiratory hypersensitivity to acid anhydrides, but whose sera tested negative in RAST assays.[29]

Irrespective of whether there exists an invariable association between respiratory allergy and IgE, there is no doubt that antibodies of this class are important mediators of sensitization.

II. REGULATION OF IgE RESPONSES: Th CELLS AND CYTOKINES

IgE responses are subject to fine immunoregulatory control. Of decisive importance is the local availability of cytokines and in particular of interleukin 4 (IL-4). It has been appreciated for some time that the initiation and maintenance of IgE antibody responses in mice is dependent upon IL-4, its importance probably being to cause a switch to, and increased expression of, C_ϵ germline transcripts.[30-32] In support of this, it has more recently been found that mice which lack the gene for IL-4 also lack IgE and fail to mount an IgE response following nematode infection, an immunogenic stimulus associated with a strong antibody response of this class in eugenic animals.[33] Conversely, mice which carry an IL-4 transgene, and which constitutively express high levels of this cytokine, exhibit increased concentrations of serum IgE and more vigorous IgE antibody responses than do wild-type animals.[34,35] In fact, differential IL-4 gene expression may, to a large extent, account for the well-known strain differences among mice in terms of IgE responses.[36] The counter-regulatory cytokine which antagonizes the production of IgE is interferon γ (IFN-γ).[37]

The regulation by cytokines of IgE production is not restricted to the mouse. In humans also, IL-4 and IFN-γ exert reciprocal antagonistic effects on IgE antibody synthesis.[38-41]

It is appropriate to consider the immunoregulatory roles of these cytokines in the context of cellular immune responses and T-helper (Th) cells. In 1986, Mosmann et al.[42] described a functional heterogeneity among clones of murine Th cells. Initially, two populations were identified, designated Th1 and Th2, which differ with respect to the spectra of cytokines they produce following immune activation.[43,44] Both populations secrete interleukin 3 (IL-3) and granulocyte/macrophage colony-stimulating factor (GM-CSF). However, only Th1 cells secrete IFN-γ, interleukin 2 (IL-2), and tumor necrosis factor β (TNF-β; lymphotoxin) and only Th2 cells secrete interleukins 4, 5, 6, and 10.[43,44] The relevance of this functional divergence is that the cytokines which are known to exert the greatest influence on the induction of IgE responses are produced by different T lymphocyte subpopulations.

Th1 and Th2 cells undoubtedly diverge and develop from common precursors. It is believed currently that the immediate progenitor is a Th cell (Th0), which is able to produce both Th1- and Th2-type cytokines. Other, and possibly less differentiated, precursors have been identified also, including Thp cells which secrete only IL-2[44,45] (Figure 3-1). The physiological significance of this functional heterogeneity is clearly to enable tailored responses to be generated which more efficiently meet the needs of the immune system in countering the inducing antigen. One theory is that differential Th cell development matures and evolves during the course of an immune response, with the appearance of selective Th1- or Th2-type responses being dependent upon the nature, strength, and duration of the immunogenic signal. Although the requirements for the expression of selective Th cell responses is not fully understood, it is apparent that the relative availability in the local microenvironment of cytokines themselves is of considerable importance. The generation of Th2 cells is favored by IL-4, while IFN-γ promotes the development of Th1 responses.[46-49] It is of interest that the cytokine products of Th1 and Th2 lymphocytes influence the development of the cells which eventually produce them. The question raised is of the cellular source, early following immune stimulation, of IL-4, IFN-γ, and other cytokines necessary for the maturation of differentiated Th cell function. One suggestion is that components of the nonadaptive immune system, natural killer (NK) cells and mast cells are responsible, respectively, for the production of IFN-γ and IL-4 which will drive Th cell differentiation.[50] In addition, the nature and concentration of the inducing antigen and the way in which it is handled, processed, and presented is likely to influence markedly the selectivity of Th cell development. It has been found in experimental systems that the allelic form of MHC class II (Ia) antigens displayed by presenting cells, the membrane density of the Ia/antigen ligand complex, the

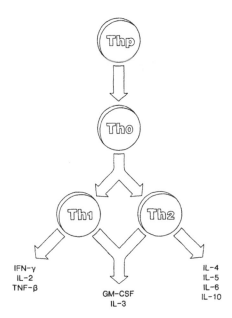

FIGURE 3-1. The development and functional diversity of T-helper (Th) cells. Th1 and Th2 cells derive from common precursors (Thp and Th0 cells). Both Th1 and Th2 cells produce interleukin 3 (IL-3) and granulocyte/macrophage colony-stimulating factor (GM-CSF). However, only Th1 cells secrete interferon γ (IFN-γ), interleukin 2 (IL-2), and tumor necrosis factor β (TNF-β; lymphotoxin) and only Th2 cells secrete interleukins 4, 5, 6, and 10 (IL-4, IL-5, IL-6, and IL-10). Th0 cells produce both Th1- and Th2-type cytokines, and Thp cells produce only IL-2.

availability of costimulatory molecules such as the cytokine interleukin 1 (IL-1), and the characteristics and functional activity of the antigen presenting cell itself, all affect the nature of Th cell responses.[51-56] However first induced, it is likely that a drive toward the selective development of either Th1- or Th2-type responses will be maintained and amplified as a result of a changing cytokine environment and the reciprocal influence of IL-4 and IFN-γ.

The relevance of divergent Th cell responses for allergic disease is clear and is emphasized further by the existence of similar subpopulations in man.[57,58] The impact of Th cell responses on allergy goes beyond the regulation by cytokines of IgE antibody production. It is known that delayed-type hypersensitivity responses are effected by Th1 cells,[59] and that IFN-γ, a product of these cells, plays an important role in this process.[60,61] Consistent with this are observations that immune responses to nickel, a potent human contact allergen, are characterized by selective Th1 cell activation. Allergen-specific T-lymphocyte clones isolated from the peripheral blood of nickel-sensitive donors were shown to secrete high levels of IFN-γ, but low or undetectable amounts of IL-4 and IL-5.[62,63]

In contrast, the selective activation of Th2 cells will favor the expression of immediate-type allergic reactions. In addition to promoting IgE responses, the soluble products of Th2 cells influence the development, distribution, and function of cells involved in the elicitation of acute allergic reactions. Interleukins 3, 4, and 10 are all mast cell growth factors or cofactors.[64,65] Interleukin 5 is an eosinophil growth and differentiation factor[66] and controls the accumulation of eosinophils at the site of allergen-induced respiratory reactions.[67-70] Moreover, it has been found recently that IL-3 and IL-4 will enhance the secretory potential of mast cells.[71,72] Accordingly, evidence has accumulated for selective Th2-type responses in human immediate hypersensitivity reactions. Clones of T lymphocytes specific for aeroallergens such as housedust mite (*Dermatophagoides pteronyssinus*) and grass pollen, which cause IgE-mediated allergic reactions in susceptible individuals, have been shown to produce Th2 cytokines, but not IFN-γ.[73] Also, Th clones derived from cells isolated from lesional biopsies of housedust mite-allergic atopic dermatitis patients were found to be of Th2 phenotype according to their pattern of cytokine secretion.[74] A predominance of Th2-type cells at the sites of skin reactions in atopic individuals has been demonstrated also by *in situ* hybridization. The cells found within lesional skin were shown to express mRNA for IL-3, IL-4, IL-5, and GM-CSF, but not for IFN-γ.[75]

Here again, during the elicitation of allergic reactions, IFN-γ and IL-4 have the potential to exert reciprocal antagonistic effects. Interferon γ inhibits the release by mast cells of serotonin,[72,76] and IL-4 has been found to inhibit contact hypersensitivity reactions induced in mice by picryl chloride.[77]

Chemicals are able to cause various types of allergic disease. Many are contact allergens, while others are associated primarily with respiratory sensitization. It has been shown that chemical contact allergens which are known or suspected not to induce respiratory hypersensitivity provoke in mice immune responses consistent with the selective activation of Th1 cells. Conversely, chemicals known to cause occupational respiratory hypersensitivity induce responses characteristic of Th2 cell activation.[78-85] Precedents exist for antigenic structure and the chemical modification of antigenic structure, influencing selectivity for different Th cell-type responses,[86-90] and it appears likely that the characteristics of the antigenic stimulus and the form in which it is recognized by the immune system will have profound effects on the quality of response induced and the type or types of allergy which may result.

It must be emphasized, of course, that in addition to the direct action of IL-4 and IFN-γ, other regulatory mechanisms serve also to influence the efficiency of IgE antibody production. Prostaglandin E_2, possibly acting via selective promotion of Th2 cell cytokine production, is known to enhance IgE synthesis.[91-94] Regulatory CD8[+] T lymphocytes may also influence IgE production secondary to alterations in the production of IL-4 and IFN-γ.[95] In addition, soluble IgE binding factors have been described which either directly or indirectly affect the production of human IgE.[96-100] Finally, it is of interest, but as yet of unknown significance, that interleukin 8 (IL-8) has been found to inhibit selectively the synthesis of IgE by IL-4 primed human B lymphocytes through an IFN-γ-independent mechanism.[101]

It is apparent, therefore, that the initiation and maintenance of IgE responses is subject to homeostasis mediated by a variety of immunoregulatory control mechanisms. For the efficient and sustained production of IgE antibody, the correct immunological microenvironment is required in which the promotional influence of IL-4 outweighs the inhibitory effects of IFN-γ. It is the balance achieved between these two cytokines which is the critical determinant of the quality of immune response generated. Of interest is the fact that some types of immunogenic stimulus and antigenic structure appear to favor the activation of Th2-type cells and induction of the quality of immune responses required for IgE production. What is not yet clear is the mechanistic basis for such selectivity and whether it is the characteristics of the antigenic structures themselves or the way in which antigen is handled and presented to the immune system which provides the initial signal for a drive toward selective Th responses.

Other factors are important in determining the vigor of induced IgE responses. Certainly genetic background is relevant. There is an increasing awareness also that external conditions, other than the presence of antigen, may play a role. Some studies have shown that cigarette smoking is associated with increased plasma levels of IgE antibody.[102,103] Further, evidence available from experimental studies has indicated that environmental pollutants such as diesel exhaust particulates and sulfur dioxide are able to enhance the production of IgE and the development of allergic sensitization in response to ovalbumin.[104-106] There is growing concern that certain environmental conditions may aggravate asthma[107] and may thereby contribute to the increasing diagnosis of this disease among urban populations. It is not clear to what extent changes in the incidence of asthma reflect the impact of environmental conditions on the immune system and on IgE production. The extrinsic factors and immune mechanisms which influence the production of IgE antibody are summarized in Figure 3-2.

III. CELL-MEDIATED IMMUNITY

There is increasing discussion regarding the possibility that other immunological effector mechanisms may act together with, or independently of, IgE in the elicitation of respiratory hypersensitivity and airway hyperreactivity. To invoke an IgE antibody-independent mechanism is particularly attractive in cases of respiratory hypersensitivity to chemical allergens where some symptomatic individuals apparently lack circulating specific IgE antibody. Recent investigations by Van Loveren and colleagues[108-110] of the role of T lymphocytes in allergic hyperresponsiveness of the respiratory tract have exploited a mouse model for measuring cellular immune function in pulmonary tissue. Mice immunized topically with picryl chloride were challenged subsequently by intranasal exposure to picryl sulphonic acid, a water soluble and antigenically cross-reactive form of the antigen. Challenge was associated with an accumulation in peribronchial and perivascular regions of mononuclear cells,

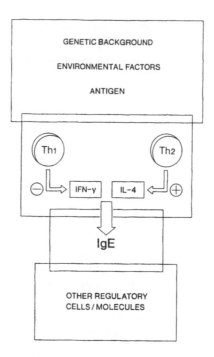

FIGURE 3-2. Factors which influence the induction of IgE antibody responses. The production of IgE is controlled by the reciprocal effects of the cytokines interferon γ (IFN-γ) and interleukin 4 (IL-4), products of activated Th1 and Th2 cells, respectively. The selective stimulation of Th cell responses will be governed by the nature of the inducing antigen and the way in which it is handled and processed and presented to the immune system. Further, IgE responses will be influenced by environmental factors, inherent genetic susceptibility, and by other immunoregulatory cells and molecules.

a response which was maximal after 48 hr and therefore reminiscent of a delayed-type hypersensitivity reaction. The inflammatory response was found to be dependent upon the release of serotonin and a cascade of cellular events leading to the infiltration of T lymphocytes and macrophages.[108] Challenge was associated also with airway hyperreactivity, a response which preceded and persisted longer than the mononuclear cell infiltrate.[109,110] A conclusion that may be drawn is that airway responses to allergenic chemicals may be induced by mechanisms other than those which are reliant upon the action of IgE antibody. Some care is necessary, however, as picryl chloride is a potent contact allergen and there is no reason to suppose that cell-mediated immune responses, characteristic of contact hypersensitivity, cannot be induced in the respiratory tract and result in local inflammation and associated changes in airway function.

Evidence of a role for cell-mediated immune mechanisms in the pathogenesis of respiratory hypersensitivity and allergic asthma does not, however, derive solely from experimental systems such as that described above.

The symptoms associated with respiratory hypersensitivity to chemical allergens are various and may include both immediate-onset and late-phase reactions.[111] Immediate-onset responses, including early asthmatic reactions, are regarded generally as being attributable to IgE antibody-dependent degranulation of local mast cells and the release of inflammatory mediators such as vasoactive amines and leukotriene B4 (LTB4) which result in vasodilation, smooth muscle contraction, and bronchoconstriction.[112] Late asthmatic responses to chemical allergens appear 2 to 8 hr following exposure and are associated with an infiltration of mononuclear cells and an increased number of leukocytes in bronchoalveolar lavage fluid. Unlike immediate-onset reactions, the symptoms associated with late phase asthma do not respond to bronchodilators but can be inhibited with anti-inflammatory steroids.[111] Chronic inflammation plays an important role in asthma and, in addition to the accumulation in the bronchial mucosa of leukocytes, is associated with the production of mucus, the destruction and sloughing of airway epithelial cells, and subepithelial fibrosis secondary to collagen deposition.[113,114] Central to the development of chronic bronchial inflammation and injury are eosinophils acting together with infiltrating T lymphocytes.[115-118] While the exact role eosinophils play in the development of bronchial hyperreactivity has yet to be defined fully, it is apparent that the eosinophilia which characterizes allergen-induced respiratory reactions is effected by cytokines and in particular by IL-5, a product of Th2 cells.[67-70] Again, a reciprocal interaction between cytokines derived from different Th cell populations is important. It has been found recently that IFN-γ downregulates antigen-induced eosinophilia in the respiratory tract secondary to an inhibition of the infiltration of T-helper cells.[119]

Thus, however the disease is first induced, there is an important role for T lymphocytes, cell-mediated immunity, and eosinophils in the pathogenesis of asthma. The argument can be developed that, irrespective of the induction of specific IgE antibody, it is CD4+ cells of Th2-type which are of primary importance. The stimulation of Th2-type responses will have a major influence on all phases of the allergic reaction. IgE antibody promoted by IL-4 will effect immediate-onset reactions in the sensitized individual and may be essential, or at least important, for triggering allergen-induced late-phase reactions and the chronic bronchial inflammation associated with asthma. The release of IL-4 and IL-5 by activated Th2 cells infiltrating the respiratory tract will augment the secretory potential of local mast cells and cause the accumulation of eosinophils. Consistent with an important role for Th2-type cells in immediate-type allergic responses is the observation that cells infiltrating lesional skin at the site of allergen-induced late-phase cutaneous reactions in atopic patients produce Th2 cytokines (IL-3, IL-4, IL-5, and GM-CSF), but not IFN-γ.[75]

An important role for T cells in the pathogenesis of allergic respiratory hypersensitivity begs questions about the local availability of accessory cells and antigen-presenting cells necessary for T lymphocyte activation. No doubt the recently described network of dendritic cells resident within the airway epithelium will be of importance.[120-122] In theory, macrophages also could play such a role. However, the available evidence suggests that pulmonary alveolar macrophages function poorly as antigen-presenting cells and may serve instead to suppress T lymphocyte responses.[123] In fact, elimination of alveolar macrophages has now been shown to enhance pulmonary immune function.[124-126] The induction and maintenance of effective T lymphocyte responses in the respiratory tract may be regulated by the equilibrium between dendritic cells and alveolar macrophages and between alveolar macrophages and infiltrating monocytes.[127-130]

IV. CONCLUDING COMMENTS

IgE antibody is an important mediator of allergic respiratory hypersensitivity. For the production of IgE antibody, the appropriate quality of immune response must be induced and this in turn is dependent largely upon the balance between Th1 cell- and Th2 cell-type responses and the relative availability of the cytokines IL-4 and IFN-γ. Cellular immune function and the action of T lymphocytes may play an important role in late-phase reactions to chemical allergens and, together with eosinophilia, is necessary for the chronic bronchial inflammation associated with asthma. The induction of IgE antibody responses, the activation of Th2 cells, and the severity of allergic reactions are all subject to various regulatory mechanisms, which together serve to control the development and expression of respiratory hypersensitivity to chemicals.

REFERENCES

1. **Meredith, S. K., Taylor, V. M., and McDonald, J. C.**, Occupational respiratory disease in the United Kingdom 1989: a report to the British Thoracic Society and the Society of Occupational Medicine by the SWORD project group, *Br. J. Ind. Med.*, 48, 292, 1991.
2. **Salvaggio, J. E.**, The impact of allergy and immunology on our expanding industrial environment, *J. Allergy Clin. Immunol.*, 85, 689, 1990.
3. **Bernstein, D. I.**, Occupational asthma, *Clin. Allergy*, 76, 917, 1992.
4. **Bernstein, J. A.**, Occupational asthma, *Postgrad. Med. J.*, 92, 109, 1992.
5. **Salvaggio, J. E., Butcher, B. T., and O'Neil, C. E.**, Occupational asthma due to chemical agents, *J. Allergy Clin. Immunol.*, 78, 1053, 1986.
6. **Karol, M. H.**, Occupational asthma and allergic reactions to inhaled compounds, in *Principles and Practice of Immunotoxicology*, Miller, K., Turk, J., and Nicklin, S., Eds., Blackwell Scientific Publications, Oxford, 1992, 228.
7. **Maccia, C. A., Bernstein, I. L., Emmett, E. A., and Brooks, S. M.**, In vitro demonstration of specific IgE in phthalic anhydride hypersensitivity, *Am. Rev. Respir. Dis.*, 113, 701, 1976.
8. **Bernstein, D. I., Patterson, R., and Zeiss, C. R.**, Clinical and immunologic evaluation of trimellitic anhydride- and phthalic anhydride-exposed workers using a questionnaire and comparative analysis of enzyme linked immunosorbent and radioimmunoassay studies, *J. Allergy Clin. Immunol.*, 69, 311, 1982.

9. **Howe, W., Venables, K. M., Topping, M. D., Dally, M. B., Hawkins, R., Law, J. S., and Taylor, A. T.,** Tetrachlorophthalic anhydride asthma: evidence for specific IgE antibody, *J. Allergy Clin. Immunol.*, 71, 5, 1983.

10. **Moller, D. R., Gallagher, J. S., Bernstein, D. I., Wilcox, T. G., Burroughs, H. E., and Bernstein, I. L.,** Detection of IgE-mediated respiratory sensitization in workers exposed to hexahydrophthalic anhydride, *J. Allergy Clin. Immunol.*, 76, 663, 1985.

11. **Tansar, A. R., Bourke, M. P., and Blandford, A. G.,** Isocyanate asthma: respiratory symptoms caused by diphenylmethane diisocyanate, *Thorax*, 28, 596, 1973.

12. **Zeiss, C. R., Kanellakes, T. M., Bellone, J. D., Levitz, D., Pruzansky, J. J., and Patterson, R.,** Immunoglobulin E-mediated asthma and hypersensitivity pneumonitis with precipitating anti-hapten antibodies due to diphenylmethane diisocyanate (MDI) exposure, *J. Allergy Clin. Immunol.*, 65, 346, 1980.

13. **Zammit-Tabona, M., Sherkin, M., Kijek, K., Chan, H., and Chan-Yeung, M.,** Asthma caused by diphenylmethane diisocyanate in foundry workers. Clinical, bronchial provocation and immunologic studies, *Am. Rev. Respir. Dis.*, 128, 226, 1983.

14. **Danks, J. M., Cromwell, O., and Buckingham, J. A., and Newman Taylor, A. J., and Davies, R. J.,** Toluene diisocyanate induced asthma: evaluation of antibodies in the serum of affected workers against a tolyl monoisocyanate protein conjugate, *Clin. Allergy*, 11, 161, 1981.

15. **Keskinen, H., Tupasela, O., Tiikkainen, U., and Nordman, H.,** Experience of specific IgE in asthma due to diisocyanates, *Clin. Allergy*, 18, 597, 1988.

16. **Alanko, K., Keskinen, H., Bjorksten, F., and Ojanen, S.,** Immediate-type hypersensitivity to reactive dyes, *Clin. Allergy*, 8, 25, 1978.

17. **Docker, A., Wattie, J. M., Topping, M. D., Luczynska, C. M., Newman Taylor, A. J., Pickering, C. A. C., Thomas, P., and Gompertz, D.,** Clinical and immunological investigations of respiratory disease in workers using reactive dyes, *Br. J. Ind. Med.*, 44, 534, 1987.

18. **Topping, M. D., Forster, H. W., Ide, C. W., Kennedy, F. M., Leach, A. M., and Sorkin, S.,** Respiratory allergy and specific immunoglobulin E and immunoglobulin G antibodies to reactive dyes in the wool industry, *J. Occup. Med.*, 31, 857, 1989.

19. **Biagini, R. E., Bernstein, I. L., Gallagher, J. S., Moorman, W. J., Brooks, S., and Gann, P. H.,** The diversity of reaginic immune responses to platinum and palladium salts, *J. Allergy Clin. Immunol.*, 76, 794, 1985.

20. **Murdoch, R. D., Pepys, J., and Hughes, E. G.,** IgE antibody responses to platinum group metals: a large scale refinery survey, *Br. J. Ind. Med.*, 43, 37, 1986.

21. **Newman Taylor, A. J.,** Occupational asthma, *Postgrad. Med. J.*, 64, 505, 1988.

22. **Burge-Sherwood, P.,** Occupational asthma, rhinitis and alveolitis due to colophony, *Clinics Immunol. Allergy*, 4, 55, 1984.

23. **Hagmar, L., Nielson, J., and Skerfving, S.,** Clinical features and epidemiology of occupational obstructive respiratory disease caused by small molecular weight organic chemicals, *Monogr. Allergy*, 21, 42, 1987.

24. **Karol, M. H.,** Respiratory effects of inhaled isocyanates, *CRC Crit. Rev. Toxicol.*, 16, 349, 1986.

25. **Karol, M. H., Hauth, B. A., Riley, E. J., and Magreni, C. M.,** Dermal contact with toluene diisocyanate (TDI) produces respiratory tract hypersensitivity in guinea pigs, *Toxicol. Appl. Pharmacol.*, 58, 221, 1981.

26. **Botham, P. A., Rattray, N. J., Woodcock, D. R., Walsh, S. T., and Hext, P. M.,** The induction of respiratory allergy in guinea-pigs following intradermal injection of trimellitic anhydride: a comparison with the response to 2,4-dinitrochlorobenzene, *Toxicol. Lett.*, 47, 25, 1989.

27. **Pauluhn, J. and Eben, A.,** Validation of a non-invasive technique to assess immediate or delayed onset of airway hypersensitivity in guinea pigs, *J. Appl. Toxicol.*, 11, 423, 1991.

28. **Newman Taylor, A. J.,** Clinical and epidemiological methods in investigating occupational asthma, *Clinics Immunol. Allergy*, 4, 3, 1984.

29. **Drexler, H., Schaller, K-H., Weber, A., Letzel, S., and Lehnert, G.,** Skin prick tests with solutions of acid anhydrides in acetone, *Int. Arch. Allergy Immunol.*, 100, 251, 1993.

30. **Finkelman, F. D., Katona, I. M., Urban, J. F., Jr., Snapper, C. M., Ohara, J., and Paul, W. E.,** Suppression of in vivo polyclonal IgE production by monoclonal antibody to the lymphokine B-cell stimulatory factor 1, *Proc. Natl. Acad. Sci. U.S.A.*, 83, 9675, 1986.

31. **Azuma, M., Hirano, T., Miyajima, H., Watanabe, N., Yagita, H., Enomoto, S., Furusawa, S., Ovary, Z., Kinashi, T., Honja, T., and Okumura, K.,** Regulation of IgE production in SJA/9 and nude mice. Potentiation of IgE production by recombinant interleukin 4, *J. Immunol.*, 139, 2538, 1987.

32. **Finkelman, F. D., Katona, I. M., Urban, J. F., Jr., Holmes, J., Ohara, J., Tung, A. S., Sample, J. G., and Paul, W. E.,** IL-4 is required to generate and sustain in vivo IgE responses, *J. Immunol.*, 141, 2335, 1988.

33. **Kuhn, R., Rajewsky, K., and Muller, W.,** Generation and analysis of interleukin-4 deficient mice, *Science*, 254, 707, 1991.

34. **Tepper, R. I., Levinson, D. A., Stanger, B. Z., Campos-Torres, J., Abbas, A. K., and Leder, P.,** IL-4 induces allergic-like inflammatory disease and alters T cell development in transgenic mice, *Cell*, 62, 457, 1990.

35. **Burstein, H. J., Tepper, R. I., Leder, P., and Abbas, A. K.,** Humoral immune functions in IL-4 transgenic mice, *J. Immunol.*, 147, 2950, 1991.

36. **Mori, A., Yamamoto, K., Suko, M., Watanabe, N., Ito, M., Miyamoto, T., and Okudaira, H.,** Interleukin-4 gene expression in high and low IgE responder mice, *Int. Arch. Allergy Appl. Immunol.*, 92, 100, 1990.

37. **Finkelman, F. D., Katona, I. M., Mosmann, T. R., and Coffman, R. L.,** IFN-γ regulates the isotypes of Ig secreted during in vivo humoral immune responses, *J. Immunol.*, 140, 1022, 1988.

38. **Del Prete, G., Maggi, E., Parronchi, P., Chretien, I., Tiri, D., Macchia, D., Ricci, M., Ansari, A. A., and Romagnani, S.,** IL-4 is an essential factor for the IgE synthesis induced in vitro by human T cell clones and their supernatants, *J. Immunol.*, 140, 1493, 1988.

39. **Pene, J., Rousset, F., Briere, F., Chretien, I., Paliard, X., Banchereau, J., Spits, H., and De Vries, J. E.,** IgE production by normal human B cells induced by alloreactive T cell clones is mediated by IL-4 and suppressed by IFN-γ, *J. Immunol.*, 141, 1218, 1988.

40. **Gauchat, J. F., Lebman, D. A., Coffman, R. L., Gascan, H., and De Vries, J. E.,** Structure and expression of germline ε transcripts in human B cells induced by interleukin 4 to switch to IgE production, *J. Exp. Med.*, 172, 463, 1990.

41. **Romagnani, S.,** Regulation and deregulation of human IgE synthesis, *Immunol. Today*, 11, 316, 1990.

42. **Mosmann, T. R., Cherwinski, H., Bond, M. W., Giedlin, M. A., and Coffman, R. L.,** Two types of murine T cell clone. I. Definition according to profiles of lymphokine activities and secreted proteins, *J. Immunol.*, 136, 2348, 1986.

43. **Mosmann, T. R. and Coffman, R. L.,** Heterogeneity of cytokine secretion patterns and functions of helper T cells, *Adv. Immunol.*, 46, 111, 1989.

44. **Mosmann, T. R., Schumacher, J. H., Street, N. F., Budd, R., O'Garra, A., Fong, T. A. T., Bond, M. W., Moore, K. W. M., Sher, A., and Fiorentino, D. F.,** Diversity of cytokine synthesis and function of mouse CD4+ T cells, *Immunol. Rev.*, 123, 209, 1991.

45. **Bendelac, A. and Schwartz, R. H.,** Th0 cells in the thymus: The question of T-helper lineages, *Immunol. Rev.*, 123, 169, 1991.

46. **Swain, S. L., Bradley, L. M., Croft, M., Tonkonogy, S., Atkins, G., Weinberg, A. D., Duncan, D. D., Hedrick, S. M., Dutten, R. W., and Huston, G.,** Helper T-cell subsets: phenotype, function and the role of lymphokines in regulating their development, *Immunol. Rev.*, 123, 115, 1991.

47. **Coffman, R. L., Varkila, K., Scott, P., and Chatelain, R.,** Role of cytokines in the differentiation of CD4+ T-cell subsets in vivo, *Immunol. Rev.*, 123, 189, 1991.

48. **Abehsira-Amar, O., Gibert, M., Joliy, M., Theze, J., and Jankovic, D. L.,** IL-4 plays a dominant role in the differential development of Th0 into Th1 and Th2 cells, *J. Immunol.,* 148, 3820, 1992.
49. **Hsieh, C-S., Heimberger, A. B., Gold, J. S., O'Garra, A., and Murphy, K. M.,** Differential regulation of T helper phenotype development by interleukins 4 and 10 in an αβ T-cell-receptor transgenic system, *Proc. Natl. Acad. Sci. U.S.A.,* 89, 6065, 1992.
50. **Romagnani, S.,** Induction of TH1 and TH2 responses: a key role for the "natural" immune response?, *Immunol. Today,* 13, 379, 1992.
51. **Weaver, C. T., Hawrylowicz, C. M., and Unanue, E. R.,** T helper cell subsets require the expression of distinct costimulatory signals by antigen-presenting cells, *Proc. Natl. Acad. Sci. U.S.A.,* 85, 8181, 1988.
52. **Chang, T. -L., Shea, C. M., Urioste, S., Thompson, R. C., Boom, W. H., and Abbas, A. K.,** Heterogeneity of helper/inducer T lymphocytes. III. Responses of IL-2 and IL-4 producing (Th1 and Th2) clones to antigens presented by different accessory cells, *J. Immunol.,* 145, 2803, 1990.
53. **Gajewski, T. F., Pinnas, M., Wong, T., and Fitch, F. W.,** Murine Th1 and Th2 clones proliferate optimally in response to distinct antigen-presenting cell populations, *J. Immunol.,* 146, 1750, 1991.
54. **Pfeiffer, C., Murray, J., Madri, J., and Bottomly, K.,** Selective activation of Th1- and Th2-like cells in vivo — response to human collagen IV, *Immunol., Rev.,* 123, 65, 1991.
55. **Marcelletti, J. F. and Katz, D. H.,** Antigen concentration determines helper T cell subset participation in antibody responses, *Cell. Immunol.,* 143, 405, 1992.
56. **Rocken, M., Muller, K. M., Saurat, J-H., Muller, I., Louis, J. A., Cerottini, J.-C., and Hauser, C.,** Central role for TCR/CD3 ligation in the differentiation of CD4⁺ T cells toward Th1 or Th2 functional phenotype, *J. Immunol.,* 148, 47, 1992.
57. **Romagnani, S.,** Human TH1 and TH2 subsets: doubt no more, *Immunol. Today,* 12, 256, 1991.
58. **Romagnani, S.,** Human TH1 and TH2 subsets: regulation of differentiation and role in protection and immunopathology, *Int. Arch. Allergy Immunol.,* 98, 279, 1992.
59. **Cher, D. J. and Mosmann, T. R.,** Two types of murine helper T cell clone. II. Delayed type hypersensitivity is mediated by Th1 clones, *J. Immunol.,* 138, 3688, 1987.
60. **Fong, T. A. T. and Mosmann, T. R.,** The role of IFN-γ in delayed-type hypersensitivity mediated by Th1 clones, *J. Immunol.,* 143, 2887, 1989.
61. **Diamanstein, T., Eckert, R., Volk, H.-D., and Kupier-Weglinski, J.-W.,** Reversal by interferon-γ of inhibition of delayed-type hypersensitivity induction by anti-CD4 or anti-interleukin 2 receptor (CD25) monoclonal antibodies. Evidence for the physiological role of CD4⁺ Th1⁺ subset in mice, *Eur. J. Immunol.,* 18, 2101, 1988.
62. **Kapsenberg, M. L., Wierenga, E. A., Stiekma, F. E. M., Tiggelman, A. M. B. C., and Bos, J. D.,** Th1 lymphokine production profiles of nickel-specific CD4⁺ T lymphocyte clones from nickel contact allergic and non-allergic individuals, *J. Invest. Dermatol.,* 98, 59, 1992.
63. **Kapsenberg, M. L., Wierenga, E. A., Bos, J. D., and Jansen, H. M.,** Functional subsets of allergen-reactive human CD4⁺ T cells, *Immunol. Today,* 12, 392, 1991.
64. **Smith, C. A. and Rennick, D. M.,** Characterization of a murine lymphokine distinct from interleukin 2 and interleukin 3 (IL-3) possessing a T-cell growth factor and a mast cell growth factor activity that synergizes with IL-3, *Proc. Natl. Acad. Sci. U.S.A.,* 83, 1857, 1986.
65. **Thompson-Snipes, L., Dhar, V., Bond, M. W., Mosmann, T. R., Moore, K. W., and Rennick, D. M.,** Interleukin-10: a novel stimulatory factor for mast cells and their progenitors, *J. Exp. Med.,* 173, 507, 1991.

66. Yokota, T., Coffman, R. L., Hagiwara, H., Rennick, D. M., Takebe, Y., Yokota, K., Gemmell, L., Schrader, B., Yang, G., Meyerson, P., Luh, J., Hoy, P., Pene, J., Briere, F., Spits, H., Banchereau, J., De Vries, J., Lee, F. D., Arai, N., and Arai, K-I., Isolation and characterization of lymphokine cDNA clones encoding mouse and human IgA-enhancing and eosinophil-colony stimulating factor activities. Relationship to interleukin 5, *Proc. Natl. Acad. Sci. U.S.A.*, 84, 7388, 1987.

67. Chand, N., Harrison, J. E., Rooney, S., Pillar, J., Jakubicki, R., Nolan, K., Diamantis, W., and Sofia, R. D., Anti-IL-5 monoclonal antibody inhibits allergic late phase bronchial eosinophilia in guinea pigs: a therapeutic approach, *Eur. J. Pharmacol.*, 211, 121, 1992.

68. Gulbenkian, A. R., Egan, R. W., Fernandez, X., Jones, H., Kreutner, W., Kung, T., Payvandi, F., Sullivan, L., Zurcher, J. A., and Watnik, A. S., Interleukin-5 modulates eosinophil accumulation in allergic guinea pig lung, *Am. Rev. Respir. Dis.*, 146, 263, 1992.

69. Iwami, T., Nagai, H., Suda, H., Tsuruoka, N., and Koda, A., Effect of murine recombinant interleukin-5 on the cell population in the guinea-pig airways, *Br. J. Pharmacol.*, 195, 19, 1992.

70. Iwami, T., Nagai, H., Tsuruoka, N., and Koda, A., Effect of murine recombinant interleukin-5 on bronchial reactivity in guinea pigs, *Clin. Exp. Allergy*, 23, 32, 1993.

71. Coleman, J. W., Holliday, M. R., Kimber, I., Zsebo, K. M., and Galli, S. J., Regulation of mouse peritoneal mast cell secretory function by stem cell factor, IL-3 or IL-4, *J. Immunol.*, 150, 556, 1993.

72. Coleman, J. W., Holliday, M. R., and Buckley, M. G., Regulation of the secretory function of mouse peritoneal mast cells by IL-3, IL-4 and IFN-γ, *Int. Arch. Allergy Immunol.*, 99, 408, 1992.

73. Parronchi, P., Macchia, D., Piccinni, M.-P., Biswas, P., Simonelli, C., Maggi, E., Ricci, M., Ansari, A. A., and Romagnani, S., Allergen and bacterial antigen-specific T-cell clones established from atopic donors show a different profile of cytokine production, *Proc. Natl. Acad. Sci. U.S.A.*, 88, 4538, 1991.

74. van der Heijden, F. L., Wierenga, E. A., Bos, J. D., and Kapsenberg, M. L., High frequency of IL-4-producing CD4+ allergen-specific T lymphocytes in atopic dermatitis lesional skin, *J. Invest. Dermatol.*, 97, 389, 1991.

75. Kay, A. B., Ying, S., Varney, V., Gaga, M., Durham, S. R., Moqbel, R., Wardlaw, A. J., and Hamid, Q., Messenger RNA expression of the cytokine gene cluster, interleukin 3 (IL-3), IL-4, IL-5 and granulocyte/macrophage colony stimulating factor, in allergen-induced late phase cutaneous reactions in atopic subjects, *J. Exp. Med.*, 173, 775, 1991.

76. Coleman, J. W., Buckley, M. G., Holliday, M. R., and Morris, A. G., Interferon-γ inhibits serotonin release from mouse peritoneal mast cells, *Eur. J. Immunol.*, 21, 2559, 1991.

77. Gautam, S. C., Chikkala, N. F., and Hamilton, T. A., Anti-inflammatory action of IL-4. Negative regulation of contact sensitivity to trinitrochlorobenzene, *J. Immunol.*, 148, 1411, 1992.

78. Dearman, R. J., Hegarty, J. M., and Kimber, I., Inhalation exposure of mice to trimellitic anhydride induces both IgG and IgE anti-hapten antibody, *Int. Arch. Allergy Appl. Immunol.*, 95, 70, 1991.

79. Dearman, R. J. and Kimber, I., Differential stimulation of immune function by respiratory and contact chemical allergens, *Immunology*, 72, 563, 1991.

80. Dearman, R. J. and Kimber, I., Divergent immune responses to respiratory and contact chemical allergens: antibody elicited by phthalic anhydride and oxazolone, *Clin. Exp. Allergy*, 22, 241, 1992.

81. Dearman, R. J., Spence, L. M., and Kimber, I., Characterization of murine immune responses to allergenic diisocyanates, *Toxicol. Appl. Pharmacol.*, 112, 190, 1992.

82. Dearman, R. J., Mitchell, J. A., Basketter, D. A., and Kimber, I., Differential ability of occupational chemical contact and respiratory allergens to cause immediate and delayed dermal hypersensitivity reactions in mice, *Int. Arch. Allergy Immunol.*, 97, 315, 1992.

83. **Dearman, R. J., Basketter, D. A., Coleman, J. W., and Kimber, I.,** The cellular and molecular basis for divergent allergic responses to chemicals, *Chem-Biol. Interact.*, 84, 1, 1992.

84. **Kimber, I. and Dearman, R. J.,** The mechanisms and evaluation of chemically induced allergy, *Toxicol. Lett.*, 64/65, 79, 1992.

85. **Kimber, I., Gerberick, G. F., Van Loveren, H., and House, R.V.,** Chemical allergy: molecular mechanisms and practical applications, *Fund. Appl. Toxicol.*, 19, 479, 1992.

86. **Baum, C. G., Szabo, P., Siskind, G. W., Becker, C. G., Firpo, A., Clarick, C. J., and Francus, T.,** Cellular control of IgE induction by a polyphenol-rich compound. Preferential activation of Th2 cells, *J. Immunol.*, 145, 779, 1990.

87. **Snapper, C. M., Pecanha, L. M. T., Levine, A. D., and Mond, J. J.,** IgE class switching is critically dependent upon the nature of the B cell activator, in addition to the presence of IL-4, *J. Immunol.*, 147, 1163, 1991.

88. **Hayglass, K. T., Gieni, R. S., and Stefura, W. P.,** Long-lived reciprocal regulation of antigen-specific IgE and IgG2a responses in mice treated with glutaraldehyde-polymerized ovalbumin, *Immunology*, 73, 407, 1991.

89. **Gieni, R. S., Yang, X., and Hayglass, K. T.,** Allergen-specific modulation of cytokine synthesis patterns and IgE responses in vivo with chemically modified allergen, *J. Immunol.*, 150, 302, 1993.

90. **Yamada, M., Nakazawa, M., and Arizono, N.,** IgE and IgG2a antibody responses are induced by different antigen groups of the nematode Nippostrongylus brasiliensis in rats, *Immunology*, 78, 298, 1993.

91. **Ohmori, H., Hikida, M., and Takai, T.,** Prostaglandin E_2 as a selective stimulator of antigen-specific IgE response in murine lymphocytes, *Eur. J. Immunol.*, 20, 2499, 1990.

92. **Roper, R. L., Conrad, D. H., Brown, D. M., Warner, G. L., and Phipps, R. P.,** Prostaglandin E_2 promotes IL-4 induced IgE and IgG1 synthesis, *J. Immunol.*, 145, 2644, 1990.

93. **Betz, M. and Fox, B. S.,** Prostaglandin E_2 inhibits production of Th1 lymphokines but not of Th2 lymphokines, *J. Immunol.*, 146, 108, 1991.

94. **Roper, R. L. and Phipps, R. P.,** Prostaglandin E_2 and cAMP inhibit B lymphocyte activation and simultaneously promote IgE and IgG1 synthesis, *J. Immunol.*, 149, 2984, 1992.

95. **Diaz-Sanchez, D., Noble, A., Staynov, D. Z., Lee, T. H., and Kemeny, D. M.,** Elimination of IgE regulatory rat $CD8^+$ T cells in vivo differentially modulates interleukin-4 and interferon-γ but not interleukin 2 production by splenic T cells, *Immunology*, 78, 513, 1993.

96. **Ishizaka, K. and Sandberg, K.,** Formation of IgE-binding factors by human T lymphocytes, *J. Immunol.*, 125, 1692, 1981.

97. **Leung, D. Y. M. and Geha, R. S.,** Regulation of the human IgE antibody response, *Int. Rev. Immunol.*, 2, 75, 1987.

98. **Pene, J.,** Regulatory role of cytokines and CD23 in the human IgE antibody synthesis, *Int. Arch. Allergy Appl. Immunol.*, 90, 32, 1989.

99. **Delespesse, G., Hofstetter, H., and Sarfati, M.,** Low-affinity receptor for IgE (FcERII, CD23) and its soluble fragments, *Int. Arch. Allergy Appl. Immunol.*, 90, 41, 1989.

100. **Bonnefoy, J.-Y., Pochon, S., Aubry, J.-P., Graber, P., Gauchat, J.-F., Jansen, K, and Flores-Romo, L.,** A new pair of surface molecules involved in human IgE regulation. *Immunol. Today*, 14, 1, 1993.

101. **Kimata, H., Yoshida, A., Ishioka, C., Lindley, I., and Mikawa, H.,** Interleukin 8 (IL-8) selectively inhibits immunoglobulin E production induced by IL-4 in human B cells, *J. Exp. Med.*, 176, 1227, 1992.

102. **Zetterstrom, O., Osterman, K., Machado, L., and Johansson, S. G.,** Another smoking hazard: raised serum IgE concentration and increased risk of occupational allergy, *Br. Med. J.*, 283, 1215, 1981.

103. **Venables, K. M., Topping, M. D., Howe, W., Luczynska, C. M., and Hawkins, R., and Newman Taylor, A. J.,** Interaction of smoking and atopy in producing specific IgE antibody against a hapten protein conjugate, *Br. Med. J.,* 290, 201, 1985.

104. **Muranaka, M., Suzuki, S., Koizumi, K., Takafuji, S., Miyamoto, T., Ikemori, R., and Tokiwa, H.,** Adjuvant activity of diesel-exhaust particulates for the production of IgE antibody in mice, *J. Allergy Clin. Immunol.,* 77, 616, 1986.

105. **Takafuji, S., Suzuki, S., Koizumi, K., Tadokoro, K., Miyamoto, M., Ikemori, R., and Muranaka, M.,** Diesel-exhaust particulates inoculated by the intranasal route have an adjuvant activity for IgE production in mice, *J. Allergy Clin. Immunol.,* 79, 639, 1987.

106. **Riedel, F., Kramer, M., Scherbenbugen, C., and Rieger, C. H. L.,** Effects of SO_2 exposure on allergic sensitization in the guinea pig, *J. Allergy Clin. Immunol.,* 82, 527, 1988.

107. **Wardlaw, A. J.,** The role of air pollution in asthma, *Clin. Exp. Allergy,* 23, 81, 1993.

108. **Garssen, J., Nijkamp, F. P., Wagenaar, S. S., Zwart, A., Askenase, P. W., and Van Loveren, H.,** Regulation of delayed-type hypersensitivity responses in the mouse lung, determined with histological procedures: serotonin, T suppressor inducer factor and high antigen dose tolerance regulate the magnitude of T cell dependent inflammatory reactions, *Immunology,* 68, 51, 1989.

109. **Garssen, J. and Nijkamp, F. P., Van Der Vliet, H., and Van Loveren, H.,** T cell-mediated induction of airway hyperreactivity in mice, *Am. Rev. Respir. Dis.,* 144, 931, 1991.

110. **Van Loveren, H., Garssen, J., and Nijkamp, F. P.,** T cell-mediated airway hyperreactivity in mice, *Eur. Respir. J.,* 4, 16s, 1991.

111. **Fabbri, L. M., Mapp, C. E., Saetta, M., and Allegra, L.,** Occupational asthma, in *Asthma as an Inflammatory Disease,* O'Byrne, P. M., Ed., Marcel Dekker, New York, 1990, 127.

112. **Chung, K. F.,** Inflammatory mediators in asthma, in *Asthma as an Inflammatory Disease,* O'Byrne P. M., Ed., Marcel Dekker, New York, 1990, 159.

113. **Roche, W. R., Beasley, R., Williams, J. H., and Holgate, S. T.,** Subepithelial fibrosis in the bronchi of asthmatics, *Lancet,* 1, 520, 1989.

114. **Beasley, R., Roche, W. R., Roberts, J. A., and Holgate, S. T.,** Cellular events in the bronchi in mild asthma and after bronchial provocation, *Am. Rev. Respir. Dis.,* 139, 806, 1989.

115. **Gleich, G. J.,** The eosinophil and bronchial asthma: current understanding, *J. Allergy Clin. Immunol.,* 85, 422, 1990.

116. **Bentley, A. M., Maestrelli, P., Saetta, M., Fabbri, L. M., Robinson, D. S., Bradley, B. L., Jeffrey, P. K., Durham, S. R., and Kay, A. B.,** Activated T-lymphocytes and eosinophils in the bronchial mucosa in isocyanate-induced asthma, *J. Allergy Clin. Immunol.,* 89, 821, 1992.

117. **Corrigan, C. J.,** Allergy of the respiratory tract, *Curr. Opinion Immunol.,* 4, 798, 1992.

118. **Corrigan, C. J. and Kay, A. B.,** T cells and eosinophils in the pathogenesis of asthma, *Immunol. Today,* 13, 501, 1992.

119. **Iwamoto, I., Nakajima, H., Endo, H., and Yoshida, S.,** Interferon γ regulates antigen-induced eosinophil recruitment into the mouse airways by inhibiting the infiltration of CD4$^+$ T cells, *J. Exp. Med.,* 177, 573, 1993.

120. **Holt, P. G., Schon-Hegrad, M. A., Phillips, M. J., and McMenamin, P. G.,** Ia-positive dendritic cells form a tightly meshed network within the human airway epithelium, *Clin. Exp. Allergy,* 19, 597, 1989.

121. **Holt, P. G., Schon-Hegrad, M. A., and McMenamin, P. G.,** Dendritic cells in the respiratory tract, *Int. Rev. Immunol.,* 6, 139, 1990.

122. **Schon-Hegrad, M. A., Oliver, J., McMenamin, P. G., and Holt, P. G.,** Studies on the density, distribution and surface phenotype of intraepithelial class II MHC antigen (Ia)-bearing dendritic cells (DC) in the conducting airways, *J. Exp. Med.,* 173, 1345, 1991.

123. **Holt, P. G.,** Down regulation of immune responses in the lower respiratory tract: role of alveolar macrophages, *Clin. Exp. Immunol.,* 63, 261, 1985.
124. **Thepen, T., Van Rooijen, N. and Kraal, G.,** Alveolar macrophage elimination in vivo is associated with an increase in pulmonary immune responses in mice, *J. Exp. Med.,* 170, 494, 1989.
125. **Thepen, T., McMenamin, C., Oliver, J., Kraal, G., and Holt, P. G.,** Regulation of immune responses to inhaled antigen by alveolar macrophages (AM): differential effects of AM elimination in vivo on the induction of tolerance versus immunity, *Eur. J. Immunol.,* 21, 2845, 1991.
126. **Thepen, T., McMenamin, C., Girn, B., Kraal, G., and Holt, P. G.,** Regulation of IgE production in presensitized animals: in vivo elimination of alveolar macrophages selectively increases IgE responses to inhaled allergen, *Clin. Exp. Allergy,* 22, 1107, 1992.
127. **Holt, P. G., Degebrodt, A., O'Leary, C., Krska, K., and Ploza, T.,** T-cell activation by antigen presenting cells from lung tissue digests: suppression by endogenous macrophages, *Clin. Exp. Immunol.,* 62, 586, 1985.
128. **Holt, P. G., Schon-Hegrad, M. A., and Oliver, J.,** MHC class II antigen-bearing dendritic cells in pulmonary tissues of the rat: regulation of antigen presentation activity by endogenous macrophage populations, *J. Exp. Med.,* 167, 262, 1988.
129. **Holt, P. G., Oliver, J., Bilyk, N., McMenamin, C., McMenamin, P. G., Kraal, G., and Thepen, T.,** Down regulation of the antigen presenting cell function(s) of pulmonary dendritic cells in vivo by resident alveolar macrophages, *J. Exp. Med.,* 177, 397, 1993.
130. **Holt, P. G.,** Macrophage: dendritic cell interaction in regulation of the IgE response in asthma, *Clin. Exp. Allergy,* 23, 4, 1993.

Chapter 4

CLINICAL AND EPIDEMIOLOGICAL ASPECTS OF RESPIRATORY ALLERGY

Karin A. Pacheco and Lanny J. Rosenwasser

CONTENTS

I. INTRODUCTION

The respiratory effects of inhaled chemicals — from urban air pollution to specific industrial exposures — are an increasing problem as more of us live in an urban and industrial setting.

The prevalence and severity of asthma is on the rise.[1-6] Numerous studies[7,8] have correlated levels of air pollution with onset and prevalence of acute respiratory symptoms. Lung capacity declines significantly more rapidly in adults living in more polluted communities.[9-11] Occupational dust and fume exposure correlate with respiratory symptoms and portend enhanced loss of lung function.[12] The list of occupational agents capable of causing asthma, hypersensitivity pneumonitis, and fibrosing interstitial lung disease is long, and growing. The cost of treating asthma in the U.S. was estimated at $6.2 billion in 1990,[13] and is likely comparable in Western European countries.

Chemicals in the environment are causing respiratory problems at an alarming rate, and at great cost. This chapter will discuss the clinical aspects and epidemiology of the problem. We will discuss the respiratory effects of inhaled pollutants, both outdoor and indoor. We will define occupational immunological lung diseases, and discuss clinical features and etiologic agents. Finally, we will review multiple chemical sensitivities, a poorly defined symptom complex which masquerades as a chemically-mediated disease.

II. RESPIRATORY EFFECTS OF AIRBORNE POLLUTANTS

A. OUTDOOR POLLUTANTS
1. General

To those who live in cities, urban air pollution is a fact of life. However, as numerous studies over the past 40 years have shown, urban air pollution can cause costly health side effects. Both acute respiratory symptoms and long-term accelerated loss of lung function have been correlated with levels of urban air pollution. Unregulated or minimally regulated urban and industrial air pollution is a form of cost shifting from the industrial sector to the health sector — with deleterious effects on the long-term health of the citizenry.

Early episodes of high particulate air pollution, such as in Donora, Pennsylvania, in 1948 or in London in 1952, were correlated with excessive daily mortality so high that a causal relationship was evident. A subsequent time series analysis of mortality and air pollution in London from 1958 to 1972, when levels of air pollution, specifically particulates and sulfur dioxide, were lower, also showed a highly significant correlation between mortality and either particulate matter or sulfur dioxide, after controlling for temperature and humidity.[14]

An early study from Japan, published in 1964,[15] compared the incidence of respiratory symptoms between the employees of a casting company in the Tokyo–Yokohama area, a heavily polluted community, and those of an oil company with branches in Tokyo and in Niigate, a less industrialized area. The residents of the Tokyo–Yokohama area had an increased incidence of chronic cough, sputum production, and throat irritation. These symptoms were more pronounced in cigarette smokers and in persons with allergies. Lung function, measured by vital capacity (VC), was lower in the Tokyo–Yokohama residents, and was lowest in those with the longest exposure. Airway obstruction in general was more prevalent in residents over the age of 45 in the Tokyo area.

Although 60% of those studied in the Tokyo–Yokohoma area had moved there from other parts of Japan, their symptoms and loss of lung function were unique to the Tokyo area, suggesting that prolonged exposure to heavily polluted air was at fault. A follow-up study in 1968[16] found a significant positive relationship between symptoms (cough, phlegm production, sneezing, throat irritation) and quantitative changes in air pollution (dust fall and SO_2) in the Tokyo area.

More recent studies confirm these findings, even when current air quality guidelines are met or exceeded. A report from Barcelona, Spain[17] found a weak but statistically significant relationship between emergency room admissions and the levels of SO_2, black smoke, and CO. Admissions increased by 0.02 for each $\mu g/m^3$ of SO_2, by 0.01 for each $\mu g/m^3$ of black smoke, and by 0.11 for each $\mu g/m^3$ of CO. Interestingly, these effects were still seen for SO_2 and black smoke even at levels below the air quality guidelines of 100 $\mu g/m^3$ recommended by the World Health Organization (WHO). This suggests that current air quality standards are not necessarily protective with a margin of safety.

Another study has shown a correlation between levels of air pollution and the duration of acute respiratory symptoms.[18] The author followed symptom diaries of first-year nursing students in Los Angeles and correlated them to daily measurements of air pollution. Photochemical oxidants were a significant predictor of cough and sore throat duration, as well as phlegm production. SO_2 was significantly associated with the duration of chest discomfort. The natural log of pollution was always a better predictor of symptom duration, and the mean pollution level during the symptom episode was a better predictor than levels during the previous 7 or 30 days, or the day the episode began.

Long-term exposure to air pollution also leads to loss of lung capacity. An important Dutch study[9] compared respiratory symptoms and spirometry over a 9-year period from residents of a nonpolluted city, Vlagtwedde, with those of Vlaardingen, a town polluted by SO_2 and black smoke from nearby oil refineries. Results showed a significantly higher prevalence of chronic cough and phlegm production in the more polluted area. Over time, exposure to moderate air pollution caused a smaller increase in FEV_1 in the 15- to 24-year age group when lung function normally increases, and a significantly greater decline in FEV_1 and VC with increasing age.

The population studies of chronic obstructive respiratory disease, conducted by the University of California at Los Angeles, were one of the largest to examine the effects of air pollution on respiratory function over time. Four Los Angeles communities with similar demographics but different pollution exposures to photochemical oxidants, SO_2, NO_2, particulates, hydrocarbons, and sulfates, were compared. Residents in the most polluted area had a significantly higher incidence of cough and sputum production, and a significantly higher percentage of persons with FEV_1 and FVC below 50% predicted.[10] Comparison of spirometry over 5 years showed residents also had a significantly more rapid decline than in those living in the less polluted areas.[11]

2. Pollution and Asthma

Statistics of asthma morbidity, mortality, and prevalence for the U.S. and the world agree: all three are on the rise.[1-6] A review of international trends in asthma mortality from 1970 to 1985[1] examined death rates from asthma in the 5- to 34-year age group from Australia, Canada, England and Wales, Finland, France, Japan, Israel, the Netherlands, New Zealand, Singapore, Sweden, Switzerland, the U.S., and West Germany. Although there was a wide variation in asthma mortality rates and trends, 11 of the 14 countries examined showed a higher average mortality from 1982 to 1984 than from 1979 to 1981; in 6 countries the increase was 20% or more, although the actual numbers were low (0.13 to 3.63 per 100,000). Specific review of the data from the U.S. showed that deaths from any form of obstructive airways disease increased since the late 1970s. Review of the U.S. National Prescription Audit from 1972 to 1985 showed a 200% increase in antiasthma prescriptions, compared to an overall 7% increase in all drug products. A report in *JAMA* published in 1990[19] reviewed U.S. vital statistics from 1968 to 1987 for patterns in asthma mortality in the 5- to 34-year age group. The authors concluded that four state economic areas: New York City, Cook County, Illinois, Maricopa County, Arizona, and Fresno County, California, had excess asthma mortality and higher rates of asthma mortality increase than the general U.S. population. New York City and Cook County in particular were felt to drive the U.S. trend. Review of asthma hospitalizations for children aged 0 to 17 from 1979 to 1987 showed a 4.5% yearly increase, compared to a 4.6% drop in total hospitalizations.[20]

Other studies[21-22] have shown increased asthma prevalence in genetically similar populations living in urban rather than rural environments. The prevalence in industrialized countries appears to have increased in the last 20 years. Both lines of evidence suggest that the major determinants of asthma prevalence in any particular population are environmental.

What is driving the rising trends in asthma? Several factors have been implicated, including changes in host susceptibility as well as environmental exposures. Early exposure to viral infections, cigarette smoke, and other indoor air pollution are associated with the development of asthma. Early allergen exposure, especially to housedust mite, cat, and cockroach, appears to be critical in the pathogenesis of new asthma. Exposure to outdoor air pollution in urbanized areas also contributes.[4] In adults, occupational sensitivities as well as increased occupational dust and fume exposure are also important.

The association of air pollution with asthma exacerbations has been well documented. A study published in 1967 examined the pattern of emergency room (ER) visits for asthma at St. Christopher's Hospital for Children in Philadelphia compared with the levels of air pollution.[23] The authors found a threefold increase in visits for asthma during days when pollutants scored in the 90th percentile for the 2-year period. In contrast, only 3%, or 8 of the high ER visit days occurred during the 280 days of low pressure and air pollution.

A study from Los Angeles published in 1981[24] again correlated increases in asthma ER visits and hospitalizations for children with increases in NO_2, coefficients of haze, hydrocarbons, Santa Ana wind conditions, and total airborne allergen counts.

Researchers in Spain compared two groups of children with extrinsic asthma, 160 living in a polluted area, and 88 living in a nonpolluted area, for differences in mean number of wheezing crises per year and incidence of severe asthma. The mean number of yearly asthma exacerbations was higher in children living in the more polluted area (10.4 vs. 7.69), as well as the incidence of more severe asthma (14.4% vs. 5.7%). However, they could not correlate wheezing episodes with levels of fumes and SO_2.[25]

Even in less polluted areas, a relation between asthma admissions and air pollution is still found. A study from Helsinki over 3 years correlated asthma admissions with levels of SO_2, NO, NO_2, CO, O_3, total suspended particles (TSP) concentrations, temperature, relative humidity, and wind speed. NO and O_3 were most strongly associated with asthma exacerbations ($p < 0.0001$). The order of significance was NO_2 ($p < 0.001$) > CO ($p < 0.001$) > O_3 ($p < 0.006$) and SO_2 (NS). Effects of air pollutants and cold air were maximal in adults if they occurred on the same day; O_3 had a more pronounced effect after a 1 day lag. The absolute levels of pollutants were fairly low, the long-term mean being $SO_2 = 19.2 \, \mu g/m^3$, $NO_2 = 38.6 \, \mu g/m^3$, $CO = 1.3 \, \mu g/m^3$. However, TSP was high (mean concentration = 76.3 $\mu g/m^3$) and mean temperature was 4.7°C. This again suggests that levels of pollutants lower than that recommended may be associated with an increased incidence of asthma attacks.[26]

3. Outdoor Specific Pollutants

Four of the six pollutants regulated by the U.S. National Primary Air Quality Standards (O_3, SO_2, TSP, and NO_2) have been shown to worsen lung function when inhaled at realistic ambient concentrations.[27] The evidence for each will be discussed separately.

a. Ozone (O_3)

Ozone is produced when nitrogen oxides react with volatile hydrocarbons, in the presence of sunlight. Levels are usually higher in the summer and fall, (e.g., during days with more sunlight). Levels peak in the late afternoon, representing maximal reactions from car exhaust, industrial production, and sunlight. U.S. Environmental Protection Agency has set standards of <0.12 ppm per hour; however, half the U.S. population lives in areas where that standard is occasionally exceeded. Most effects of ozone have been assessed at concentrations moderately greater than 0.12 ppm under conditions of moderate exercise. Lung function acutely decreases with falls in FEV_1, FVC, and TV and increases in RR and specific airways resistance. Nonspecific bronchial hyperresponsiveness (BHR) shows a small increase in sensitivity (one doubling dose of histamine), and sensitivity to allergen also increases by 1.5

doubling dose dilutions.[28] A comparison of 218 children living in a high ozone area (45.39% of time >60 ppb over 2 years) to 281 children living in a low ozone area (0.33% of time >60 ppb) showed increased BHR in the ozone exposed children (29.4% vs. 19.9%, $p < 0.02$). BHR was somewhat more severe (PD_{20} MC 2.1 µg vs. 2.35 µg, $p < 0.05$), although there was no difference in subjective respiratory symptoms between the two groups.[29] Changes in BHR are reflected in bronchoalveolar lavage (BAL) fluid. After a 2-hour exposure to 0.4 ppm of O_3 followed by exercise in 11 healthy nonsmoking normals, BAL fluid showed an 8.2-fold increase in neutrophils, evidence of increased vascular permeability, and increased levels of proteolytic enzymes.[30]

b. Sulfur Dioxide (SO₂)

Sulfur dioxide is generated from the combustion of sulfur-containing fossil fuels, e.g., coal and oil, and it is a ubiquitous source of urban air pollution. Common large-scale sources are factories, mines, and smelters. In the U.S., the annual mean of 0.05 ppm, and 0.14 ppm highest daily level for SO_2 is exceeded approximately one-third of the time in urban areas, and urban levels are on average two to three times higher than in rural areas.[31]

Multiple studies have documented the effect of inhaling sulfur dioxide, usually during light to moderate exercise. The degree of bronchoconstriction induced depends on the concentration of SO_2 inhaled, on the amount and rate of work performed, and on the degree of underlying bronchial hyperreactivity.[32] A study of ten nonsmoking volunteers with mild to moderate asthma, not requiring chronic medication, examined the effect of free breathing 0.50 ppm SO_2 in filtered air on airways resistance. Subjects performed moderately heavy exercise, and airways resistance increased significantly. A more recent study in nine adult nonallergic asthmatics showed a statistically significant dose response curve to freely inhaled SO_2 at 0.5 ppm and 1.0 ppm. Subjects performed light to moderate exercise 20 minutes after exposure, and subsequent spirometry showed significant drops in FEV_1 ($p < 0.008$), specific total respiratory resistance ($p = 0.033$), and maximal expiratory flow rates at 50% ($p = 0.017$) and 75% ($p = 0.048$) compared to placebo.[33]

Sulfuric acid exposure can cause bronchoconstriction in both normals and asthmatics. A 4-hr exposure to 450 µg/m³ followed by exercise, caused increased airways reactivity in normals as measured by carbachol challenges 24 hr after exposure. Asthmatics were more sensitive, and showed increased bronchial hyperreactivity after a 16-min exposure to 450 µg/m³ without subsequent exercise, and decreased FEV_1 after a 2-hr exposure to 75 µg/m³ followed by exercise. The latter concentration approaches actual high ambient concentrations achieved outdoors whereas 1,000 µg/m³ is the maximum level adopted by the Occupational Safety and Health Administration (OSHA) for workplace exposures.[34] Changes in lung function and airways reactivity are reflected in cellular changes in bronchial lavage fluid. Both macrophages and lymphocytes were significantly elevated in BAL fluid obtained from healthy subjects 24 hr after exposure to 8 ppm of SO_2.[35]

c. Nitrogen Dioxide (NO₂)

Nitrogen dioxide is generated when nitric oxide (NO) is oxidized by the sun. Afternoon rush-hour traffic is a common outdoor source; kitchen gas stoves are a common indoor source. NO_2 has many of the same biological effects as ozone, but at somewhat higher concentrations (1.6 to 2.0 ppm). Previous studies have documented an increased incidence of respiratory illness and pulmonary function decrements in children exposed to a kitchen gas stove at home,[36] as well as in school children living near an industrial plant.[37] Numerous studies have documented increased bronchial hyperreactivity in asthmatics after exposure to 0.10 to 0.20 ppm of NO_2.[38-40] A study in 1986 examined the effect of low level NO_2 inhalation on exercise-induced bronchospasm in moderate asthmatics.[41] Exposure to 0.30 ppm NO_2 for 20 min, followed by 10 min of exercise, resulted in statistically significant falls in FEV_1 ($p < 0.01$) and $FEF_{60\%}$ ($p < 0.05$), although these returned to baseline 1 hr following exercise. After the recovery period though, subjects showed increased bronchial hyperreactivity as measured by isocapnic cold air hyperventilation. The levels of NO_2 used in the study were felt to represent peak urban values and thus true ambient exposures. No effects were seen on respiratory function in asthmatics at lower doses, e.g., 0.12 to 0.18 ppm.[42]

Several studies have documented a dose-response relationship to NO_2, where those with lung disease had a lower threshold dose effect than normals. A summary of several studies on the effect of NO_2 noted that whereas subjects with asthma developed increased airways responsiveness to carbachol after exposure to NO_2 at concentrations of 0.1 ppm, those with chronic bronchitis only increased airways resistance after exposure to 1.5 ppm, and normals after exposure to 2.5 ppm. Other studies found that exposure levels of 5 to 7.5 ppm were required to increase BHR to acetylcholine in normals.[43]

d. Total Suspended Particulates (TSP)

Fine particulate pollution also correlates with acute and chronic loss of lung function. Fine particulates are defined as those with an aerodynamic diameter of less than or equal to 10 μm (PM_{10}). They are of special concern as they contain a higher percentage of toxic metals and acidic sulfur oxides, and are able to penetrate to the lung periphery because of their small size.

Several recent studies have documented their respiratory effect. Results from the six cities study of air pollution and health documented increased reported rates of chronic cough, bronchitis, and chest illness with increased levels of all particulate pollution. Prevalence of symptoms in children with a history of wheeze or asthma was much higher. However, no association was found in this study with decreased measures of pulmonary function.[44] A more recent study found an acute 3 to 6% loss in peak expiratory flow (PEF) when PM_{10} levels were 150 μg/m³, and smaller, but correlated fluctuations in PEF with smaller increments of PM_{10}. Elevated levels of PM_{10} were also associated with increased respiratory complaints and use of asthma medication in both adults and children. Interestingly, these associations were seen even at PM_{10}

levels below the U.S. 24-hr ambient standard of 150 µg/m³.[45] A report from Hong Kong[46] found a highly significant correlation between TSP levels and asthma hospitalization rates for 1 to 4 year olds, suggesting that children are especially vulnerable to particulate pollution.

B. INDOOR POLLUTANTS

As more of us spend the majority of our time indoors — at home, at work, or in transit, indoor air pollution has an increasingly important effect on our respiratory health. Outdoor pollutants may enter through the ventilation system or through natural inlets; indoor pollutants may be generated through unvented combustion, solvent evaporation, grinding, or abrasion. Biologic sources such as pets, mold, fungal, or bacterial growth may also contribute, but will not be discussed further here. The reduced ventilation rates of the new energy efficient buildings may lead to higher pollutant concentrations.

1. Cigarette Smoke

Cigarette smoke is probably the most important indoor pollutant, and one whose effects have been best characterized. In children, exposure to parental side-stream smoke has been well documented to increase the risk of early childhood respiratory illness as well as the prevalence of respiratory symptoms.[47] Tobacco smoke exposure also increases bronchial hyperreactivity in children, especially in those with a family history of asthma or allergies. A study in children as young as $4^{1}/_{2}$ weeks showed enhanced responsiveness to inhaled histamine, as measured by a 40% fall of the maximal flow rates at FRC, in infants with a family history of asthma, exposure to parental smoking, or both.[48] Similar results were found in 9-year-old boys, with increased BHR to carbachol in those children with smoking parents. Bronchial responsiveness was significantly correlated with atopy, and was more pronounced in asthmatics, although the increased BHR was still present in nonasthmatics.[49] The risk of developing asthma in childhood is increased if the child is exposed to parental smoking. A population-based questionnaire survey of households in Michigan revealed that if mothers smoked, the prevalence of parent-reported asthma in their children increased from 5 to 7.7% (RR = 1.5) and the prevalence of functionally impairing asthma increased from 1.1 to 2.2% (RR = 2).[50] Several studies have documented that maternal smoking is a worse risk factor than paternal smoking for asthma and respiratory symptoms in children, probably related to increased exposure to the mother. If a child already has asthma, exposure to secondhand cigarette smoke will worsen symptoms, decrease flow rates,[51] and increase bronchial hyperreactivity. There appears to be a dose–response relationship between number of cigarettes smoked, and decrease in lung function tests.[52]

Adults show a similar response to cigarette smoke exposure. A study published in the *New England Journal of Medicine* in 1980 showed that nonsmokers chronically exposed to cigarette smoke (>20 years) had a 13 to 15%

lower mid-expiratory flow rate, FEF_{25-75}, than nonexposed cohorts ($p < 0.005$). The decrement in nonsmokers was similar to light smokers (1 to 10 cigarettes a day) and to smokers who did not inhale.[53] The effects of passive smoke in asthmatics is more pronounced. An hour exposure to passive smoke causes a 20% decrement in FEV_1, FEF_{25-75}, and FVC in asthmatics as compared to normals.[54,55] Exposure to passive cigarette smoke increases bronchial hyperreactivity in asthmatics, as measured by both histamine challenge,[56] and by methacholine challenge.[57]

In summary, exposure to cigarette smoke has been well documented to cause increased respiratory symptoms, respiratory illness, worsened asthma symptoms, increased bronchial hyperreactivity, and loss of lung function in both asthmatic and nonasthmatic exposed persons.

2. Nitrogen Dioxide (NO_2)

Indoor sources of NO_2 include gas cooking stoves, kerosene, and gas space heaters. The effect on respiratory function appears to be variable, and small at best.[43,47] Large questionnaire-based studies seem to be evenly split between those finding increased respiratory symptoms and illness in children, although with low odds ratios (1.12 to 1.97), and those finding no effect. It appears that indoor NO_2 exposure is only a minor factor in causing respiratory impairment.

3. Formaldehyde

Formaldehyde is a highly water soluble volatile gas with a characteristic pungent odor. At high concentrations (5 to 30 ppm) it has lower airway and pulmonary effects. However, most formaldehyde never reaches the peripheral airways, as it is quickly solubilized and metabolized in the upper airway.[47] Mean conventional home levels are estimated to be 0.03 ppm; office and day care centers have levels ranging from 0.37 to 0.55 ppm, and many occupational settings have levels of 1 ppm.[58] Formaldehyde is generated in the production of textiles, floor coverings, insulations, plywood, ordinary and carbonless papers, particleboard, embalming fluid, fungicides, bactericides, air fresheners, cosmetics, and toothpaste. Incomplete combustion of wood (e.g., wood-burning stoves) or of fuels including gasoline, alcohol, and refuse (e.g., incinerators) is a major outdoor source. Cigarette smoke has been estimated to contain between 25 to 250 ppm formaldehyde.

Specific respiratory effects at current ambient concentrations have been hard to document despite numerous studies. Most have assessed symptom prevalence not correlated to formaldehyde concentration. A recent exhaustive review of studies examining the respiratory effects of formaldehyde[59] concluded that although formaldehyde was capable of acting as a respiratory irritant, there was no coexistent evidence that it could act as a respiratory sensitizer. Most bronchoconstrictor responses after inhalation challenge, with an average dose of 3 to 6 ppm, were small and reversed after 1 to 3 hr. Late reactions were not observed, and transient or permanent bronchial hyperresponsiveness was never induced. Most studies were flawed by concomitant industrial exposures, undocumented formaldehyde levels at work, and cigarette smoking as a confounding variable. Thus, although formaldehyde is a ubiquitous chemical, at present it has not been documented to cause significant respiratory disease at commonly encountered levels.

4. Woodsmoke

Domestic woodsmoke exposure may contribute to respiratory complaints, but the effect is probably minor. Woodsmoke is a complex mixture of aldehydes, including formaldehyde, polycyclic aromatic hydrocarbons, and fine particulates. Most studies have compared the prevalence of respiratory symptoms and respiratory disease with the extent of domestic smoke exposure. Results have been mixed, with some studies noting a correlation[60-62] and some finding no association.[63]

5. Sick-Building Syndrome

Since the introduction in the 1970s of new, more energy-efficient buildings with central air handling systems, there has been an outbreak of work-related symptoms in office workers. In a classic study published in 1984, workers were interviewed from nine buildings, of which three were naturally ventilated and six had mechanical ventilation. Of the latter six, five had humidified air and four had air recirculation. Excess work-related symptoms were found in the humidified and mechanically ventilated buildings.[64] A symptom complex emerged affecting the nose, eyes, and mucous membranes. Workers complained of chest tightness, headache, dry skin, and afternoon lethargy out of proportion to the work done.[65] These symptoms are, in general, nonspecific, and there may be no objective findings. The diagnosis requires a cluster of cases from the same workplace.

Evaluation for the source of the problem is generally frustrating, unrewarding, and time consuming. A specific source of pollutants is found in less than 40% of the cases. In a review of 446 U.S. National Institute for Occupational Safety and Health (NIOSH) health hazard evaluations of indoor air complaints,[65] inadequate ventilation was found to be the primary problem in 52% of the cases. This includes not enough outdoor air, poor air distribution and mixing, temperature and humidity extremes, and inadequate maintenance of air filtration devices. Often, increasing the ventilation flow rates was sufficient to fix the problem. However, a recent study found no impact of increasing flow rates or the amount of outdoor air delivery on office workers' ratings of the office environment or in symptom reporting.[66]

III. OCCUPATIONAL EXPOSURES

A. OCCUPATIONAL IMMUNOLOGIC LUNG DISEASE

Occupational lung disease leads the list of the ten most important categories of occupational disease as compiled by NIOSH. The lung is both a common port of entry and target organ for inhaled chemicals. The potential for exposure and for injury is vast. NIOSH has estimated that over 1.2 million U.S. workers are potentially exposed to silica dust[67]; over 27.5 million U.S. workers were potentially exposed to asbestos from 1940 to 1979.[68] By 1986, occupational asthma was the most prevalent occupational lung disease in Canada, surpassing both asbestosis and silicosis.[69] In the U.S., a population-based assessment of the prevalence of asthma found a 7.7% incidence of asthma, of which 1.2%, or 15.4% of those with asthma, attributed it to workplace exposures. Relative risk for occupational asthma was higher in industrial and agricultural workers than in white collar or service occupations.[70] Although this is self-reported asthma, with probable overlap into emphysema, chronic bronchitis, hypersensitivity pneumonitis, and pneumoconiosis, it does suggest the extent of the problem.

Occupational immunologic lung disease may be broadly defined as a set of lung diseases characterized by an etiologic agent inhaled in the work environment and a defined immunologic response in the worker.[71]

Who gets occupational lung disease? Risk factors include both agent and host. The specific agent, its physical characteristics, e.g., particle size, chemical reactivity, etc., and exposure dose are all important in determining risk of developing disease. Host factors, including host ability to recognize the antigen, and host-specific patterns of response, are critical as well. Not all exposed workers develop lung disease, and the prevalence varies among agents. Atopy and cigarette smoke are also important variables.

For example, development of the pneumoconioses appears principally due to the size and exposure dose of the implicated dust: the smaller the particle and the higher the dose, the higher the incidence of disease. Cigarette smoke exacerbates the disease process. By comparison, host factors appear more critical in occupational asthma, since only a portion of those receiving the same dose develop disease. Between 10 to 45% of workers exposed to proteolytic enzymes develop asthma.[69] However, only 6.8% of workers exposed to trimellitic anhydride (TMA) develop a TMA immunologic syndrome.[72] The incidence of asthma in workers exposed to Western red cedar or to isocyanates is estimated at 5%, although after high dose exposure to isocyanates, many will develop symptoms of asthma. In both cases, atopy is not a factor and nonsmoking workers are more affected than smokers. Many of the asthma-causing agents are also capable of causing hypersensitivity pneumonitis in a smaller cohort of exposed workers, again suggesting that host factors are critical in determining the nature of the immune response to the offending agent.

B. OCCUPATIONAL ASTHMA

1. Definition

Occupational asthma is defined as variable air flow limitation caused by exposure to a specific agent in the workplace. Controversy exists as to the definition of asthma and the nature of the effect of the specific agent. In addition to airflow limitation, occupational asthma usually is associated with BHR to nonspecific challenges (e.g., methacholine, histamine, etc.). BHR may be considered a marker of the extent of lung inflammation. Occasionally, BHR is not demonstrated in conjunction with asthma elicited by specific agent exposure, but this is probably related to the chronicity of exposure. That is, BHR may not be present early in the process of sensitization, before chronic lung inflammation is established. It may be lost late in the process, once the worker is no longer exposed to the inciting agent. A specific sensitizing agent in the workplace, to which unexposed workers are not sensitive, is generally required as part of the definition. A period of sensitization, defined as latency between initial exposure and the development of symptoms, is important in establishing medical-legal probability of causation. Sensitization to the agent can be established by specific agent challenges causing a 20% drop in FEV_1 at exposure levels lower than that causing irritant reactions in normals, or bronchoconstriction in nonexposed asthmatics.[73] However, specific agent challenge may revert to negative in those no longer exposed.

Pre-existing asthma may be exacerbated by workplace exposure. Asthma may worsen by sensitization to an inciting agent, but it may also worsen with nonspecific irritant exposure. In many states, asthma worsened by workplace exposure, whether or not a sensitizing agent can be demonstrated, is compensable as work-related asthma.

2. Types

Occupational asthma has been traditionally divided by different types of inciting agents. These include pharmacologic bronchoconstriction, allergic asthma due to high molecular weight compounds or low molecular weight agents, and dust exposure. The distinction between high molecular weight and low molecular weight compounds may in fact reflect our relative ignorance of the pathogenesis of asthma. The initial pathways may be similar, with divergent T-cell responses responsible for the different clinical manifestations. Since chronic dust exposure may lead to BHR and chronic airflow limitation, a chronic inflammatory response appears to be involved, although it is poorly understood.

Pharmacologic bronchoconstriction is due to the direct bronchoconstrictor properties of the inhaled agent. Organophosphate insecticides, which act as cholinesterase inhibitors and therefore enhance cholinergic tone in the lung, are a good example. Since they are direct bronchoconstrictors, a dose–response relationship means that all exposed workers will develop bronchoconstriction if the dose is high enough. Hypersensitivity and a chronic inflammatory response is *not* involved, and therefore eosinophilia or BHR are not associated.

3. Allergic Occupational Asthma

A defined allergic or immune response to an inhaled allergen underlies most occupational asthma. Allergens have been subdivided, perhaps artificially, into high molecular weight and low molecular weight substances which elicit different clinical patterns of response.

High molecular weight (HMW) allergens tend to be protein derivatives of plant, animal, or bacterial sources. Examples include mammals (veterinarians, laboratory workers), crustacea or mollusks (crab, lobster, oyster workers), plant parts such as coffee beans, cocoa beans, costar beans, tea leaves, tobacco leaves, flour dust (Baker's asthma), and psyllium (hospital workers), arachnids (grain mites), and bacterial enzymes, such as papain, trypsin, and pepsin (detergent workers). Positive skin tests with IgE to the inciting agent can be demonstrated, and the symptomatic workers are by definition atopic. Specific IgG may occasionally be demonstrated. Other allergic symptoms besides asthma, such as allergic rhinitis or atopic dermatitis, may also be present or precede the development of asthma. Inhalation challenges show a typical early bronchoconstrictor response, or in about 50%, a dual early and late response. Cigarette smoking appears to be a risk factor for the development of asthma to HMW compounds. Smoking increases epithelial permeability, and may permit greater penetration of antigens. Enhanced production of IgE has been shown in smoking workers exposed to green coffee beans, prawns, humidifier antigens, and acid anhydrides. Onset of symptoms is shorter in smoking workers exposed to ethylene diamine or to platinum.[74]

More than 140 low molecular weight (LMW) species have been identified to cause occupational asthma.[75] They include copolymerizing or hardening agents for plastics, varnish and paints (isocyanates and anhydrides), exotic wood dust (western red cedar is the prototype), reactive metals (platinum, nickel, chromium, etc.), fluxes (colophony), drugs (penicillin, cephalosporins, psyllium), and other chemicals (urea, freon, furan resins). In some agents, such as toluene diisocyanate, trimellitic anhydride, Western red cedar, and platinum salts, specific IgE has been demonstrated. IgE may be directed to the unmetabolized compound, or to the reactive compound covalently linked to a carrier protein. In many cases, however, IgE has not been demonstrated. This may reflect a different pathway of immune activation, without the generation of IgE. Or it may simply reflect our inability to generate the relevant hapten-carrier protein of these highly reactive chemicals. A review of the structure of many of the LMW species suggests that the presence of multiple reactive groups capable of reacting with human macromolecules is a prerequisite.[76] Only 5 to 10% of those exposed develop asthma. The clinical pattern of asthma induced by low molecular weight compounds is often one of late-phase bronchoconstriction after challenge — either an isolated late reaction in 50%, or a biphasic pattern in the rest. Other, more atypical reactions have been described for isocyanates, including a progressive pattern of bronchoconstriction peaking 5 to 6 hr later, and a square-waved pattern similar to a biphasic pattern but with <10% recovery in the interim.[77] Atypical patterns of bronchoconstriction may apply to other LMW compounds as well, and should be looked for during provocation studies, or a positive reaction may be missed. Neither atopy nor cigarette smoking appears to be important risk factors for developing asthma to LMW compounds. Review of worker characteristics of those developing asthma to isocyanates,[78,79] or to Western red cedar,[80,81] show that the affected are predominantly nonatopic and nonsmokers. Only in colophony workers has smoking been shown to increase the risk for work-related disease.[82] Once removed from exposure, the majority of affected workers fail to recover completely. Most continue to be symptomatic, with evidence of persistent BHR. Resolution of symptoms is associated with shorter duration, near normal lung function, and less BHR at the time of diagnosis. This suggests that early diagnosis and removal from exposure is critical.

4. RADS

Reactive airways dysfunction syndrome (RADS) was best described in 1985 in a group of ten patients after a single, high level exposure to an irritating vapor, fume or smoke. The majority were exposed to a toxic spill at work. Symptoms developed minutes to hours later, with a mean of 8.9 hr. Although none had previous respiratory complaints, all demonstrated BHR to methacholine after exposure, and symptoms and BHR persisted for several years after. Trans-bronchial biopsies from two patients showed features of chronic inflammation and epithelial damage. A lymphocytic and plasma cell infiltrate was present in bronchial and bronchiolar walls, with desquamation of respiratory epithelium in one patient, and goblet cell hyperplasia in the other. Two of the patients were atopic and 6 of the 10 smoked cigarettes; neither of these appears to be a risk factor for the development of RADS. Clinical criteria for the diagnosis of RADS include:[83]

1. A documented absence of preceding respiratory complaints.
2. The onset of symptoms occurred after a single specific exposure incident or accident.
3. The exposure was to a gas, smoke, fume, or vapor which was present in very high concentrations and had irritant qualities to its nature.
4. The onset of symptoms occurred within 24 hr after the exposure and persisted for at least 3 months.
5. Symptoms simulated asthma with cough, wheezing, and dyspnea predominating.
6. Pulmonary function tests may show airflow obstruction.
7. Methacholine challenge testing was positive.
8. Other types of pulmonary diseases were ruled out.

RADS does not fit the classic description of occupational asthma induced by exposure to a sensitizing agent over time. However, it is persistent asthma induced by an acute toxic exposure, often in the workplace. As such, we feel it fits the other criteria for the diagnosis of occupational asthma.

C. HYPERSENSITIVITY PNEUMONITIS/ EXTRINSIC ALLERGIC ALVEOLITIS

Hypersensitivity pneumonitis (HP), also known as extrinsic allergic alveolitis, is an immune-mediated lung disease caused by exposure and subsequent sensitization to inhaled organic dusts and chemicals. It is commonly associated with occupational exposure to small organic dust particles <10 μg in diameter. Farmer's Lung, caused by inhalation of thermophilic actinomycetes in moldy hay, is the prototype described in 1932.[84] Many other agents have since been described, including Bird Fancier's Lung to avian droppings or feathers, Chemical Worker's Lung to isocyanates, Woodworker's Lung to wood dust, Coffee Worker's Lung to coffee beans, Detergent Worker's Disease to bacillus subtilis enzymes, Laboratory Worker's HP to male rat urine, etc. Clearly, antigens known to cause occupational asthma may cause HP in a subset of sensitive workers. Disease prevalence is generally low, but may vary from 10 to 70% of those exposed.

Traditionally, HP has been divided into an acute, subacute, and chronic form.[85-87] It has been postulated, but not proven, that dose exposure predicts the form of disease. The acute form, thought to occur after high dose intermittent exposure, resembles an acute viral or bacterial illness. Symptoms occur 2 to 9 hr after exposure, peak at 6 to 24 hr, and resolve over hours to days. The subacute form, felt to reflect a more chronic low level exposure, resembles progressive chronic bronchitis. However, one study of 287 Scottish bird fanciers examined at a major national convention found that intensity or duration of pigeon exposure did not discriminate between those with and without chronic bronchitis. However, the prevalence of chronic bronchitis increased with higher serum levels of pigeon antibodies,[88] suggesting that host susceptibility, rather than exposure dose alone, determined the development of disease. The chronic form may present as end-stage pulmonary fibrosis.

In fact, a dose–response association has never been proven, and other variables may be important.[89] Concentrations of organic dusts or chemicals, duration and frequency of exposure, particle size, antigen solubility and reactivity, as well as host factors, may all influence the presentation and course of disease.

Precipitating (IgG) antibodies, once felt to be a hallmark of HP, now are felt to reflect exposure: 40 to 50% of other exposed, but asymptomatic workers, may also have serum precipitins. Conversely, patients with the disease may lack precipitating antibodies,[90] which may reflect our inability to characterize the antigen, either as a single entity or a complex mixture from multiple exposures.

If the relevant antigen has been characterized, skin testing may be helpful, suggesting that IgE is also involved in the disease. In one of the original descriptions of pigeon breeders disease, 10 of 13 patients had positive immediate wheal and flare reactions to intradermal skin testing with nonirritating extracts of pigeon serum or droppings. Interestingly, all patients, even those with negative immediate reactions, had delayed reactions at 4 hr.[85] Once away from exposure, the intensity of the skin test response and titers of precipitating antibody decreased.[91]

The T-cell is critical to the pathogenesis of HP. In animal models, athymic nude mice exposed to thermoactinonyces vulgans will not develop the lung lesions of HP until sensitized spleen-derived T-cells are adoptively transferred.[92,93]

The long-term prognosis of HP is variable, and appears to be determined by the nature of the inciting antigen and individual patient characteristics rather than by continued exposure alone. Early studies of Farmer's Lung suggested that recurrent episodes of interstitial pneumonitis could lead to interstitial fibrosis, and recommended removal of farmers from the farm environment.[94,95] However, subsequent reports have suggested that not all continuously exposed farmers will have progressive disease. Barbee's study of 50 farmers with Farmer's Lung found a mortality rate of 10% after a follow-up of 6 years. Of the survivors, approximately 30% were symptomatic with dyspnea. All symptomatic patients showed varying degrees of fibrosis on chest radiograph, and lower lung function as compared to assymptomatic farmers. Importantly, symptoms did not correlate with the presence of serum precipitins, or with continued exposure, since half of both the symptomatic and asymptomatic farmers remained on the farm. While persistent symptoms correlated with evidence of persistent disease, it could not be explained by persistent exposure, and there was no evidence that ongoing exposure was causing permanent pulmonary damage.[96] Another study of 141 patients with Farmer's Lung concurred. The authors found no significant relationship between continued farming or length of disease and reduced lung function. The authors concluded that symptomatic recurrences, rather than continued exposure, correlated with progression of disease.[91]

Conversely, studies of bird breeders' hypersensitivity pneumonitis suggest that continued exposure to bird antigens can lead to progressive disease.[97-99] One study of 9 patients exposed to birds from 1 to 20 years found that only younger patients exposed to antigen for 6 months or less after symptoms developed were able to recover completely. Older patients with longer exposure had residual impairments, indicating persistent interstitial damage. In some instances, the changes were progressive, even despite removal from exposure. Neither the nature of the clinical presentation, whether acute or chronic, nor the degree or type of lung function abnormality, predicted outcome. Immediate removal from this antigen exposure was recommended.[100]

D. INORGANIC DUST OCCUPATIONAL LUNG DISEASE

Occupational and environmental exposures, especially the inorganic dusts, are the most common causes of interstitial lung disease. These include the mining, refining, and industrial use of silica, silicates (e.g., asbestos, talc, kaolin or "china clay"), aluminum, carbon (e.g., coal dust or graphite), beryllium, and hard metal dusts. Inhalation of metal fumes may also cause lung disease; these include the oxides of zinc, copper, manganese, cadmium, iron, magnesium, nickel, brass, selenium, tin, and antimony.[101]

Work in the dusty trades leads to chronic air flow limitation, as documented in multiple studies. Community-based studies, which compare exposed to nonexposed persons, longitudinal studies of lung function in workers in a specific occupation, pathology studies comparing exposure levels to development of emphysema, and cohort mortality studies, all support similar conclusions.[102,103] Exposed workers have a significantly elevated prevalence of chronic cough, chronic phlegm, persistent wheeze, and breathlessness, compared to the unexposed. Loss of lung function and obstructive pulmonary changes are more prevalent in exposed workers.[104-106] Smoking has been shown to exacerbate the effects of dust exposure in all studies. In some studies, the effect of gas and fumes appears synergistic with dust exposure,[106] in others, it was not.[104] Longitudinal workforce-based studies have shown significant annual lung function loss associated with exposure to organic and inorganic dusts.[107] Support for causality is shown by the effect of job changes; when this change was from heavy to lighter exposure, the annual loss of lung function was diminished.[107] Pathology studies in coal miners and South African hard rock gold miners have shown significant associations between mining exposure and the presence of emphysema at autopsy. Occupational dust exposure has also been shown to be a significant predictor of mortality, independent of the effects of smoking.[103]

One interesting study in hard rock miners has suggested that the effects of smoking and dust exposure are not additive. Rather, dust effects differ in never-smokers compared to smokers. In smokers, dust exposure was associated with obstructive changes: increased lung volume, lower flow rates, and lower DL_{co}/VA than that accounted for by smoking alone. In contrast, dust exposure in never-smokers was associated with more restrictive changes: decreased lung volume, increased flow rates, and increased DL_{co}/VA. These respiratory effects were observed several months after cessation of mining exposures, suggesting that dust-induced physiologic changes may be irreversible.[108]

Several other important conclusions emerged from these studies. Acute respiratory responses to occupational exposures, measured as cross-shift changes in FEV_1 or nonspecific BHR, were predictive of long-term airflow limitation and excessive annual loss of lung fuction. The development of acute and chronic respiratory changes appears to be based on individual susceptibility, since not all workers who are equally exposed are equally affected: this is the "Dutch hypothesis."[109]

In contrast, many of the classic pneumoconioses such as silicosis, asbestosis, and Coal Workers' Lung, appear to be nonspecific inflammatory reactions to irritant dusts. Development of disease is largely dependent on dose exposure to respirable particles (<10 μm). Workers with the highest exposures are the most likely to develop disease,[110-112] and individual susceptibility appears less important. Although certain autoimmune phenomena are observed in association with the pneumoconioses, including hypergammaglobulinemia, antinuclear antibodies, rheumatoid factor, and circulating immune complexes, these are felt to be nonspecific markers of chronic inflammation. A hypersensitivity response to silica, asbestos, or carbon is not thought to be involved in pathogenesis, and these chemicals will not be considered further here.

However, a hypersensitivity mechanism is felt to be causative in the lung diseases induced by beryllium and hard metal dust exposure, where the responsible agent has been shown to be cobalt. Both cobalt and beryllium exposure may cause dermatitis, conjunctivitis, irritation of the upper respiratory tract, and chronic bronchitis in cobalt-exposed workers. Whereas both can cause a granulomatous interstitial lung disease leading to interstitial fibrosis, cobalt has also been associated with occupational asthma documented by specific inhalation challenges. The incidence of disease in exposed workers is low. The prevalence of hard metal occupational asthma is reported at 5.6%; hard metal interstitial lung disease affects only 0.7% of exposed workers.[113] The prevalence of chronic beryllium lung disease is reported to be from 1 to 3%.[114] Specific immune mechanisms are implicated in the pathogenesis of both cobalt- and beryllium-induced diseases. Specific IgE antibodies to cobalt have been demonstrated by radioallergosorbent test (RAST) and skin testing in workers with hard metal asthma,[115] and not in exposed controls.[116] Bronchoalveolar lavage from patients with hard metal interstitial lung disease shows unusual multinucleated giant cells, numerous mononuclear macrophages, giant cells phagocytizing macrophages, and peri-bronchiolar chronic inflammation and fibrosis on biopsy.[113,117] Chronic beryllium disease is characterized by an abnormal T-helper lymphocytosis on bronchial lavage, and noncaseating granulomas or a diffuse mononuclear interstitial infiltrate on biopsy. T-lymphocyte proliferation in response to beryllium is seen in both blood and lavage cells in sensitized individuals. Thus, development of disease in only a subset of exposed workers, and the demonstration of specific immune changes, suggests that hypersensitivity mechanisms are critical to the pathogenesis of beryllium and hard metal lung disease.

IV. MULTIPLE CHEMICAL SENSITIVITIES

Multiple chemical sensitivities (MCS) has been defined by Cullen in 1987 as "an acquired disorder characterized by recurrent symptoms, referable to multiple organ systems, occurring in response to demonstrable exposure to many chemically unrelated compounds at doses far below those established in the general population to cause harmful effects. No single widely accepted test of physiologic function can be shown to correlate with symptoms."[118] MCS is derived from earlier concepts of environmental or ecologic illness (EI) proposed by Randolph and later by Rea, in which exposure to environmental incitants such as common foods and chemicals triggers a variety of symptoms. The journal *Clinical Ecology* defines ecologic illness as a "polysymptomatic, multisystem chronic disorder manifested by adverse reactions to environmental excitants as they are modified by individual susceptibility in terms of specific adaptations. The excitants are present in air, water, drugs, and our habitats."[119] Other syndromes considered to overlap with MCS and EI include cerebral allergy, chemically-induced immune dysregulation, 20th-century disease, and total allergy syndrome.

The key components of the disease include (1) multiple symptoms affecting many organ systems, (2) adverse reactions to ubiquitous environmental agents or chemicals, (3) reactions that may occur at very low exposure doses,

(4) reactions that are idiosyncratic, and (5) no objective measures that correlate with symptoms. The most frequent complaints involve the upper and lower respiratory tracts, including cough, hoarseness, sore throat, recurrent otitis, rhinitis, eye irritation, recurrent respiratory infection, aphonia, or asthma. Multiple other nonspecific complaints are also common, including headache, fatigue, malaise, insomnia, weakness, weight change, memory loss, and dark circles under the eyes. Cutaneous complaints include rash, facial swelling, burning, pruritis, easy bruising, eczema, or peripheral edema. Neuromuscular symptoms including numbness, poor coordination, lack of concentration, tremor, visual disturbances, syncope, muscle aches and spasms, and even paralysis have been described. The cardiovascular system may be involved, with complaints of chest pain, tachycardia, arrhythmias, palpitations, or hypertension. Gastrointestinal complaints include abdominal pain and bloating, nausea, vomiting, increased appetite, or anorexia. Psychologic complaints are common, such as confusion, anxiety, agitation, fright, panic, anger, hyperventilation, or depression.[120,121]

The list of putative causative agents is endless, although synthetic chemicals are considered to be major offenders. The most frequently cited chemicals include formaldehyde, phenol, ethanol, ammonia, hydrocarbons, and petrochemicals. Environmental exposures at home or work include cleaning solvents, paints, perfumes, synthetic clothing, pesticides, structural plastics, building construction materials, new carpeting, smoke, gasoline, vehicle exhaust fumes, and fumes from office machines. In general, most patients first develop symptoms at work, and are more symptomatic in stores, shopping malls, and public buildings than at home. Other causes of illness include exposure to natural gas, electromagnetic radiation, yeast, viruses, fungi, dust, and especially foods. Most patients with MCS usually attribute their symptoms to a combination of environmental chemicals, foods, and drugs.

Once the symptom complex has been initiated, often by a single or sustained toxic exposure, symptoms are felt to be triggered by very low doses of the offending agent, as well as by many other, unrelated chemical exposures. Clinical ecologists postulate that the inciting chemical exposure has caused dysregulation and dysfunction of the patient's immune system. Reactions tend to be idiosyncratic, such that the symptoms of one patient to a series of chemicals are not referable to the possible reactions of another. Case series from the clinical ecology literature, therefore, are anecdotal. Comparison of the results of immune testing by clinical ecologists show conflicting results. Testing includes serum immunoglobulin levels, complement components, T-cell subsets, B-cell numbers, and inflammatory mediators such as histamine, prostaglandins, etc. Values are often in the normal range, or conflict between studies (e.g., high in one, low in another). Changes in immune markers, such as total eosinophil counts, total complement, IgG levels, or T-lymphocyte counts have been reported after environmental challenge or time in environmental control.[122-124] However, there are no consistent patterns of immune responses described in any series of patients. There are no comparison controls of immune markers between challenged and unchallenged patients, or between affected and unaffected patients, and double-blind controlled studies of any abnormality have not been done.

The diagnosis of MCS is established by an extensive environmental history, and is usually followed by a series of provocation-neutralization tests. The history emphasizes diet, food preferences, medications, previous use of corti-costeroids or antibiotics, and potential exposure to chemicals at home and at work. The historian elicits the patient's assessment of symptoms and their relationship to possible chemical exposure. Patients may then be admitted to special, environmentally "clean" units and placed on a 4- to 5-day detoxifying fast. Once the patient feels his or her symptoms have cleared, they then undergo food and chemical challenges to document incriminating substances. End-points are the reproduction of previously reported symptoms. Alternatively, patients may undergo provocation-neutralization tests, as initially described by Willoughby[125] for foods, which have been expanded to include a variety of chemicals as well. Doses are administered subcutaneously or sublingually; the dose which elicits symptoms is considered the provoking dose. Lower or higher doses are given until symptoms disappear; this is considered the neutralizing dose which the patient may subsequently use for self-treatment.

Treatment primarily consists of scrupulous avoidance of the implicated chemicals and foods. Drugs are avoided as well, since they are also chemicals, with the exception of nystatin, mycostatin, ketokonazole, or amphotericin for the treatment of candida hypersensitivity. Megadose vitamin therapy, mineral and amino acid supplements, and antioxidant treatments may also be prescribed. Neutralization therapy may be self-administered. The end-point of all therapies is the elimination of symptoms. However, none of these treatments has been shown to be effective, utilizing blinded trials of therapy.[126]

A number of theories have been presented in the clinical ecology literature to explain the phenomenon of MCS. Randolph suggested that "maladaptation to specific materials to which one is regularly exposed and susceptible presents as both physical and mental illnesses ... there are also highly significant variations in the same susceptible individual at different times, depending principally on the frequency of dosage and the stage of specific adaptation." He suggested that humans may adapt more satisfactorily to foods and chemicals to which they have been exposed, as a race, for a longer time, than to newer foods such as coffee, cereal grains, potato, citrus, etc., and to new aspects of the chemical environment, especially fossil fuels and their combustion products and derivatives.[127] Other theories suggest that environmental chemicals may act as haptens to trigger an immune response, or that MCS is an autoimmune phenomenon. Others have suggested that MCS is due to a relative deficiency in antioxidants, such that environmental chemicals are not detoxi-fied properly but cause toxic lipid peroxidation of cell membranes, and subsequent release of cell mediators. Total body load and chemical overload suggest that the immune system is unable to handle environmental toxins once its load has been exceeded. Masked food sensitivity is a phenomenon whereby a food which has been avoided for several days causes symptoms, but when eaten continuously, such symptoms disappear. The concept of a spreading phenomenon suggests that sensitivity to one environmental chemical may spread to involve immune sensitivity to a wide variety of chemicals.[126] However, none of these theories has been documented in any scientific system.

Established medicine has rejected the tenets of clinical ecology and MCS on several grounds. Firstly, MCS is a diagnosis based on the patient's self-reported symptom complex only. Any organ system may be involved, and any reported symptom is considered valid. There are no characteristic signs or symptoms, or physiologic and/or laboratory measurement of disease or immune function that has been consistently found in the majority of patients. Secondly, the diagnostic procedures used to confirm the diagnosis have serious flaws. The diagnostic end-point is the elicitation of symptoms, which is subjective. Double-blinded studies of the efficacy of sublingual or subcutaneous food testing have shown similar responses to food or to placebo.[128,129] Further, the tests were not reproducible from month to month. Treatment is based on symptom relief only, without changes in any objective parameters. The best of these critical studies was performed by Jewett et al., using the offices and technicians of seven experienced clinical ecologists. He comments that "a system of treatment based on the relief of symptoms alone may come to be dominated by placebo reactions. That the field of clinical ecology may have developed disease and treatment concepts based on placebo responses is suggested by the increase in the number of placebo responses that are considered to be symptoms of disease."[130] Lastly, the theories suggesting the pathogenesis of MCS lack scientific validation. Changes in immune system markers, if they occur at all, e.g., changes in white blood cell counts, complement levels, eosinophil counts, immunoglobulin levels, etc., before and after environmental challenge are neither reproducible nor predictable, and do not constitute causality. Further, none has been compared to placebo-challenged patients, or to unaffected but exposed subjects.

Despite these objections, ecological illness and MCS remain widespread diagnoses and clinical ecologists are busy. Because the practice of environmental medicine cannot be considered harmless, it is important to understand its appeal. Ecological illness speaks directly to a public already concerned about the health consequences of living in an increasingly polluted environment. Modern medicine is poor at addressing patient complaints that are vague, multiple, and with a psychological component. A recent editorial in the *Annals of Internal Medicine* suggests that clinical ecology has the characteristics of a cult which would appeal to such patients neglected by modern medicine. "Accusations made by clinical ecologists that organized medicine is prejudiced against them certainly represent one such feature. Another is self-portrayal as embattled fighters for new truths against an entrenched establishment. Most characteristic is reliance on testimonials and anecdotes as evidence and the lack of training in scientific methodology among the chief proponents of clinical ecology."[131] Whereas modern medicine needs to be open to new ideas and theories, it has the right to demand that such new theories be subject to the same scrutiny and scientific validation to which traditional medicine is subjected.

The practice of clinical ecology has a cost. Severe limitations in terms of food, work exposure, and lifestyle are imposed on patients, who often are reinforced in their isolation from family, work, and society at large. More importantly, the investment in MCS and its trappings and treatments often neglects addressing the real underlying issues that may be important in the pathogenesis of the symptoms perceived by these patients. Often MCS deflects diagnosis of true physical or pyschological problems. Furthermore, treatment is not cheap and imposes a severe financial burden on patients, health insurers, and workers' compensation systems. The use of diagnostic and treatment plans that have not been shown to be either safe or efficacious poses serious ethical problems, which traditional medicine has the obligation to address.

REFERENCES

1. **Jackson, R., Sears, M. R., Beaglehole, R., and Rea, H. H.,** International trends in asthma mortality: 1970 to 1985, *Chest,* 94:914–919, 1988.
2. **Buist, A. S.,** Postgraduate course. Asthma mortality: What have we learned? *J. Allergy Clin. Immunol.,* 84:275–283, 1989.
3. **Woolcock, A. J.,** Worldwide trends in asthma morbidity and mortality. Explanation of trends, *Bull. Int. Union Tuber. Lung Dis.,* 66:85–89, 1991.
4. **Gergen, P. J. and Weiss, K. B.,** The increasing problem of asthma in the United States, *Am. Rev. Respir. Dis.,* 146:823–824, 1992.
5. **Juel, K. and Pedersen, P. A.,** Increasing asthma mortality in Denmark 1969–88 not a result of a changed coding practice, *Ann. Allergy,* 68:180–182, 1992.
6. **Caraballo, L., Cadavid, A., and Mendoza, J.,** Prevalence of asthma in a tropical city of Colombia, *Ann. Allergy,* 68:525–529, 1992.
7. **Pierson, W. E. and Koenig, J. Q.,** Postgraduate course. Respiratory effects of air pollution on allergic disease, *J. Allergy Clin. Immunol.,* 90:557–566, 1992.
8. **Molfino, N. A., Slutsky, A. S., and Zamel, N.,** The effects of air pollution on allergic bronchial responsiveness, *Clin. Exp. Allergy,* 22:667–672, 1992.
9. **van der Lende, R., Kok, T., Peset, R., Quanjer, P. H., Schouten, J. P., and Orie, N. G. M.,** Long-term exposure to air pollution and decline in VC and FEV$_1$. Recent results from a longitudinal study in the Netherlands, *Chest,* 80:23S–26S, 1981.
10. **Detels, R., Sayre, J. W., Coulson, A. H., Rokow, S. N., and Massey, F. J., et al.,** Respiratory effect of longterm exposure to two mixes of air pollutants in Los Angeles county, *Chest,* 80:27S–29S, 1981.
11. **Detels, R., Tashkin, D. P., Sayre, J. W., Rokow, S. N., and Coulson, A. H., et al.,** The UCLA population study of chronic obstructive respiratory disease. Lung function changes associated with chronic exposure to photochemical oxidants; a cohort study among never smokers, *Chest,* 92:594–603, 1987.

12. **Xu, X., Christiani, D. C., Dockery, D. W., and Wang, L.,** Exposure-response relationship between occupational exposures and chronic respiratory illness: a community-based study, *Am. Rev. Respir. Dis.,* 146:413–418, 1992.

13. **Weiss, K. B., Gergen, P. J., and Hodgson, T. A.,** An economic evaluation of asthma in the United States, *N. Engl. J. Med.,* 326:862–866, 1992.

14. **Schwartz, J. and Marcus, A.,** Mortality and air pollution in London: a time series analysis, *Am. J. Epidem.,* 131:185–194, 1990.

15. **Oshima, Y., Ishizaki, T., Miyamoto, T., Shimizu, T., Shida, T., and Kabe, J.,** Air pollution and respiratory diseases in the Tokyo–Yokohama area, *Am. Rev. Respir. Dis.,* 90:572–581, 1964.

16. **Ishizaki, T., Makino, S., Miyamoto, T., and Kodama, T.,** A follow-up study of the daily incidence of respiratory symptoms among a group having chronic bronchitis in connection with air pollution of the Tokyo–Yokohama area, *Asian Med. J.,* 11:607–618, 1968.

17. **Sunyer, J., Anto, J. M., Murillo, C., and Saez, M.,** Effects of urban air pollution on emergency room admissions for chronic obstructive pulmonary disease, *Am. J. Epidem.,* 134:277–286, 1991.

18. **Schwartz, J.,** Air pollution and the duration of acute respiratory symptoms, *Arch. Environ. Health,* 47:116–122, 1992.

19. **Weiss, K. B. and Wagener, D. K.,** Changing patterns of asthma mortality. Identifying target populations at high risk, *JAMA,* 264:1683–1687, 1990.

20. **Gergen, P. J. and Weiss, K. B.,** Changing patterns of asthma hospitalizations among children: 1979–1987, *JAMA,* 264:1688–1692, 1990.

21. **Barry, D. M., Burr, M. L., and Limb, E. S.,** Prevalence of asthma among 12 year old children in New Zealand and South Wales: a comparative survey, *Thorax,* 46:405–409, 1991.

22. **Waite, D. A., Eyles, E. F., Tonkin, S. L., and O'Donnell, T. V.,** Asthma prevalence in Tokelauan children in two environments, *Clin. Allergy,* 10:71–75, 1980.

23. **Girsh, L. S., Shubin, E., Dick, C., and Schulaner, F. A.,** A study on the epidemiology of asthma in children in Philadelphia. The relation of weather and air pollution to peak incidence of asthmatic attacks, *J. Allergy,* 39:347–357, 1967.

24. **Richards, W., Azen, S. P., Weiss, J., Stocking, S., and Church, J.,** Los Angeles air pollution and asthma in children, *Ann. Allergy,* 47:348–354, 1981.

25. **Berciano, F. A., Dominguez, J., and Alvarez, F. V.,** Influence of air pollution on extrinsic childhood asthma, *Ann. Allergy,* 62:135–141, 1989.

26. **Ponka, A.,** Asthma and low level air pollution in Helsinki, *Arch. Environ. Health,* 46:262–270, 1991.

27. **Koenig, J. Q. and Pierson, W. E.,** Air pollutants and the respiratory system: toxicity and pharmacologic interventions, *Clin. Toxicol.,* 29:401–411, 1991.

28. **Anon. Editorial,** Ozone: too much in the wrong place, *Lancet,* 338:221–222, 1991.

29. **Zwick, H., Popp, W., Wagner, C., Reiser, K., Schmoger, J., Bock, A., Herkner, K., and Radunsky, K.,** Effects of ozone on the respiratory health, allergic sensitization and cellular immune system in children, *Am. Rev. Respir. Dis.,* 144:1075–1079, 1991.

30. **Koren, H. S., Devlin, R. B., Graham, D. E., Mann, R., and McGee, M. P., et al.,** Ozone-induced inflammation in the lower airways of human subjects, *Am. Rev. Respir. Dis.,* 139:407–415, 1989.

31. **Weiss, S. T.,** Indoor and outdoor air pollution and airway disease, *Mediguide Pulm. Med.,* 3, 1986.

32. **Bethel, R. A., Epstein, J., Sheppard, D., Nadal, J. A., and Boushey, H. A.,** Sulfur dioxide-induced bronchoconstriction in freely breathing, exercising, asthmatic subjects, *Am. Rev. Respir. Dis.,* 128:987–990, 1983.

33. **McManus, M. S., Koenig, J. Q., Altman, L. C., and Pierson, W. E.,** Pulmonary effects of sulfur dioxide exposure and ipratropium bromide pretreatment in adults with non-allergic asthma, *J. Allergy Clin. Immunol.,* 83:619–626, 1989.

34. **Utell, M. J., Frampton, M. W., and Morrow, P. E.,** Air pollution and asthma: clinical studies with sulfuric acid aerosols, *Allergy Proc.,* 12:385–388, 1991.
35. **Sandstrom, T., Stjernberg, N., Andersson, M-C., Hedman, B. K., Lundgren, R., Rosenhall, L., and Angstrom, T.,** Cell response in bronchoalveolar lavage fluid after exposure to sulfur dioxide: a time response study, *Am. Rev. Respir. Dis.,* 140:1828–1831, 1989.
36. **Spiezer, F. E., Ferris, B., Bishop, Y. M. M., and Spengler, J. D.,** Respiratory disease rates and pulmonary function in children associated with NO₂ exposure, *Am. Rev. Respir. Dis.,* 121:3–10, 1980.
37. **Mostardi, R. A., Woebkenberg, N. R., Ely, D. L., Conlon, M., and Atwood, G.,** The University of Akron study on air pollution and human health effects II. Effects on acute respiratory illness, *Arch. Environ. Health,* 36:250–255, 1981.
38. **Orchek, J., Massari, J. P., Gaynard, P., Grimaud, C., and Charpin, J.,** Effect of short-term, low-level nitrogen dioxide exposure on bronchial sensitivity of asthmatic patients, *J. Clin. Invest.,* 57:301–307, 1976.
39. **Ahmed, T., Marchette, B., and Dauta, I., et al.,** Effect of 0.1 ppm NO₂ on bronchial reactivity in normals and subjects with bronchial asthma, *Am. Rev. Respir. Dis.,* 125:152S, 1982.
40. **Kleinman, M. T., Baily, R. M., and Linn, W. S., et al.,** Effect of 0.2 ppm nitrogen dioxide on pulmonary function and response to bronchoprovocation in asthmatics, *J. Toxicol. Environ. Health,* 12:815–826, 1983.
41. **Bauer, M. A., Utell, M. J., Morrow, P. E., Speers, D. M., and Gibb, F. R.,** Inhalation of 0.30 ppm nitrogen dioxide potentiates exercise-induced bronchospasm in asthmatics, *Am. Rev. Respir. Dis.,* 134:1203–1208, 1986.
42. **Koenig, J. Q., Covert, D. S., Marshall, S. G., van Belle, G., and Pierson, W. E.,** The effects of ozone and nitrogen dioxide on pulmonary function in healthy and in asthmatic adolescents, *Am. Rev. Respir. Dis.,* 136:1152–1157, 1987.
43. **Magnussen, H.,** Experimental exposures to nitrogen dioxide, *Eur. Resp. J.,* 5:1040–1042, 1992.
44. **Dockery, D. W., Speizer, F. E., Stram, D. O., Ware, J. H., Spengler, J. D., and Ferris, B. G.,** Effects of inhalable particles on respiratory health of children, *Am. Rev. Respir. Dis.,* 139:587–594, 1989.
45. **Pope, C. A., Dockery, D. W., Spengler, J. D., and Raizenne, M. E.,** Respiratory health and PM₁₀ pollution. A daily time series analysis, *Am. Rev. Respir. Dis.,* 144:668–674, 1991.
46. **Tseng, R. Y. M., Li, C. K., and Spinks, J. A.,** Particulate air pollution and hospitalization for asthma, *Ann. Allergy,* 68:425–432, 1992.
47. **Samet, J. M., Marburg, M. C., and Spengler, J. D.,** Respiratory effects of indoor air pollution, Fourth Ann. Aspen Allergy Conf., *J. Allergy Clin. Immunol.,* 79:685–700, 1987.
48. **Young, S., LeSouef, P. N., Geelhoed, G. C., Stick, S. M., Turner, K. J., and Landau, L. I.,** The influence of a family history of asthma and parental smoking on airway responsiveness in early infancy, *N. Engl. J. Med.,* 324:1168–1173, 1991.
49. **Martinez, F. D., Antognoni, G., Macri, F., Bonci, E., Midulla, F., deCastro, G., and Ronchetti, R.,** Parental smoking enhances bronchial responsiveness in nine-year-old children, *Am. Rev. Respir. Dis.,* 138:518–523, 1988.
50. **Gortmaker, S. L., Walker, D. K., Jacobs, F. H., and Ruch-Ross, H.,** Parental smoking and the risk of childhood asthma, *Am. J. Public Health,* 72:574–579, 1982.
51. **Section of Allergy, Canadian Paediatric Society,** Secondhand cigarette smoke worsens symptoms in children with asthma, *Can. Med. Assoc. J.,* 135:321–323, 1986.
52. **Murray, A. B. and Morrison, B. J.,** Passive smoking and the seasonal difference of severity of asthma in children, *Chest,* 94:701–708, 1988.
53. **White, J. R. and Froeb, H. F.,** Small-airways dysfunction in nonsmokers chronically exposed to tobacco smoke, *N. Engl. J. Med.,* 302:702–703, 1980.
54. **Dahms, T. E., Bolin, J. E., and Slavin, R. G.,** Passive smoking. Effects on bronchial asthma, *Chest,* 80:530–534, 1981.

55. **Stankus, R. P., Menon, P., Rando, R., Glindmeyer, H., Salvaggio, J., and Lehrer, S.,** Cigarette smoke-sensitive asthma: challenge studies, *J. Allergy Clin. Immunol.,* 82:331–338, 1988.

56. **Knight, A. and Breslin, A. B. X.,** Passive cigarette smoking and patients with asthma, *Med. J. Austr.,* 142:194–195, 1985.

57. **Menon, P., Rando, R. J., Stankus, R. P., Salvaggio, J. E., and Lehrer, S. B.,** Passive cigarette smoke-challenge studies: increase in bronchial hyperreactivity, *J. Allergy Clin. Immunol.,* 89:560–566, 1992.

58. **Imbus, H. R.,** Clinical evaluation of patients with complaints related to formaldehyde exposure, *J. Allergy Clin. Immunol.,* 76:831–840, 1985.

59. **Bardana, E. J. and Montanero, A.,** Formaldehyde: an analysis of its respiratory, cutaneous, and immunologic effects, *Ann. Allergy,* 66:441–452, 1991.

60. **Honicky, R. E., Akpom, C. A., and Osborne, J. S.,** Infant respiratory illness and indoor air pollution from a woodburning stove, *Pediatrics,* 71:126–128, 1983.

61. **Honicky, R. E., Osborne, J. S., and Akpom, C. A.,** Symptoms of respiratory illness in young children and the use of woodburning stoves of indoor heating, *Pediatrics,* 75:587–593, 1985.

62. **Morris, K., Morganlander, M., Coulehau, J. L., Gahagan, S., and Arena, V. C.,** Woodburning stoves and lower respiratory tract infection in American indian children, *Am. J. Dis. Child,* 144:105–108, 1990.

63. **Tuthill, R. W.,** Woodstoves, formaldehyde, and respiratory disease, *Am. J. Epidemiol.,* 120:952–955, 1984.

64. **Finnegan, M. J., Pickering, C. A. C., and Burge, P. S.,** The sick building syndrome: prevalence studies, *Br. Med. J.,* 289:1573–1575, 1984.

65. **Letz, G. A.,** Sick building syndrome: acute illness among office workers — the role of building ventilation, airborne contaminants and work stress, *Allergy Proc.,* 11:109–116, 1990.

66. **Menzies, R., Tamblyn, R., Farant, J.-P., Hanley, J., Nunes, F., and Tamblyn, R.,** The effect of varying levels of outdoor-air supply on the symptoms of sick building syndrome, *N. Engl. J. Med.,* 328:821–827, 1993.

67. **National Institute for Occupational Safety and Health,** National Occupational Hazard Survey 1972–1974. DHEW Publ. No. 78-114. Cincinnati, NIOSH, 1978.

68. **Nicholson, W. J., Perkel, G., and Selikoff, I. J.,** Occupational exposure to asbestos: population at risk and projected mortality — 1820–2030, *Am. J. Ind. Med.,* 3:259–312, 1982.

69. **Chan-Yeung, M.,** Occupational asthma, *Chest,* 98:148S–161S, 1990.

70. **Blanc, P.,** Occupational asthma in a national disability survey, *Chest,* 92:613–617, 1987.

71. **Zeiss, C. R.,** Occupational immunologic lung disease, *Allergy Proc.,* 11:175, 1990.

72. **Zeiss, C. R., Mitchell, J. H., Van Peenen, P. F. D., Kavich, D., et al.,** A clinical and immunologic study of employees in a facility manufacturing trimellitic anhydride, *Allergy Proc.,* 13:193–198, 1992.

73. **Smith, D. D.,** Medical-legal definition of occupational asthma, *Chest,* 98:1007–1011, 1990.

74. **Burge, P. S.,** New developments in occupational asthma, *Br. Med. Bull.,* 48:221–230, 1991.

75. **Butcher, B. T. and Salvaggio, J. E.,** Occupational asthma, *J. Allergy Clin. Immunol.,* 78:547–556, 1986.

76. **Agius, R. M., Nee, J., McGovern, B., and Robertson, A.,** Structure activity hypothesis in occupational asthma caused by low molecular weight substances, *Ann. Occup. Hyg.,* 35:129–137, 1991.

77. **Perrin, B., Cartier, A., Ghezzo, H., Grammer, L., Harris, K., Chan, H., Chan-Yeung, M., and Malo, J.-L.,** Reassessment of the temporal patterns of bronchial obstruction after exposure to occupational sensitizing agents, *J. Allergy Clin. Immunol.,* 87:630–639, 1991.

78. **Brugsch, H. G. and Elkins, H. B.,** Toluene di-isocyanate (TDI) toxicity, *N. Engl. J. Med.,* 268:353–357, 1963.

79. **Banks, D. E., Sastre, J., Butcher, B. T., Ellis, E., Rando, R. J., et al.,** Role of inhalation challenge testing in the diagnosis of isocyanate-induced asthma, *Chest,* 95:414–423, 1989.

80. **Chan-Yeung, M., Barton, G. M., MacLean, L., and Grzybowski, S.,** Occupational asthma and rhinitis due to western red cedar (Thuja plicata), *Am. Rev. Respir. Dis.,* 108:1094–1102, 1973.

81. **Chan-Yeung, M., MacLean, L., and Paggiaro, P. L.,** Follow-up study of 232 patients with occupational asthma caused by western red cedar (Thuja plicata), *J. Allergy Clin. Immunol.,* 79:792–796, 1987.

82. **Burge, P. S., Perks, W. H., O'Brien, I. M., Burge, A., Hawkins, R., Brown, D., and Green, M.,** Occupational asthma in an electronics factory: a case control study to evaluate etiological factors, *Thorax,* 34:300–307, 1979.

83. **Brooks, S. M., Weiss, M. A., and Bernstein, I. L.,** Reactive airway dysfunction syndrome (RADS). Persistent asthma syndrome after high level irritant exposures, *Chest,* 88:376–384, 1985.

84. **Campbell, J. M.,** Acute symptoms following work with hay, *Br. Med. J.,* 12:1143–1166, 1932.

85. **Fink, J. N., Sosman, A. J., Barboriak, J. J., Schlueter, D. P., and Holmes, R. A.,** Pigeon breeders' disease. A clinical study of a hypersensitivity pneumonitis, *Ann. Int. Med.,* 68:1205–1219, 1968.

86. **Fink, J. N.,** Clinical features of hypersensitivity pneumonitis, *Chest,* 89:193S–195S, 1986.

87. **Richerson, H. B., Fink, J. N., Reed, C. E., and Schwartz, H. J.,** Guidelines for the clinical evaluation of hypersensitivity pneumonitis, *J. Allergy Clin. Immunol.,* 84:839–844, 1989.

88. **Bourke, S., Anderson, K., Lynch, P., Boyd, J., King, S., Banham, S., and Boyd, G.,** Chronic simple bronchitis in pigeon fanciers. Relationship of cough with expectoration to avian exposure and pigeon breeders' disease, *Chest,* 95:598–601, 1989.

89. **Rose, C. and King, T. E.,** Controversies in hypersensitivity pneumonitis, *Am. Rev. Respir. Dis.,* 145:1–2, 1992.

90. **Sennekamp, J., Niese, D., Stroehmann, I., and Rittner, C.,** Pigeon breeders' lung lacking detectable antibodies, *Clin. Allergy,* 8:305–310, 1978.

91. **Fink, J. N., Barboriak, J. J., and Sosman, A. J.,** Immunologic studies of pigeon breeder's disease, *J. Allergy,* 39:214–221, 1967.

92. **Salvaggio, J. E.,** Recent advances in pathogenesis of allergic alveolitis, *Clin. Exp. Allergy,* 20:137–144, 1990.

93. **Takizawa, H., Ohta, K., Horiuchi, T., Suzuki, N., Ueda, T., et al.,** Hypersensitivity pneumonitis in athymic nude mice. Additional evidence of T cell dependency, *Am. Rev. Respir. Dis.,* 146:479–484, 1992.

94. **Dickie, H. A. and Rankin, J.,** Farmer's lung: an acute granulomatous disease occurring in agricultural workers, *JAMA,* 167:1069, 1958.

95. **Emanuel, D. A., Wenzel, F. J., Bowerman, C. F., and Lawton, B. R.,** Farmer's lung: clinical, pathologic, and immunologic study of twenty-four patients, *Am. J. Med.,* 37:392–401, 1964.

96. **Barbee, R. A., Callies, Q., Dickie, H. A., and Rankin, J.,** The long-term prognosis in farmer's lung, *Am. Rev. Respir. Dis.,* 97:223–231, 1968.

97. **Schlueter, D. P., Fink, J. N., and Sosman, A. J.,** Pulmonary function in pigeon breeders' disease. A hypersensitivity pneumonitis, *Ann. Int. Med.,* 70:457–470, 1969.

98. **Dinda, P., Chatterjee, S. S., and Riding, W. D.,** Pulmonary function studies in bird breeder's lung, *Thorax,* 24:374–378, 1969.

99. **Riley, D. J. and Saldana, M.,** Pigeon breeder's lung. Subacute course and the importance of indirect exposure, *Am. Rev. Respir. Dis.,* 107:456–460, 1973.

100. **Allen, D. H., Williams, G. V., and Woolcock, A. J.,** Bird breeder's hypersensitivity pneumonitis: progress studies of lung function after cessation of exposure to the provoking antigen. *Am. Rev. Respir. Dis.,* 114:555–566, 1976.

101. **Schwarz, M. I. and King, T. E.,** Eds. *Interstitial Lung Disease.* 2nd ed., Mosby-Year Book, St. Louis, 1993.
102. **Becklake, M. R.,** Chronic airflow limitation: its relationship to work in dusty occupations, *Chest,* 88:608–617, 1985.
103. **Becklake, M. R.,** Occupational exposures: evidence for a causal association with chronic obstructive pulmonary disease, *Am. Rev. Respir. Dis.,* 140:S85–S91, 1989.
104. **Korn, R. J., Dockery, D. W., Speizer, F. E., Ware, J. H., and Ferris, B. G.,** Occupational exposures and chronic respiratory symptoms. A population-based study, *Am. Rev. Respir. Dis.,* 136:298–304, 1987.
105. **Viegi, G., Prediletto, R., Paoletti, P., Carrozzi, L., diPede, F., et al.,** Respiratory effects of occupational exposure in a general population sample in North Italy, *Am. Rev. Respir. Dis.,* 143:510–515, 1991.
106. **Xu, X., Christiani, D. C., Dockery, D. W., and Wang, L.,** Exposure-response relationships between occupational exposures and chronic respiratory illness: a community-based study, *Am. Rev. Respir. Dis.,* 146:413–418, 1992.
107. **Kauffmann, F., Drouet, D., Lellouch, J., and Brille, D.,** Occupational exposure and 12-year spirometric changes among Paris area workers, *Br. J. Ind. Med.,* 39:221–232, 1982.
108. **Kreiss, K., Greenberg, L. M., Kogut, S. J. H., Lezotte, D. C., Irvin, C. G., and Cherniack, R. M.,** Hard-rock mining exposures affect smokers and nonsmokers differently. Results of a community prevalence study, *Am. Rev. Respir. Dis.,* 139:1487–1493, 1989.
109. **Van der Lende, R.,** A critical analysis of three field surveys of CNSLD carried out in the Netherlands, in: *The Epidemiology of Chronic Nonspecific Lung Disease,* Vol. 1, Van Gorcum & Co., Assen, 1969, Charles C. Thomas, USA, 1969.
110. **Official statement of the American Thoracic Society.** The diagnosis of nonmalignant diseases related to asbestos, *Am. Rev. Respir. Dis.,* 134:363–368, 1986.
111. **Ng, T.-P., Chan, S.-L., and Lam, K.-P.,** Radiological progression and lung function in silicosis: a ten year follow up study, *Br. Med. J.,* 295:164–168, 1987.
112. **Reisner, M. T. R.,** Results of epidemiologic studies on the progression of coal workers' pneumoconiosis, *Chest,* 78:406S–407S, 1980.
113. **Austenfeld, J. and Colby, T. V.,** Recognizing lung diseases induced by hard metal exposure, *J. Respir. Dis.,* 10:65–75, 1989.
114. **Newman, L. S., Kreiss, K., King, T. E., Seay, S., and Campbell, P. A.,** Pathologic and immunologic alterations in early stages of beryllium disease. Re-examination of disease definition and natural history, *Am. Rev. Respir. Dis.,* 139:1479–1486, 1989.
115. **Shirakawa, T., Kusaka, Y., Fujimura, N., Goto, S., and Morimoto, K.,** The existence of specific antibodies to cobalt in hard metal disease, *Clin. Allergy,* 18:451–460, 1988.
116. **Shirakawa, T., Kusaka, Y., Fujimura, N., Goto, S., Kato, M., et al.,** Occupational asthma from cobalt sensitivity in workers exposed to hard metal dust, *Chest,* 95:29–37, 1989.
117. **Davison, A., Haslam, P. L., Corrin, B., Coutts, I., Dewar, A., Riding, W. D., Studdy, P. R., and Newman-Taylor, A. J.,** Interstitial lung disease and asthma in hard-metal workers: bronchoalveolar lavage, ultrastructural, and analytical findings and results of bronchial provocation tests, *Thorax,* 38:119–128, 1983.
118. **Cullen, M. R.,** The worker with multiple chemical hypersensitivities: an overview, *State Art. Rev. Occup. Med.,* 2:655–661, 1987.
119. *Clinical Ecology,* Definition in each issue.
120. **Black, D. W., Rathe, A., and Goldstein, R. B.,** Environmental illness: a controlled study of 26 subjects with '20th century disease,' *JAMA,* 264:3166–3170, 1990.
121. **Terr, A. I.,** Clinical ecology in the workplace, *J. Occ. Med.,* 31:257–261, 1989.
122. **Rea, W. J.,** Environmentally triggered small vessel vasculitis, *Ann. Allergy,* 38:245–251, 1977.
123. **Rea, W. J.,** Environmentally triggered cardiac disease, *Ann. Allergy,* 40:243–251, 1978.

124. **Rea, W. J., Bell, I. R., Suits, C. W., and Smiley, R. E.,** Food and chemical susceptibility after environmental chemical overexposure: case histories, *Ann. Allergy,* 41:101–110, 1978.

125. **Willoughby, J. W.,** Provocative food test technique, *Ann. Allergy,* 23:543–554, 1965.

126. **American College of Physicians,** Clinical ecology, *Ann. Int. Med.,* 111:168–178, 1989.

127. **Randolph, T. G.,** Ecologic orientation in medicine: comprehensive environmental control in diagnosis and therapy, *Ann. Allergy,* 23:7–22, 1965.

128. **Crawford, L. V., Lieberman, P., Harfi, H. A., Hale, R., Nelson, H., Selner, J., Wittig, H., Postman, M., and Zietz, H.,** A double-blind study of subcutaneous food testing sponsored by the Food Committee of the American Academy of Allergy (abstr. 114), *J. Allergy Clin. Immunol.,* 57:236, 1976.

129. **Lehman, C. W.,** A double-blind study of sublingual provocative food testing: a study of its efficacy, *Ann. Allergy,* 45:144-149, 1980.

130. **Jewett, D. L., Fein, G., and Greenberg, M. H.,** A double-blind study of symptom provocation to determine food sensitivity, *N. Engl. J. Med.,* 323:429–433, 1990.

131. **Kahn, E. and LeFe, G.,** Clinical ecology: environmental medicine or unsubstantiated theory? *Ann. Int. Med.,* 111:104–106, 1989.

Chapter 5

DIAGNOSIS AND TREATMENT
OF RESPIRATORY ALLERGY

Karin A. Pacheco and Lanny J. Rosenwasser

CONTENTS

I. DIAGNOSIS

A. ASTHMA

Asthma has been defined as a clinical syndrome characterized by symptoms of dyspnea, wheezing, and coughing, and bronchial hyperresponsiveness manifested by variable airflow obstruction.[1] Asthma may initially present as chest tightness or heaviness. The patient may complain of a nonproductive cough, associated with wheezing or dyspnea. The cough may become productive of small amounts of viscous, mucoid sputum which may appear purulent. Purulence may in fact represent eosinophilia rather than the neutrophilia of infection. Mild asthma may have wheezing only on forced expiration. The more severe asthmatic may display wheezing on quiet expiration and inspiration. In severe asthma, increasing airways obstruction and decreased airflow may result in the abatement of wheezing. Other signs of severe airways obstruction, including restlessness, agitation, orthopnea, tachypnea, pursed lip breathing with a prolonged expiratory phase, use of accessory muscles of respiration, and difficulty speaking may be present. At that point, a widened pulse pressure, and pulsus paradoxus > 10 mmHg may be seen. The arterial blood gas is usually normal in controlled asthma. Moderate bronchospasm is associated with a respiratory alkalosis, a normal PO_2, and a widened (alveolar–arterial) gradient. Severe bronchospasm is associated with hypercapnia and a respiratory acidosis, when hypoxia may become apparent.

Airflow obstruction may be documented by peak flow measurements, spirometry, response to bronchodilators, or measurement of nonspecific bronchial hyperreactivity.[2] Normal mean diurnal variation in PEFR is 8%, with highest values occurring between 12 to 5 p.m., and lowest values at 12 to 5 a.m. A mean diurnal variation of 20% or greater is indicative of active asthma, and correlates with bronchial hyperreactivity.[3] PEFR should be recorded at least four times a day to document asthma. Daily variability may be calculated as:

$$\frac{\text{highest PEFR} - \text{lowest PEFR}}{\text{highest PEFR}} \times 100$$

Obstruction on spirometry may be defined as a disproportionate reduction in FEV_1 compared to FVC or vital capacity (VC) due to early airtrapping. Current ATS guidelines recommend against using fixed FEV_1/FVC ratios to define the lower limits of normal, since limits are variable and depend on patient characteristics. They suggest using FEV_1 instead as a measure of severity of obstruction.[4] Decreases in FEF_{25-75} are not felt to be specific enough to diagnose small airways disease in the individual patient. Mild obstruction is defined as an FEV_1 between 70% and 100% of predicted; moderate obstruction is an FEV_1 between 60% and 70% of predicted. Older references define obstruction as an FEV_1/FVC ratio < 70% predicted or a TLC > 120% predicted, RV > 120% predicted or FRC (TGV) > 120% predicted.[5] A significant bronchodilator response has most recently been defined as a 12% improvement in FEV_1 or FVC over baseline and an absolute change of 200 ml.[4]

Bronchial challenge with a variety of agents can document nonspecific bronchial hyperresponsiveness. Methacholine or histamine are most commonly used; isocapnic cold air inhalation is an alternative. Methacholine and histamine may measure different phenomena, and some authors feel the two methods cannot be used interchangeably.[6] Methacholine has been shown to be a better predictor of asthma severity and steroid requirements, and a better discriminator between asthmatics and normals. However, if methacholine hyperresponsiveness is lacking in a patient with a clinical picture of asthma, it is reasonable to repeat the challenge with histamine. A review of the specific methodology of bronchial inhalation challenge may be found in *Chest*.[7] Some form of objective measurement is important to establish the diagnosis of asthma, since the symptom complex alone is not sensitive enough.

1. Occupational Asthma

Occupational asthma has been defined as variable airways narrowing causally related to exposure in the working environment to airborne dusts, gases, vapors, or fumes.[8] Other suggested criteria include exposure to a specific sensitizing agent in the workplace to which unexposed individuals are not ordinarily sensitive. A variable symptomless period of sensitization should precede the development of symptoms. Workplace exposure resulting in an abnormal state of sensitization to a specific agent must be proven. A drop of 20% or greater in FEV_1 after exposure to the workplace agent, at a concentration lower than that producing a nonspecific irritant response in normals or in unexposed asthmatics, is indicative of sensitization.[9] Controversy with these suggested criteria abound, as they exclude those with preexisting asthma exacerbated by irritant exposure in the workplace. Also excluded are those developing bronchial hyperreactivity in response to high level toxic or irritant exposures at work (e.g., reactive airways dysfunction syndrome, RADS), where the agent cannot be considered a sensitizer.

Defining occupational asthma requires: (1) diagnosing asthma by both clinical and objective criteria; (2) establishing work-relatedness of symptoms and (3) determining the etiologic agent involved.

The clinical diagnosis of occupational asthma is established by a careful history of symptom patterns, home and work exposures, and a detailed assessment of the job duties and work environment. An important clue to occupational sensitization may be the preceding or concurrent development of upper respiratory tract (ocular and nasal) or skin sensitization, e.g., watery, burning eyes, rhinitis, hives, or dermatitis. The work-relatedness of symptoms may be difficult to ascertain. Asthma symptoms often initially occur late, several hours after coming home from work or at night. Only later will symptoms occur soon after exposure at work. Symptoms may improve on weekends away from work but may not; there should be some improvement after at least 2 weeks away from exposure. A patient history of atopy or smoking may have predisposing or confounding effects.

However, a clinical history alone is not sensitive enough to diagnose occupational asthma. One study of 162 workers compared the sensitivity of diagnosing occupational asthma by symptom questionnaire alone, with concurrent confirmation by specific inhalation challenge or PEFR monitoring, or both. The predictive value of a positive history was only 63%, whereas the predictive value of a negative history was 83%.[10] Clearly, objective criteria are also needed to make the diagnosis.

Objective tests to establish the work-relatedness of asthma include on- and off-the-job serial peak flow monitoring, serial measurements of nonspecific bronchial hyperreactivity, or specific agent inhalation challenge.

Serial peak flow monitoring is cheap and easy to do. When performed correctly, PEFR can reveal a work-related pattern of bronchoconstriction. For best results, they must be done every 2 hr while awake, with at least 3 blows each time with <10% variability. PEFR should be recorded for at least three working periods separated by periods away from work. At least one of the periods away should be 1 week or greater. Several studies have documented the validity of peak flow monitoring in establishing the diagnosis of asthma, with

93 to 100% sensitivity, and 45 to 77% specificity.[11-14] Drawbacks are that the compliance and honesty of the subject may affect the reading, 4 to 6 weeks may be needed to obtain valid results, and re-exposing the worker to the workplace may be dangerous. Work-related patterns may not be readily apparent, especially if the time away from work is not long enough. However, PEFR monitoring is a simple, direct, and inexpensive way of documenting a work-related pattern to asthma.

Documenting nonspecific bronchial hyperreactivity (BHR) as related to work exposure may also establish work-related asthma. BHR may increase while the subject is at work and revert to normal when the subject is removed from exposure.[15] However, normal BHR does not refute the diagnosis of occupational asthma, but may simply reflect the removal from a sensitizing exposure. One review of four cases of occupational asthma noted negative methacholine challenges in subjects away from work, with reversion to positive after exposure at work or through specific agent challenge.[16] Although the method is reliable and reproducible, drawbacks include nonspecificity, e.g., other agents such as allergen exposure or viral infection may also cause increased BHR. A lack of changes with exposure does not rule out occupational asthma, and the test does not define the etiologic agent.

Inhalation challenge testing is regarded as the gold standard to establish the presence of occupational asthma and to determine the etiologic agent. Pepys and Hutchcroft first proposed detailed protocols and described specific challenges in 1975.[17] They suggested that the challenges should be safe and reproducible, and that they should be performed under controlled settings. The chemical exposure should never exceed the recommended workplace exposure limits. This means that challenges should be performed in a hospital setting with resuscitative equipment available. Patients should ideally stay overnight to monitor and treat possible late reactions. Chemicals may be inhaled by (1) simulating the work environment (e.g., painting, sanding, soldering, etc.), (2) in nebulized solutions, or (3) mixed with dried lactose powder and tipped from one receiver to another such that the subject inhales the dust. Some dusts, such as flour or sawdust, can be used directly without dilution. Direct nebulization inhalation challenges with toxic volatile chemicals such as isocyanates are currently considered unethical.[18] In inhalation challenges simulating the work environment, the amount of chemical inhaled is difficult to quantitate, especially compared to nebulized solutions. Nevertheless, when they are performed for the same amount of time in the same space under the same conditions, the results have been reproducible.

Challenges should be performed in patients with stable pulmonary function, and away from work long enough to allow nonspecific BHR to stabilize as well. If possible, the patient should stop all medication which might mask a late response for several weeks, and suspend bronchodilator use for 6 to 8 hr prior to the challenge. A drawback is that occupational sensitization may no longer be readily apparent in a patient stabilized away from exposure.

If possible, the patient should be blinded as to the investigator's suspicion of the causative agent, and as to the actual exposures of the challenge. In practice, this may be difficult to do. A baseline exposure to the vehicle without the implicated agent should be performed on the control day, for example, to paint without the isocyanate hardener, to solder without colophony containing flux, to dried lactose powder without the platinum salt, etc. On the second challenge day, exposures mimicking the work environment should consist of short exposures with 10-min intervals between exposures. Nebulized inhalations should contain the minimum amount of material possible, which, where applicable, may be based on the amount eliciting a positive skin test. Inhalations should initially be brief, 30 to 60 seconds, and followed by longer periods of 1 to 2 min, up to 5 min, at 10-min intervals. Spirometry should be obtained between each challenge, then at 10 to 15 min after challenge for the first hour, and hourly for the next 6 to 7 hr for a late response. A fall in FEV_1 from 10 to 20%, for an average 15%, is considered a positive result. Where a late reaction is predicted on the basis of the patient's history or the chemical's toxicity, the patient should be hospitalized and monitored overnight. Challenges should not be oriented toward eliciting a positive immediate response, since these are often absent in occupational asthma, and the approach may result in an unusually severe late response. The researcher must keep an open mind in interpreting results, as patterns other than the typical early, late, or dual response may appear. A comparison of responses to IgE sensitizing agents (flour, psyllium, and guar gum), Western red cedar, and isocyanates showed that isocyanates often induced an atypical response. Alternate patterns included a progressive decline peaking at 5 hr, a square-waved response starting 10 min after challenge but persisting for 7 hr, and a prolonged immediate reaction occurring after 10 min and gradually resolving over 7 hr.[19] If possible, nonspecific BHR should be evaluated before and 8 hr after challenge to look for an increase.[17,18,20]

If positive, inhalation challenges reliably document the presence of occupational asthma and the etiologic agent involved. However, they are time consuming, expensive, inconvenient for the patient, and not devoid of risk of severe bronchospasm, and subsequent sensitization with increased BHR. If negative, inhalation challenges do not exclude the diagnosis. The agent selected may have been wrong, the dosage may have been too low, the exposure time may have been too short, current medication may have masked the response, or removal from exposure may have caused loss of responsiveness.

Jointly monitoring PEFR and nonspecific BHR by histamine or methacholine in relationship to workplace exposures has been suggested as an alternative to inhalation challenge.[15] However, several other studies have suggested that PEFR monitoring alone, in conjunction with a positive clinical history, was 100% sensitive. Additional monitoring of changes in BHR did not add to the power of the analysis. If both PEFR and clinical history were negative, specific inhalation challenge was not indicated. If only one were positive, then specific agent challenge should be performed.[11,14]

Workplace sensitization may also be evaluated by skin prick testing. A positive skin test in the presence of appropriate negative irritant controls, reflects specific tissue bound IgE to the implicated agent. Positive skin tests have been found to a variety of high molecular weight agents. Negative skin tests to not necessarily rule out the presence of disease. They may reflect an inadequate dose or an inappropriate form of the allergen, e.g., nonmetabolized, or bound to the wrong carrier. Alternatively, IgE may not be a feature of the sensitization. The dose eliciting a positive skin test reaction may be an appropriate starting point for inhalation challenges.[21]

B. HYPERSENSITIVITY PNEUMONITIS/ EXTRINSIC ALLERGIC ALVEOLITIS

Hypersensitivity pneumonitis (HP) is an immunologically-mediated lung disease which occurs after repeated exposure and sensitization to a wide variety of organic dusts and chemicals. Diagnosis is based on the congruence of clinical features and diagnostic tests, although no single feature is pathognomonic.

Typically, patients with acute disease present with complaints of dyspnea, dry cough, malaise, headache, nausea, myalgias, fever and chills occurring 4 to 6 hr after exposure. On exam, the patient may be febrile, tachypneic, or tachycardic. Fine rales may be noted diffusely or bibasally, and may take weeks to resolve. Subacute or chronic exposure causes more insidious symptoms. Fever and chills are often absent, and the patient may complain of a productive cough, dyspnea, fatigue, anorexia, and occasionally weight loss. Some patients may present with simple chronic bronchitis.[22] HP may also present as end-stage fibrosis, with pulmonary hypertension, right heart failure, or cyanosis; digital clubbing is usually absent.

The chest X-ray may be normal if the exposure is intermittent. A fine reticular nodular pattern is typical with accentuation of bronchovesicular markings. During acute episodes, patchy, poorly-defined parenchymal infiltrates may appear which may coalesce, consistent with an acute pneumonitis and secondary alveolar involvement. End-stage fibrosis may present with volume loss or honeycombing.

Serum precipitins are indicative of exposure, not disease, and thus are nondiagnostic. One study has suggested a correlation between prevalence of symptoms and level of precipitating antibodies.[22] Another found a positive correlation between persistently positive precipitins and decreased DL_{co}.[23] Absence of antibodies does not rule out disease, since other studies have shown progressive loss of antibodies over time from an acute episode of pneumonitis.[24] Technical factors may be responsible for negative results, for example, the wrong kind or dose of antigen was used in the assay. If the appropriate antigen can be used, skin testing may be useful in documenting exposure. Using nonirritating concentrations of pigeon serum or pigeon dropping extracts, one study found that 10 of 13 symptomatic patients had positive immediate intradermal reactions, and all had late induration at 4 hr.[25] Whether this represents exposure, sensitization, or disease is not known, since exposed but asymptomatic subjects were not tested as a comparison. Another study compared one sensitive pigeon breeder studied at the height of his clinical illness and one year after pigeon avoidance, to another sensitive pigeon breeder 3 years after complete recovery, though with continued intermittent exposure. The size of the skin test response correlated with disease activity. It was largest in the active patient and smaller one year away from pigeon contact, although the delayed reaction was still positive. In the inactive patient 3 years after the acute episode, the immediate skin test was positive but the delayed reaction was absent. Precipitin titers were also highest in the symptomatic patient and fell away from exposure.[26] These results suggest that the degree of skin test positivity, the presence of a late reaction, and the titer of serum precipitins may give an indirect picture of disease severity and may be followed over time to document improvement or relapse.

Pulmonary function testing may show a restrictive defect, an obstructive defect, or features of both.[25,27-29] The kind of defect does not appear to correlate with progression of disease, since obstructive defects have been noted in both acute episodes and chronic disease. Exercise testing is useful in exposing subtle restrictive defects and in documenting extent and progress of disease. Abnormalities may include an abnormal rise in respiratory rate with exercise, a rise in the alveolar-arterial oxygen difference of >20 mmHg, a drop in the PaO_2, or a dead space to tidal volume ratio (V_D/V_T) that rises or fails to fall with exercise. This shunt-like effect seen with exercise is due to maldistribution of perfusion to diseased lung units.[30]

Bronchoprovocation may be useful in documenting disease and identifying the inciting antigen,[31] but because it may exacerbate disease, it is no longer used for diagnosis.

Bronchoscopy has proved to be a valuable tool in the diagnosis of hypersensitivity pneumonitis. Bronchoalveolar lavage cells from patients with hypersensitivity pneumonitis show an increase in total cells recovered up to 70×10^6 compared to 10×10^6 in controls. A relative lymphocytosis is seen, 65% compared to 16% of the total count in controls, and in established disease T8 suppressor cells predominate, 42% compared to 24% in controls. The T4:T8 ratio is thus reversed, to an average of 0.8, compared to 1.8 in controls. Interestingly, asymptomatic exposed pigeon breeders showed similar changes, though less pronounced. This suggests there is a spectrum of disease among the exposed, and that cellular changes found on bronchoscopy are not necessarily pathogenetic but may reflect exposure.[32] In another study, when subjects with hypersensitivity pneumonitis were removed from exposure, the helper:suppressor ratio was found to revert toward normal.[33] A follow-up study of 14 farmers with diagnosed Farmer's Lung who remained working at their farm found that the abnormal cell counts found on the initial lavage remained stable in 13 of the 14. There was no evidence of disease progression despite continued exposure, although biopsies were not taken.[34]

An analysis of lavage cells examined during and after an acute episode showed that neutrophils are prominant early, to be replaced after a few hours, and for several weeks, by cytotoxic T-cells, both CD8 suppressor and NK cells. After a few weeks, and for several months, CD4 helper cells accumulate and granulomas are formed in the interstitium.[35] Bronchial biopsies are characterized by granulomatous mononuclear cell infiltrates, both alveolar and interstitial. Pulmonary vasculitis has been reported early in the disease, but it is rare. Activated alveolar macrophages and T-cell infiltrates are prominent. Noncaseating, loosely organized granulomas are seen, and there may be marked involvement and narrowing of bronchial walls, characteristic of bronchiolitis obliterans.[36] An abnormal peribronchial cellular infiltrate may also be present, consisting of plasma cells, lymphocytes, large histiocytes with foamy cytoplasm, or occasional eosinophils. Fibrotic changes can occur, with moderate to severe interstitial thickening, and fibrotic intra-alveolar septae lined with lymphocytes.

The diagnosis of occupational hypersensitivity pneumonitis requires an appropriate clinical picture supported by objective diagnostic findings. Terho listed three major and six minor criteria and suggested that the diagnosis could be confirmed if the patient fulfilled all major and at least two minor criteria. The major criteria include: (1) exposure to offending antigens, established by history, industrial hygiene evaluation, or precipitating antibodies; (2) typical symptoms occurring or worsening several hours after antigen exposure; and (3) lung infiltrates on chest X-ray. Minor criteria are: (1) basal crepitant rales on lung exam; (2) impaired DL_{co}; (3) decreased PO_2 or O_2 saturation at rest or with exercise; (4) restrictive ventilatory defect on spirometry; (5) typical histology on bronchial biopsy; or (6) positive inhalation challenge. If the chest X-ray is normal, the biopsy specimen should be confirmatory.[37]

C. FIBROSING INTERSTITIAL LUNG DISEASE

The pneumoconioses are interstitial lung diseases (ILDs) resulting from the inhalation of inorganic dusts, and are associated with chronic inflammation in the lower respiratory tract. The most commonly associated dusts are asbestos, coal, and silica. Cobalt and beryllium disease have been shown to be due to an immune-mediated hypersensitivity response.

In the usual clinical setting, the diagnosis of a pneumoconiosis can be made on the basis of typical clinical findings and chest X-ray changes in the context of occupational exposures, without the confirmation of a lung biopsy. The earliest symptoms are cough, sputum production, and dyspnea, usually with an insidious onset and gradual progression over years. Asbestosis has been associated with fine bibasilar crackles and with clubbing of the fingers. The chest X-ray in asbestosis is characterized by irregular, fine-to-coarse opacifications that are initially basilar but subsequently spread, irregular linear shadows, and evidence of pleural disease, e.g., plaques or extensive pleural thickening.[38] The chest X-ray in silicosis typically shows small, rounded, uniform nodules, more prominent in the upper lung fields, and which may coalesce. Pulmonary lymph nodes may calcify in an "eggshell" pattern. Silicosis is often complicated by the coexistence of tuberculosis. Mixed exposures to silica and asbestos are common, and the chest X-ray may reflect both patterns. The chest X-ray may be normal in up to 10% of patients,[39] and has been shown to correlate poorly with the stage of disease. High resolution chest computed tomography has been suggested as a more sensitive marker of early interstitial disease in asbestos-exposed workers.[40] Both the symptomatic patient with a normal chest X-ray, and the asymptomatic patient with radiographic evidence of ILD, should be thoroughly evaluated.

The pneumoconioses are characterized by restrictive pulmonary physiology. Lung volumes are reduced, primarily the inspiratory capacity (IC) and VC. FRC and RV tend to be less affected. In contrast, although flow rates (FEV$_1$ and FVC) are decreased, the FEV$_1$/FVC ratio is usually preserved in pure disease. Lung compliance is reduced, and may be the only abnormality in a symptomatic patient with a normal chest X-ray. However, since many workers are cigarette smokers, pulmonary function tests may be confounded by obstructive defects as well. Diffusing capacity may drop, reflecting a loss of alveolar–capillary units as well as a V/Q mismatch, where damaged alveolar units are still perfused. However, DL$_{co}$ does not correlate with extent of disease, and may be only minimally reduced in the face of severely reduced lung volumes and hypoxemia.

In many patients with ILD, gas exchange may be normal at rest due to compensatory changes in breathing patterns. However, exercise testing will reveal the ventilatory defect in most, and correlates best with the severity of disease.[30] Characteristic changes with exercise include a fall in PaO$_2$, a rise in the (alveolar–arterial) oxygen gradient, V$_d$/V$_t$ (dead space to tidal volume) which rises or fails to fall, reflecting an inability to recruit more lung ventilation units with exercise, and an excessive rise in the respiratory rate, reflecting stiff, poorly compliant lungs.

Bronchoscopy is not necessary to establish the diagnosis. It may be useful to document mineral dust exposure, but is not evidence of disease. For example, asbestos bodies of short uncoated fibers are commonly found in patients with asbestosis, but a minority of patients with asbestosis may not demonstrate asbestos bodies in their BAL fluid, and a small amount is commonly found in city dwellers. The exact chemical composition of the particles may be established by electron microscopy using energy dispersive X-ray analysis (EDXA), or with neutron activation analysis, which has been especially useful in detecting trace metals including tungsten, tantalum, and cobalt in hard metal lung disease.[35,41] Broncoalveolar lavage is also useful in excluding other causes of ILD.

Cell differentials from BAL fluid are occasionally useful in establishing the extent of the associated alveolitis. In general, the cell count in patients with ILD from mineral dust exposure is normal or only slightly increased, and should be corrected for the effect of smoking. Cell differentials are normal or show slightly increased lymphocytes, neutrophils, and eosinophils. Bronchoalveolar lavage studies in asbestosis have shown a small increase in lymphocyte counts from 17 to 30%, and a specific increase in CD4+ helper counts. Higher lymphocyte percentages were associated with less severe lung impairment. Conversely, higher neutrophil counts correlate with more lung function impairment, the presence of inspiratory crackles, hypoxemia, increased alveolar–arterial oxygen differences, and a longer history of disease. In silicosis, suppressor CD8+ lymphocytes predominate. In coal workers' pneumoconiosis, neutrophilia is associated with progressive massive fibrosis.[35]

The BAL cell differential for hard metal lung disease is not consistent, some studies showing a mild lymphocytosis up to 36% and others reporting increases in neutrophil counts. An increased percentage of giant cells has been reported. Although giant cells are also found in other ILDs, e.g., sarcoid, the presence of metallic inclusions is specific for hard metal ILD. BAL cytology in chronic beryllium disease shows an activated T-helper lymphocytosis. Chronic beryllium disease can be distinguished from sarcoidosis by a positive blood or BAL lymphocyte transformation test for beryllium.

In general, the diagnostic or prognostic significance of BAL in mineral dust occupational lung disease remains unclear except in those diseases with a defined immune mechanism. The demonstration of a specific dust in BAL fluid or cells is helpful in establishing exposure, but is not indication of disease. At present, no threshold values have been established, above which development of disease is inevitable. A specific diagnosis cannot be made on the basis of cell differentials, nor are they predictive for progression to fibrosis or response to therapy.[35]

In summary, the diagnosis of mineral dust occupational lung disease is based on a typical constellation of clinical findings and radiographic and pulmonary function tests in the setting of the appropriate exposure.[38,42] In the diagnosis of hard metal lung disease, BAL may prove useful in documenting the presence of multinucleated giant cells and in demonstrating increased levels of tungsten, tantalum, and cobalt. The diagnosis of beryllium disease is confirmed by typical noncaseating granulomas or mononuclear cell infiltrates on bronchial biopsy and a positive blood or BAL lymphocyte transformation test.

D. OCCUPATIONAL RHINITIS

Occupational rhinitis is defined as the episodic occurrence of sneezing, nasal discharge, and nasal obstruction which develops with a work-related pattern. It may occur alone, or may precede the development of lower respiratory tract symptoms. Similar to asthma, the etiology may be irritative, such as dust, fumes, cold air exposure, or allergic. Interestingly, most of the causative agents for occupational asthma have been shown to cause rhinitis in both a separate and an overlapping cohort of patients. Both high molecular weight and low molecular weight agents have been implicated.

In one survey of a university-based occupational medicine clinic, upper respiratory irritation was the third most common occupational disease diagnosed, and slightly more common even than occupational asthma.[43]

Rhinitis and upper respiratory tract irritation have been described as the earliest clinical signs of toluene diisocyanate (TDI) sensitization.[44-47] Exposure to trimellitic anhydride (TMA) is associated with both irritant and allergic rhinitis[48] documented by positive skin tests to TMA and significantly elevated TMA-IgE levels not seen in other TMA-induced syndromes.[49] Workers exposed to colophony (abeitic acid) in soldering fluxes or hot-melt glue commonly develop rhinitis (48% of cases) before the onset of respiratory symptoms.[50]

Allergic rhinitis is common in laboratory animal workers[51,52] and in those exposed to other animal products at work.[53-56]

Grain mill workers[57] and bakers[58] have significant rates of occupational rhinitis. Rhinitis is a common complaint in wood workers, present in 49 to 64%.[59,60] Nasal irritation and rhinorrhea are common in Western red cedar workers (55% in one study[61]), and often precede the onset of bronchial symptoms.

Latex sensitization may initially present as rhinitis and upper respiratory tract symptoms, or a dermatitis, before progressing to asthma or anaphylaxis.[62]

Exposure to psyllium-based powdered laxatives in nurses[63,64] and pharmaceutical workers processing psyllium[65] has been documented to cause rhinitis, sneezing, and nasal congestion. Guar gum used in carpet dying, insulation, and as a food thickener, causes allergic rhinitis in exposed workers, with positive skin tests, guar-specific IgE by RAST, and positive nasal provocation tests.[66,67]

Highly reactive metals, including platinum,[68] and aluminum[69] can also cause occupational rhinitis.

The diagnosis of occupational rhinitis is based on history and physical exam, and substantiated by nasal challenge and/or demonstration of positive skin tests or RAST where specific IgE is suspected. Symptoms include clear rhinorrhea, nasal itching, sneezing, and nasal congestion, and are often but not invariably associated with conjunctival complaints of watery, itchy eyes. Onset should be work-related, occurring after the start of a new job, or introduction of a new process or product. Symptoms may occur during the weekday, but may also have a later onset, in the evening after work. Symptoms often improve on the weekends but may require longer periods of time away from work to remit. Physical exam may reveal erythema, swollen nasal turbinates, or increased nasal secretions.

Work-related exposures should be as detailed as possible. Exposure to agents known to cause occupational asthma should alert the clinician to the possibility of occupational rhinitis, though other agents may be involved. The history should include home and hobby exposures as well.

If skin tests or RAST are available to implicated agents, they may document IgE sensitization. Nasal challenge is the most direct way to establish a work-related diagnosis, although irritant mechanisms cannot be excluded. Nasal exposure can be done directly at the workplace, or through duplicating work conditions in the challenge laboratory. Alternatively, the nose can be directly exposed to a cotton swab impregnated with the suspected agent, or an aqueous solution can be blown into the nose with a syringe. A positive response can be monitored by symptom score, or increasing nasal resistance measured by rhinomanometry.[70]

Occupational rhinitis may be a harbinger of occupational asthma. Many of the studies of asthma-causing occupational agents have documented previous or coexistant nasal symptoms. A patient who presents with occupational rhinitis, therefore, and remains in the workplace, should be closely monitored for the development of asthma.

II. TREATMENT

The treatment of established cases of occupational immunologic lung disease is largely symptomatic and rarely curative. Once the worker is sensitized, continued exposure to even low levels of the sensitizing agent leads to ongoing symptoms, and often to progressive disease.

A. ASTHMA

The treatment of asthma has been well documented in the NHLB Institute Guidelines from August 1991.[71] Although many of the early bronchoprovocation studies showed ablation of the bronchoconstrictor response by pretreatment with inhaled cromolyn or corticosteroids, most sensitized workers need to be completely removed from exposure. Continued low level exposure can lead to progressive loss in pulmonary function. Deaths in sensitized isocyanate workers from minute exposure, even with respiratory protection, have occurred.[72] The distinction between respiratory sensitization to a specific agent, as compared to a RADS-type syndrome from high level irritants, becomes important. The worker who develops RADS or has nonspecific exacerbation of pre-existing disease may continue to work in the same environment with adequate respiratory protection. The specifically sensitized worker may not. Even away from exposure, many studies have shown persistence of asthma.[73-76]

B. OCCUPATIONAL RHINITIS

The treatment of occupational rhinitis is similar to that of other types of rhinitis. Pharmacotherapy includes antihistamines, decongestants, topical saline washes, and cromolyn or corticosteroid nasal sprays. Since many of the same occupational agents which cause rhinitis can also cause asthma, occupational rhinitis may be considered a sentinel event for excess occupational exposure and the possible development of more severe disease. Once the offending agent has been identified, environmental controls should be instituted. Both the patient and other exposed workers should be carefully monitored for the progression of disease and the development of asthma.

C. HYPERSENSITIVITY PNEUMONITIS/
EXTRINSIC ALLERGIC ALVEOLITIS

Hypersensitivity pneumonitis in general has shown a good response to steroids in conjunction with removal from exposure. In some, progressive loss of lung function occurred despite these precautions.[29] In others, continued exposure has not necessarily led to progression of disease.[23,24] A recent 5-year study of 36 farmers with Farmers' Lung, of whom half received a 2-month course of prednisolone and half received placebo, showed no difference in lung function at the end of the study period. The majority in each group had continued farming, with continued exposure. Interestingly, lung function as assessed by FEV_1, FVC, DL_{co}, and PO_2, had improved equally in both the treated and untreated groups, although it improved faster in the treated group. However, the recurrence of Farmers' Lung was more common in the steroid treated group, although the numbers did not reach statistical significance.[77] The decision, then, to remove a worker with HP from exposure is dependent on the characteristics both of the inciting antigen and of the host response. The patient who continues to develop multiple symptomatic recurrences should be removed from exposure.

D. FIBROSING INTERSTITIAL LUNG DISEASE

The development of the simple pneumoconioses is primarily due to a dose–response relationship. The higher the dust exposure, the more severe the disease. Limiting dust exposure should protect workers from further health damage. However, studies of both coal workers' pneumoconiosis and silicosis have demonstrated progression of radiologic changes and decline of lung function even with removal from exposure.[78,79]

Treatment of the hypersensitivity interstitial lung diseases, including beryllium and cobalt, requires immunosuppression as well as removal from exposure. The mainstay of treatment of chronic beryllium disease is low-dose prednisone. Both symptoms and objective parameters including chest X-ray and pulmonary function tests improve on therapy, but often regress on withdrawal of therapy. This suggests the immune response to beryllium may be suppressed but not ablated.

Removal from exposure of the patient with cobalt interstitial lung disease is the first therapeutic step. Corticosteroids or other immunosuppressives, such as cyclosporine or azathioprine, are often used in the patient refractory to steroids. The subsequent clinical course is difficult to predict. Some patients improve dramatically, while others may show progressive interstitial fibrosis despite intensive therapeutic efforts.[80]

The treatment of occupational respiratory disease, therefore, requires control of exposure prior to sensitization or initiation of disease. Once the worker is sensitized, treatment and removal from exposure are often ineffective in preventing disease progression.

III. SURVEILLANCE

Surveillance of disease patterns is the cornerstone for prevention, diagnosis, and treatment of respiratory disease due to chemicals in the environment. Surveillance data are necessary to portray the ongoing pattern of disease occurrence, detect unusual disease patterns, and trigger disease control and prevention efforts. In addition, these data can be used to allocate resources in public health planning and to evaluate control and prevention measures. Moreover, unusual events detected from surveillance data are often a stimulus for health-related research, and provide an important archive of disease activity.[81]

The conceptual framework of any surveillance system will determine its long-term impact. The U.K. and the U.S. have both instituted surveillance programs to monitor the occurrence of occupational lung disease, each with somewhat different goals.

A. SWORD

In 1989, a voluntary program for the reporting of occupational respiratory disease, SWORD (Surveillance of Work-related and Occupational Respiratory Diseases), was established conjointly by the British Thoracic Society, the Society of Occupational Medicine, and the Health and Safety Executive. The objectives of the program were (1) to monitor the frequency of work-related respiratory disease; (2) to promote the early recognition, investigation, and control of new problems; (3) to provide rapid feedback and information to participants; and (4) to undertake collaborative investigations where indicated. A report of the methods used and the results obtained[82] in the first 2 years of operation was published in 1991. Close to national coverage was achieved by chest physicians, with increasing participation by occupational medicine specialists. There were 2101 cases reported of work-related respiratory disease in 1989. Diagnostic categories included inhalation accidents (including acute pulmonary edema), allergic alveolitis, asthma, building-related illness, byssinosis, infectious disease, lung cancer (with or without pulmonary fibrosis), malignant mesothelioma, pneumoconiosis, bronchitis, benign pleural disease, and other diseases. The data gave valuable information for the incidence of each diagnosis, suspected agents, sex and age distribution, occupational rates, and regional rates. Asthma was the most frequently reported single respiratory disease (26%), with mineral dust diseases accounting for 34% of the cases. Isocyanates, flour/grain dusts, and wood dusts accounted for 22%, 8%, and 6%, respectively, of the reported occupational asthma cases. The incidence of occupational asthma and asbestos-related disease was analyzed by high-risk occupations. Regional rates of occupational asthma were compared to expected rates, and used to calculate the ratios of relative risk. The value of the SWORD program has been to provide anonymous national information of value in setting priorities for research and regulation by government or within companies.[83]

B. SENSOR

In 1987, the U.S. National Institute for Occupational Safety and Health (NIOSH) initiated an occupational surveillance program SENSOR (Sentinel Event Notification System for Occupational Risk) in conjunction with selected state health departments. Several health conditions were targeted for surveillance, including occupational asthma, but not other occupationally-related respiratory diseases. Occupational asthma was selected as a sentinel event because the prevalence of asthma can be high in industries where occupational asthma has been identified. One affected worker may indicate many others are also affected or at risk.[84] A case definition of occupational asthma with reporting and surveillance guidelines was developed by NIOSH.[85]

The program is voluntary and depends on a network of health care providers such as physicians, clinics, or nurse practitioners to report suspected cases to the state health department. Case reports are analyzed by the state health department, which can then direct worksite investigations, and disseminate information to health care providers about implicated workplaces and processes. Although a single case may not provide enough information to implement specific intervention, it may target a specific workplace or industry for further industrial hygiene or epidemiologic studies.

The direction of SENSOR, then, is somewhat different from SWORD, and involves more direct worksite intervention using the expertise of the state industrial hygiene department. The goals are to generalize the information provided by the individual case report to serve others affected or at risk. Worksite investigations and/or recommendations attempt to limit futher exposure to known sensitizing agents in already sensitized individuals. Other workers at risk may then be identified, and their exposures limited as well. Further exposures in other similar work environments may be prevented, and new processes using known sensitizing agents identified. New causes of occupational asthma may also be identified. In each state, reporting physicians receive letters summarizing the results of worksite investigations. Summaries of reported cases and noteworthy case investigations are disseminated in state health department newsletters, in state medical journals, at professional meetings, and at hospital grand rounds. Case follow-up and provider education is critical to the success of the program.

Results of the first year of SENSOR reporting in different states review different exposure settings, including cases that led to the recognition of a new setting for occupational asthma. Clusters of cases have been identified, with the subsequent identification of other affected workers. Worksite investigations in New Jersey assessed the adequacy of air monitoring and engineering controls, as well as identifying other symptomatic workers.[85]

The limitations of SENSOR have in part been related to inadequate physician documentation of disease. Asthma was diagnosed by symptom complex only, and objective data or work-relatedness, other than by history, was lacking.[86] Since the reporting physicians in the SWORD program are respiratory or occupational specialists only, the diagnostic accuracy of the reported cases was not felt to be in question. The U.S. system accepts the reports of all health care providers, which necessarily increases sensitivity but decreases the specificity of reported cases. Since both systems are voluntary, under-reporting likely occurs in both. Limitations in industrial hygiene techniques have also hampered efforts at documentation and control of exposures. Often the sensitizing agent is unknown, or cannot be demonstrated to exceed recommended limits. The ability of the state to intervene with feasible suggestions is thus also limited.

C. FUTURE DEVELOPMENTS

Both SWORD and SENSOR are useful models for the surveillance of occupational respiratory allergies. Both could be easily expanded to include other occupational respiratory diseases. Surveillance data would be useful not only in documenting the incidence and prevalence of disease, but also in identifying other affected workers or those at risk for developing disease. High-risk industries, as well as new agents or processes causing respiratory disease, can be identified. Finally, worksite education and intervention are the critical links between surveillance and disease prevention. Future development and expansion of surveillance and intervention programs will lead to the better diagnosis, treatment, and ultimate prevention of respiratory disease from chemicals in the workplace and the environment.

ACKNOWLEDGMENTS

Participation of Drs. Pacheco and Rosenwasser in this consultation and monograph was supported by the International Life Science Institute (ILSI) — Allergy-Immunology Institute and Health and Environmental Sciences Institute.

REFERENCES

1. **American Thoracic Society,** Standards for the diagnosis and care of patients with chronic obstructive pulmonary disease (COPD) and asthma, *Am. Rev. Respir. Dis.,* 136:225–231, 1987.
2. **Subcommittee on "Occupational Allergy" of the European Academy of Allergology and Clinical Immunology,** Guidelines for the diagnosis of occupational asthma, *Clin. Exp. Allergy,* 22:103–108, 1992.
3. **Neukirch, F., Liard, R., Segala, C., Korobaeff, M., Henry, C., and Cooreman, J.,** Peak expiratory flow variability and bronchial responsiveness to methacholine, *Am. Rev. Respir. Dis.,* 146:71–75, 1992.
4. **American Thoracic Society,** Lung function testing: selection of reference values and intepretive strategies, *Am. Rev. Respir. Dis.,* 144:1202–1218, 1991.

5. **Kanner, R. E. and Morris, A. H.,** Eds., Intermountain Thoracic Society, Pulmonary Function Standardization Task Force, Clinical Pulmonary Function Testing 1975, 1984.

6. **Connolly, M. J., Avery, A. J., Walters, E. H., and Hendrick, D. J.,** The relationship between bronchial responsiveness to methacholine and bronchial responsiveness to histamine in asthmatic subjects, *Pulm. Pharm.,* 1:53–58, 1988.

7. **Pickering, C. A. C.,** Clinical assessment of bronchial hyperresponsiveness due to nonspecific and specific agents, *Chest,* 98(S):202–205S, 1990.

8. **Newman Taylor, A. J.,** Occupational asthma, *Thorax,* 35:241–245, 1980.

9. **Smith, D. D.,** Medical-legal definition of occupational asthma, *Chest,* 98:1007–1011, 1990.

10. **Malo, J.-L., Ghezzo, H., L'Archeveque, J., Lagier, F., Perrin, B., and Cartier, A.,** Is the clinical history a satisfactory means of diagnosing occupational asthma? *Am. Rev. Respir. Dis.,* 145:528–532, 1991.

11. **Cote, J., Kennedy, S., and Chan-Yeung, M.,** Sensitivity and specificity of PC_{20} and peak expiratory flow rate in cedar asthma, *J. Allergy Clin. Immunol.,* 85:592–598, 1990.

12. **Liss, G. M. and Tarlo, S.,** Peak expiratory flow rates in possible occupational asthma, *Chest,* 100:63–69, 1991.

13. **Henneberger, P. K., Stanbury, M. J., Trimbath, L. S., and Kipen, H. M.,** The use of portable peak flowmeters in the surveillance of occupational asthma, *Chest,* 100:1515–1521, 1991.

14. **Perrin, B., Lagier, F., L'Archevegue, J., Cartier, A., Boulet, L.-P., Cote, J., and Malo, J.-L.,** Occupational asthma: validity of monitoring of peak expiratory flow rates and nonallergic bronchial responsiveness as compared to specific inhalation challenge, *Eur Respir. J.,* 5:40–48, 1992.

15. **Cartier, A., Pineau, L., and Malo, J.-L.,** Monitoring of maximum expiratory peak flow rates and histamine inhalation tests in the investigation of occupational asthma, *Clin. Allergy,* 14:193–196, 1984.

16. **McNutt, G. M., Schlueter, D. P., and Fink, J. N.,** Screening for occupational asthma: a word of caution, *J. Occup. Med.,* 33:19–22, 1991.

17. **Pepys, J. and Hutchcroft, B. J.,** State of the art: bronchial provocation tests in etiologic diagnosis and analysis of asthma, *Am. Rev. Respir. Dis.,* 112:829–859, 1975.

18. **Cockcroft, D. W.,** Bronchial inhalation tests II. Measurement of allergic (and occupational) bronchial responsiveness, *Ann. Allergy,* 59:89–98, 1987.

19. **Perrin, B., Cartier, A., Ghezzo, H., Grammer, L., Harris, K., Chan, H., Chan-Yeung, M., and Malo, J.-L.,** Reassessment of the temporal patterns of bronchial obstructive after exposure to occupational sensitizing agents, *J. Allergy Clin. Immunol.,* 87:630–639, 1991.

20. **Cartier, A., Bernstein, I. L., Burge, P. S., Cohn, J. R., Fabbri, L. M., Hargreave, F. E., Malo, J.-L., McKay, R. T., and Salvaggio, J. E.,** Guidelines for bronchoprovocation on the investigation of occupational asthma, *J. Allergy Clin. Immunol.,* 84:823–829, 1989.

21. **Grammer, L. C., Patterson, R., and Zeiss, C. R.,** Guidelines for the immunologic evaluation of occupational lung disease, *J. Allergy Clin. Immunol.,* 84:805–814, 1989.

22. **Bourke, S., Anderson, K., Lynch, P., Boyd, J., King, S., Banham, S., and Boyd, G.,** Chronic simple bronchitis in pidgeon fanciers, *Chest,* 95:598–601, 1989.

23. **Braun, S. R., doPico, G. A., Tsiatis, A., Horvath, E., Dickie, H. A., and Rankin, J.,** Farmer's lung disease: long-term clinical and physiologic outcome, *Am. Rev. Respir. Dis.,* 119:185–191, 1979.

24. **Barbee, R. A., Callies, Q., Dickie, H. A., and Rankin, J.,** The long-term prognosis in farmer's lung, *Am. Rev. Respir. Dis.,* 97:223–231, 1968.

25. **Fink, J. A., Sosman, A. J., Barboriak, J. J., Schlueter, D., and Holmes, R. A.,** Pidgeon breeders' disease. A clinical study of a hypersensitivity pneumonitis, *Ann. Int. Med.,* 68:1205–1219, 1968.

26. **Fink, J. A., Barboriak, J. J., and Sosman, A. J.,** Immunologic studies of pigeon breeders' disease, *J. Allergy,* 39:214–221, 1962.

27. **Schlueter, D. P., Fink, J. A., and Sosman, A. J.,** Pulmonary function in pigeon breeders' disease. A hypersensitivity pneumonitis, *Ann. Int. Med.,* 70:457–470, 1969.
28. **Dinda, P., Chatterjee, S. S., and Riding, W. D.,** Pulmonary function studies in bird breeder's lung, *Thorax,* 24:374–378, 1969.
29. **Allen, D. H., Williams, G. V., and Woolcock, A. J.,** Bird breeder's hypersensitivity pneumonitis: progress studies of lung function after cessation of exposure to the provoking antigen, *Am. Rev. Respir. Dis.,* 114:555–566, 1976.
30. **Cherniack, R. M.,** Physiologic alterations in interstitial lung disease, in *Interstitial Lung Disease,* Schwarz, M. I., King, T. E., Eds, 2nd ed., Mosby-Year Book, St. Louis, 1993.
31. **Hendrick, D. J.,** Bronchopulmonary disease in the workplace. Challenge testing with occupational agents, *Ann. Allergy,* 51:179–184, 1983.
32. **Leatherman, J. W., Michael, A. F., Schwartz, B. A., and Hoidal, J. R.,** Lung T cells in hypersensitivity pneumonitis, *Ann. Int. Med.,* 100:390–392, 1984.
33. **Costobel, U., Bross, K. J., Marxen, J., and Matthys, H.,** T-lymphocytosis in bronchoalveolar lavage fluid of hypersensitivity pneumonitis: changes in profile of T-cell subsets during the course of disease, *Chest,* 85:514–518, 1984.
34. **Cormier, Y., Belanger, J., and Laviolette, M.,** Prognostic significance of bronchoalveolar lymphocytosis in farmer's lung, *Am. Rev. Respir. Dis.,* 135:692–695, 1987.
35. **Semenzato, G., Bjermer, L., Costabel, U., Haslam, P. L., Olivieri, D., and Trentin, L.,** Clinical role of bronchoalveolar lavage in extrinsic allergic alveolitis, *Eur. Resp. Rev.,* 2:8,69–74, 1992.
36. **Salvaggio, J. E.,** Recent advances in pathogenesis of allergic alveolitis, *Clin. Exp. Allergy,* 20:137–144, 1990.
37. **Terho, E. O.,** Diagnostic criteria for farmer's lung disease, *Am. J. Ind. Med.,* 10:329, 1986.
38. **Official Statement of the American Thoracic Society,** The diagnosis of nonmalignant diseases related to asbestos, *Am. Rev. Respir. Dis.,* 134:363–368, 1986.
39. **Epler, G. R., McLoud, T. C., Gaensler, E. A., et al.,** Normal chest roentgenograms in chronic diffuse infiltrative lung disease, *N. Engl. J. Med.,* 298:934–939, 1978.
40. **Dujic, Z., Tocilj, T., and Saric, M.,** Early detection of interstitial lung disease in asbestos exposed non-smoking workers by mid-expiratory flow rate and high resolution computed tomography, *Br. J. Ind. Med.,* 48:663–664, 1991.
41. **Rizzato, G., LoCicero, S., Barberis, M., Torre, M., Pietra, R., and Sabbioni, E.,** Trace metals in human lung as determined by neutron activation analysis of bronchoalveolar lavage, *Chest,* 90:101–106, 1986.
42. **Ziskind, M., Jones, R. N., and Weill, H.,** State of the art: silicosis, *Am. Rev. Respir. Dis.,* 113:643–665, 1976.
43. **Cullen, M. R., Cherniack, M. G., and Rosenstock, L. R.,** Medical progress: occupational medicine, *N. Engl. J. Med.,* 322:594–601, 1990.
44. **Brugsch, H. G. and Elkins, H. B.,** Toluene di-isocyanate (TDI) toxicity, *N. Engl. J. Med.,* 268:353–357, 1963.
45. **Butcher, B. T., O'Neil, C. E., Reed, M. A., Salvaggio, J. E., and Weill, H.,** Development and loss of toluene diisocyanate reactivity: immunologic, pharmacologic, and provocative challenge studies, *J. Allergy Clin. Immunol.,* 70:231–235, 1982.
46. **Hargreave, F. E.,** Occupational asthma without bronchial hyper-reactivity, *Am. Rev. Respir. Dis.,* 130:513–515, 1984.
47. **Banks, D. E., Sastre, J., Butcher, B., Ellis, E., Rando, R. J., Barkman, H. W., Hammad, Y. Y., Glindmeyer, H. W., and Weill, H.,** Role of inhalation challenge testing in the diagnosis of isocyanate-induced asthma, *Chest,* 95:414–423, 1989.
48. **Zeiss, C. R., Patterson, R., Pruzansky, J. J., Miller, M. M., Rosenberg, M., and Levitz, D.,** Trimellitic anhydride-induced airway syndromes: clinical and immunologic studies, *J. Allergy Clin. Immunol.,* 60:96–103, 1977.
49. **Zeiss, C. R., Mitchell, J. H., VanPeenen, P. F. D., Harris, J., Levitz, D.,** A 12 year clinical and immunologic evaluation of workers involved in the manufacture of trimellitic anhydride (TMA), *Allergy Proc.,* 11:71–77, 1990.

50. **Burge, P. S., Perks, W. H., O'Brien, I. M., Burge, A., Hawkins, R., Brown, D., and Green, M.,** Occupational asthma in an electronics factory: a case control study to evaluate aetiological factors, *Thorax,* 34:300–307, 1979.

51. **Cockcroft, A., McCarthy, P., Edwards, J., and Andersson, N.,** Allergy in laboratory animal workers, *Lancet,* 11:827–830, 1981.

52. **Aoyama, K., Ueda, A., Manda, F., Matsushita, T., Ueda, T., and Yamauchi, C.,** Allergy to laboratory animals: an epidemiological study, *Br. J. Ind. Med.,* 49:41–47, 1992.

53. **Colten, H. R., Polakoff, P. L., Weinstein, S. F., and Streider, D. J.,** Immediate hypersensitivity to hog trypsin resulting from industrial exposure, *N. Engl. J. Med.,* 292:1050–1053, 1975.

54. **Dolan, T. F. and Meyers, A.,** Bronchial asthma and allergic rhinitis associated with inhalation of pancreatic extracts, *Am. Rev. Respir. Dis.,* 110:812–813, 1974.

55. **Pepys, J., Longbottom, J. L., Hargreave, F. E., and Faux, J.,** Allergic reactions of the lungs to enzymes of bacillus subtilis, *Lancet,* i:1181–1184, 1969.

56. **Terho, F. O., Husman, K., Vohlonen, I., Rautalahti, M., and Tukiainen, H.,** Allergy to storage mites or cow dander as a cause of rhinitis among Finnish dairy farmers, *Allergy,* 40:23–26, 1985.

57. **Yach, D., Myers, J., Bradshaw, D., and Benatar, S. R.,** A respiratory epidemiologic survey of grain mill workers in South Africa, *Am. Rev. Respir. Dis.,* 131:505–510, 1985.

58. **Thiel, H. and Ulmer, W. T.,** Bakers' asthma: development and possibility for treatment, *Chest,* 78:400S–405S, 1980.

59. **Shamssain, M. H.,** Pulmonary function and symptoms in workers exposed to wood dust, *Thorax,* 47:84–87, 1992.

60. **Holness, D. L., Sass-Kortsak, A. M., Pilger, C. W., and Nethercott, J. R.,** Respiratory function and exposure effect relationships in wood dust-exposed and control workers, *J. Occ. Med.,* 27:501–506, 1985.

61. **Chan-Yeung, M., Barton, G. M., MacLean, L., and Grzybowski, S.,** Occupational asthma and rhinitis due to western red cedar (Thuja plicata), *Am. Rev. Respir. Dis.,* 108:1094–1102, 1973.

62. **Tarlo, S. M., Wong, L., Roos, J., and Booth, N.,** Occupational asthma caused by latex in a surgical glove manufacturing plant, *J. Allergy Clin. Immunol.,* 85:626–631, 1990.

63. **Nelson, W. L.,** Allergic events among health care workers exposed to psyllium laxatives in the workplace, *J. Occ. Med.,* 29:497–499, 1987.

64. **Cartier, A., Malo, J.-L., and Dolovich, J.,** Occupational asthma in nurses handling psyllium, *Clin. Allergy,* 17:1–6, 1987.

65. **Bardy, J.-D., Malo, J.-L., Seguin, P., Ghezzo, H., DesJardins, J., Dolovich, J., and Cartier, A.,** Occupational asthma and IgE sensitization in a pharmaceutical company processing psyllum, *Am. Rev. Respir. Dis.,* 135:1033–1036, 1987.

66. **Karervz, L., Tupasela, O., Jolanki, R., Vaheri, E., Estlander, T., and Keskinen, H.,** Occupational allergic rhinitis from guar gum, *Clin. Allergy,* 18:245–252, 1988.

67. **Malo, J.-L., Cartier, A., L'Archeveque, J., Ghezzo, H., Soucy, F., Somen, J., and Dolovich, J.,** Prevalence of occupational asthma and immunologic sensitization to guar gum among employees at a carpet-manufacturing plant, *J. Allergy Clin. Immunol.,* 86:562–569, 1990.

68. **Pepys, J., Pickering, C. A. C., and Hughes, E. G.,** Asthma due to inhaled chemical agents — complex salts of platinum, *Clin. Allergy,* 2:391–396, 1972.

69. **Field, G. B.,** Pulmonary function in aluminum smelters, *Thorax,* 39:743–751, 1984.

70. **Slavin, R. G.,** Occupational rhinitis, *Immunol. Allergy Clinics N. Am.,* 12(4):769–777, 1992.

71. Guidelines for the diagnosis and management of asthma. National Asthma Education Program expert panel report. NIH publication no. 91–3042, Department of Health and Human Services, Bethesda, MD, 1991.

72. **Fabbri, L. M., Danieli, D., Crescioli, S., Bevilacqua, P., Meli, S., Saelta, M., et al.,** Fatal asthma in a subject sensitized to toluene diisocyanate, *Am. Rev. Respir. Dis.,* 137:1494–1498, 1988.
73. **Chan-Yeung, M., MacLean, L., and Paggiaro, P. L.,** A follow-up of 232 patients with occupational asthma due to western red cedar (Thuja plicata), *J. Allergy Clin. Immunol.,* 80:279–284, 1987.
74. **Paggiaro, P. L., Loi, A. M., Rossi, O., Ferrante, B., Pavdi, F., Roselli, M. G., et al.,** Follow up study of patients with respiratory disease due to toluene diisoyanate (TDI), *Clin. Allergy,* 14:463–469, 1984.
75. **Losewicz, S., Assoufi, B. I. C., Hawkins, R., and Newman-Taylor, A. J.,** Outcome of asthma induced by isocyanates, *Br. J. Dis. Chest.,* 81:14–22, 1987.
76. **Burge, P. S.,** Occupational asthma in electronics workers caused by colophony fumes: follow-up of affected workers, *Thorax,* 37:348–353, 1982.
77. **Kokkarinen, J. I., Tukiainen, H. O., and Terho, E. O.,** Effect of corticosteroid treatment on the recovery of pulmonary function in farmer's lung, *Am. Rev. Respir. Dis.,* 145:3–5, 1992.
78. **Reisner, M. T. R.,** Results of epidemiologic studies on the progression of coal workers' pneumoconiosis, *Chest,* 78:406–407, 1980.
79. **Ng, T.-P., Chan, S.-L., and Lam, K.-P.,** Radiologic progression and lung function in silicosis: a ten year follow-up study, *Br. Med. J.,* 295:164–168, 1987.
80. **Austenfeld, J. L. and Colby, T. V.,** Recognizing lung diseases induced by hard metal exposure, *J. Respir. Dis.,* 10:65–75, 1989.
81. **Thacker, S. B., Choi, K., and Brachman, P. S.,** The surveillance of infectious diseases, *JAMA,* 249:1181–1185, 1983.
82. **Meredith, S. K., Taylor, V. M., and McDonald, J. C.,** Occupational respiratory disease in the United Kingdom 1989: a report to the British Thoracic Society and the Society of Occupational Medicine by the SWORD project group, *Br. J. Ind. Med.,* 48:292–298, 1991.
83. **Venables, K. M.,** Preventing occupational asthma, *Br. J. Ind. Med.,* 49:817–819, 1992.
84. **Matte, T. D., Hoffman, R. E., Rosenman, K. D., and Stanbury, M.,** Surveillance of occupational asthma under the SENSOR model, *Chest,* 98:173S–178S, 1990.
85. Occupational disease surveillance: occupational asthma, *MMWR,* 39:119–23, 1990.
86. **Klees, J. E., Alexander, M., Rempel, D., Beckett, W., Rubin, R., Barnhart, S., and Balmes, J. R.,** Evaluation of a proposed NIOSH surveillance case definition for occupational asthma, *Chest,* 98:212S–215S, 1990.

Chapter 6

PREDICTIVE TESTING FOR RESPIRATORY ALLERGY

Meryl H. Karol

CONTENTS

ABSTRACT

Respiratory allergy to chemicals is a substantial environmental and industrial concern since there exists the potential for exposure of large numbers of individuals. Remarkably, the number of chemicals causing such problems appears to be limited. The mechanism(s) underlying the disorder is unknown but has been assumed to include immunologic, pharmacologic, and neurologic involvement. Predictive testing would be helpful to prevent the problem. Several approaches have been used for predictive purposes including: structure-activity studies of known chemical allergens, assessment of the ability of the chemical to undergo *in vitro* reactivity with proteins, and development of animal models of respiratory sensitization. With the latter approach, numerous species have been employed utilizing a variety of endpoints as markers of respiratory sensitization. Such markers include both physiologic endpoints, such as airflow limitation, and serologic changes such as increases in IgE or in specific antibody formation. All methods require further validation using positive and negative agents, as well as calibration to human systems to provide guidance for regulatory purposes.

I. INTRODUCTION

Respiratory allergy has been recognized in association with numerous agents. Most frequently these agents are of high molecular weight (HMW), i.e., proteins larger than 10 kDaltons. Many of these allergens possess biological activity, for example, proteolytic and lipolytic enzymes are recognized to be potent respiratory allergens.[1] In contrast to the situation with high molecular weight allergens, only a limited number of low molecular weight (LMW) chemicals have been associated with respiratory allergy. These chemicals are listed in Table 6-1. The situation contrasts also with that of chemicals causing dermal sensitization where a greater number of chemicals and chemical classes are associated with dermal activity (Table 6-2).

A further distinction between HMW and LMW allergens is apparent when one considers the characteristics of the physiologic responses to these agents. Respiratory sensitivity can be manifested by responses which occur immediately upon exposure to the offending allergen, i.e., immediate-onset responses (IAR), or those which occur hours later, i.e., late-onset reactions (LAR). HMW allergens elicit predominantly IAR, whereas LMW allergens frequently produce LAR.[2] This distinction has implications for genetic evaluations and for predictive testing since the mechanism underlying IAR has traditionally been assumed to be IgE mediated, whereas the mechanism of the LAR is uncertain. Clearly, information regarding structural functionalities associated with respiratory sensitivity, and mechanistic information for both IAR and LAR should guide future development of predictive tests for respiratory allergens.

A. CHEMICAL CHARACTERISTICS OF RESPIRATORY ALLERGENS

Considering the large number and wide diversity of environmental and industrial chemicals, it is surprising that so few have been associated with respiratory sensitivity. Table 6-1 provides a listing of the chemicals which have either been recognized clinically as agents of respiratory sensitivity, or have been shown to have respiratory sensitizing capability in animal systems.

A visual examination of the chemicals which comprise this list reveals some common features. Many of the chemicals in Table 6-1 possess electrophilic functionalities such as diisocyanates and acid anhydrides. This property would enable the chemicals to react covalently with nucleophilic moieties such as sulfhydryl, hydroxyl, or amino groups on biologic molecules. Since it is generally accepted that covalent (or high affinity) binding of chemicals to macromolecules is essential to confer the property of immunogenicity on small molecules, electrophilicity would be a highly desirable property for immunogenic activity of small chemicals.

However, it is also apparent from consideration of Table 6-2, that electrophilicity of itself is not sufficient to endow a chemical with respiratory sensitizing ability. Many of the chemicals which are dermal sensitizers, without

TABLE 6-1
Low Molecular Weight Respiratory Allergens

2-hydroxy-1,4-naphthaquinone
Abietic acid
Cyanuric chloride
Hexamethylene diisocyanate
Diphenyl methane-4,4'-diisocyanate (MDI)
Phthalic anhydride
Piperazine
Plicatic acid
Toluene 2,4 diisocyanate (2,4 TDI)
Toluene 2,6 diisocyanate (2,6 TDI)
Trimellitic anhydride
Platinum salts
Reactive dyes (remazol black B dye)

recognized respiratory sensitizing ability, also have electrophilic functionalities. For example, dicyclohexyl methane-4,4'-diisocyanate (HMDI) contains a pair of electrophilic isocyanate functionalities yet is a contact sensitizer, not a respiratory sensitizing agent.[3,4] Its aromatic diisocyanate analog, diphenyl methane-4,4'-diisocyanate (MDI), is a respiratory allergen.[5] Moreover, HMDI is a dermal sensitizer even though industrial exposure is usually via the inhalation route with workers coming into contact with atmospheres containing vapors of the diisocyanate. Similarly, animal experiments have demonstrated the propensity of this chemical to cause dermal, rather than pulmonary sensitivity irrespective of the route of contact with the chemical.[6] Studies in which guinea pigs were exposed to HMDI by the inhalation route resulted in contact, not respiratory, sensitivity of animals.[7]

TABLE 6-2
Low Molecular Weight Contact Allergens

Benzocaine	Geraniol
1-Chloromethylpyrene	Hydroquinone monobenzylether
1 Chloro,2,4,5 trinitrobenzene	2 Hydroxy-1,4-naphthaquinone
Cinnamic aldehyde	Isoeugenol
1,4 Dichloro-2,5-dinitrobenzene	N-nitroso-dimethylaniline
1,3 Dichloro-4,5-dinitrobenzene	P-nitroso-n,n-dimethylaniline
Diethanolamine	Oxazolone
Diethyl fumarate	Penicillin G
Diethylene triamine	3-Pentadecyl ortho quinone
Dimethylbenzanthracene	P-phenylenediamine
2,4-Dinitrochlorobenzene	Picryl chloride
Ethanolamine	Streptomycin
Eugenol	Tetrachlorosalicylanilide
Fluorescein	Turpentine
Furacin	4-Vinylpyridine

Whereas some electrophiles are respiratory allergens and others are contact allergens, a third group of electrophilic chemicals have neither respiratory nor dermal sensitizing activity. Accordingly, electrophilicity, although undoubtedly important in contributing to the immunogenicity of small chemicals, is neither definitive nor sufficient for conferring on chemicals respiratory sensitizing activity.

II. MECHANISMS OF RESPIRATORY RESPONSES

It is generally accepted that IAR to respiratory allergens is mediated by IgE antibody.[8] Evidence supporting this hypothesis has been obtained from studies employing HMW allergens. Passive transfer to naive animals of serum rich in IgE followed by provocation challenge with specific allergen has consistently produced IAR in recipient animals.[9] Passive transfer studies undertaken with purified immunoglobulins have produced similar results.[10] Using a guinea pig animal model in which both IAR and LAR have been demonstrated, purified IgG_1 from sensitized animals was administered intravenously in increasing doses to naive recipients.[11] Subsequent provocative challenge of animals with an aerosol of the HMW antigen (ovalbumin) resulted only in immediate-onset responses. No LAR was produced even though extremely severe IAR resulted from administration of high doses of immunoglobulin. The failure of the purified immunoglobulins to induce LAR implied either that LAR is not mediated by immunoglobulins, or that other factors in combination with immunoglobulins were required to produce a LAR.

The mechanism(s) underlying respiratory responses to chemical allergens are uncertain. Evidence of an association of IgE with IAR to chemical allergens has been reported.[12] Two individuals who demonstrated IAR upon inhalation of MDI-containing atmospheres had elevation of MDI-specific IgE and total IgE at the time of symptomatic exposure. In both individuals, titers of specific and of total IgE decreased during the ensuing 16 months when exposure had ceased. The kinetics of the decrease indicated a half life of approximately 9 months for total and specific IgE populations.

Recently, an association of serum IgE with IAR to toluene diisocyanate (TDI) provocation challenge was noted.[13] The association was found with IAR and was not present in those responding to TDI with LAR.

For predictive purposes, testing sera for total and hapten-specific IgE is appropriate to identify individuals with the potential for IAR to respiratory chemical allergens.[14,15] The value of such tests is less certain for identification of individuals who typically respond to these allergens with LAR. Since LAR is a frequent response to respiratory chemical allergens,[2] additional predictive procedures are needed.

Although the mechanism(s) underlying LAR is uncertain, it has been hypothesized that LAR is analogous to delayed-type hypersensitivity (DTH) reactions, and is mediated by CD4+ T-cells.[16] These specific T-cells recognize antigen in association with class II major histocompatibility complex (MHC) determinants. It is believed that the resulting interaction leads to the release of particular cytokines which in turn results in recruitment of effector cells to the sites of T-cell interaction.

Factors which support recruitment of CD4+ T-cells into antigenated sites are uncertain. In mice, local serotonin release has been detected upon contact with antigen. Evidence has been obtained for a role of both IgE antibody, and for a non-IgE factor binding to mast cells and other cells.

T-cells have been observed in pulmonary tissue of mice in association with late-phase respiratory responses to chemicals sensitizers.[17] A mild mononuclear cell influx was observed around bronchioli and other airways of mice sensitized to picryl chloride and challenged by intranasal exposure to the chemical. This histologic change was accompanied by physiologic changes, most notably airway hyperreactivity. Inhibition of the latter response by nedocromil suggested involvement of mast cells in the process. These results have been interpreted as supporting a role of the Th1 subset of T-cells in respiratory hypersensitivity reactions to chemicals, and specifically in late-phase respiratory reactions.

III. ROLE OF ANTIGEN CONCENTRATION IN INDUCTION AND ELICITATION OF SENSITIZATION

Industrial experience with both HMW and LMW allergens has demonstrated a central role of exposure concentration in both the induction and elicitation phases of respiratory sensitization. Regulations requiring reduction of allowable airborne concentrations of detergent enzymes, and diisocyanates in the workplace environments have resulted in dramatic decreases in the incidence of workplace sensitization episodes.[18]

A more thorough examination of the dose–response relationship in sensitization has been achieved through the use of animal studies. Using a guinea pig model of sensitization to the LMW allergen, TDI, the need for a threshold concentration of allergen was identified for both induction and elicitation of respiratory sensitivity responses.[19] Regarding TDI-specific antibody production, a log-linear dose–response relationship was observed between the initial exposure concentration of TDI and the titer of TDI-specific allergic antibody.[19]

Antigen concentration has also been shown to play an important role in the induction and elicitation phases of allergic responses in mice.[20] The concentration of antigen influenced the amount of interleukin 4 (IL-4) produced by primed T-cells in BALB/c mice. IL-4 is critical to the generation and regulation of allergic responses, especially those responses which result in IgE or IgG_1 synthesis. High concentrations of antigen resulted in increased production of IL-4, but also caused increased interferon gamma (IFN-γ) production. The latter is known to inhibit IgE production. Low concentrations of antigen stimulated release of IL-4 from primed T-cells without concomitant IFN-γ production. It is believed that B-cells rather than macrophage preferentially bind antigen under conditions of low antigen concentration. The result is the production of IL-4 with little IFN-γ, and consequently in substantial IgE production. Thus, a sensitization response depends on a balance between opposing activities.

The continual production of IL-4 is of interest for allergic disease. Continuous IL-4 release would imply continual production of IgE. In the absence of continuous antigen stimulation, T-cells are rapidly down-regulated and cytokine release is diminished.

The application of these findings to understanding human allergic disease must be done cautiously since substantial differences exist between humans and numerous species. Indeed, significant differences have been noted between strains of mice[21] with regard to patterns of cytokine synthesis upon exposure to various antigens.[21] These differences may occur at the level of the T-cell, the antigen presenting cell, or in IL-4 regulation. Extrapolation from results in one particular murine strain to human responses must be done with caution.

IV. METHODS FOR IDENTIFYING RESPIRATORY SENSITIZING CHEMICALS

Much effort has been devoted to developing methodology for identifying chemicals which have pulmonary sensitizing activity and to estimating the respiratory sensitizing potencies of such chemicals. Methods include: (a) determination of *in vitro* reactivity of suspect chemicals with proteins,[18] (b) injection of the chemical into animals with subsequent determination of increased production of total IgE, or specific antibody,[18] and (c) inhalation exposure of animals to the agent with determination of IAR, LAR, fever, antibody formation, pulmonary histopathology, cytokine production, and airway hyperreactivity.[18,22]

A. *IN VITRO* METHODS

Several *in vitro* methods for predicting respiratory sensitizing properties of chemicals have been described. Wass and Belin[23] employed high-pressure liquid chromatography to monitor reactivity of suspect chemical allergens toward a lysine-containing peptide. Reaction conditions included aqueous solution at neutral pH and 37°C. Binding was obtained with three chemicals recognized as respiratory sensitizers. However, positive results were also obtained with chemicals which have not been classified as respiratory sensitizers (i.e., with isobutyl chloroformate). Theoretically, it is unclear how this test could distinguish contact from respiratory sensitizers since both groups of agents are known to be chemically reactive toward nucleophiles.

An *in vitro* assay is fundamental to the tier approach proposed by Sarlo and Clark[24] to evaluate the potential of low molecular weight chemicals for causing respiratory hypersensitivity. The approach utilizes both *in vitro* and *in vivo* assessment in a four-tier evaluation.

The first level of evaluation (tier 1) is an examination of the structure of the chemical to assess its likely ability to react covalently with protein. This procedure includes a search of the literature for information on immunogenic effects reported in humans or animals for the chemical, or for structurally-related compounds. The second tier is an *in vitro* test of the chemical's ability to react with protein. A variety of conditions are used to effect coupling including a range of concentrations of the chemical, variations in pH, in temperature and in time of incubation. If positive results are obtained in tiers 1 and 2, the chemical is evaluated in tier 3, which involves animal experimentation.

Tier 3 assesses the ability of the chemical to invoke immunogenic changes *in vivo*. Procedures in this tier were designed to bypass concerns of penetration and absorption of the chemical into the respiratory tract. The agent is injected subcutaneously into guinea pigs, employing two injections per week for 4 weeks, followed by a further injection 1 week later. Assessment is made of antibody titer and of respiratory reactivity toward a protein conjugate of the LMW allergen. Specific antibody formation is assessed on sera drawn 7 days after the last injection. Both total IgG and hypersensitivity antibody are assessed. In the guinea pig, the latter is composed of IgG_1 and IgE. A positive ELISA is taken as evidence that the chemical has stimulated the immune system. A positive passive cutaneous anaphylaxis test for hypersensitivity antibodies, and assessment of antibody titer, provides an indication of the allergenic potency of the agent.

Respiratory reactivity is also assessed in tier 3. A chemical–protein conjugate is intratracheally instilled in guinea pigs and IAR is investigated. Animals are monitored for 10 min following conjugate administration and diaphragmatic contractions are noted. Severity of responses provides indication of the potency of the chemical as a respiratory sensitizing agent.

If positive results are obtained in tier 3, the chemical is advanced to tier 4 evaluation. In this last tier, guinea pigs are exposed to the chemical by the inhalation route to assess the chemical's ability to cause respiratory hypersensitivity via a "relevant" route of exposure. Response is assessed by a change in the respiratory rate of animals and by determination of airway constriction by observed diaphragmatic contractions. Again, only IAR is monitored.

The validity of this tier method was evaluated by testing three known human respiratory tract sensitizers, specifically, phthalic anhydride, TDI, and reactive black B dye. Each of the chemicals was found to be positive in tiers 1 to 4. From dose-response information, TDI and phthalic anhydride were identified as strong sensitizers, and the dye was scored as a moderately potent sensitizer. Phthalic acid, a nonsensitizer, was correctly identified as such in the procedure.

This approach is promising. There is a need for testing additional chemicals using this approach to determine the latter's ability to distinguish contact from pulmonary sensitizers. It would be expected that both types of sensitizers would yield positive results in tiers 1 and 2, but would be distinguished in tier 3.[26]

TABLE 6-3
Representative Animal Models of Pulmonary Sensitivity

Species	Sensitization	Challenge	Response
Rat			
Donryu	OA, i.p., B. pertus.	OA, i.v.	IAR (R_L)
Brown Norway	OA, s.c., B. pertus., alum	OA, aerosol	IAR (R_L); LAR (R_L); AHR: Ach, Mch; IgE
Guinea Pig	OA, i.p.	OA, aerosol	IAR (f, airway constrict, t); LAR (f, airway constrict, t); AHR: histamine; IgE, IgG_1
Rabbit	A. tenius, i.p., alum, neonate	A. tenius, aerosol	IAR (R_L, C_{dyn}, f,v,vt); LAR (R_L, C_{dyn}, f,v, vt); AHR: Hist; IgE
Dog			
Mongrel	Ragweed, grasses, injection	Ragweed, aerosol	IAR (R_L); AHR: Ach; IgE
Basenji-greyhound	Natural, A. suum	A. suum, aerosol	IAR (R_L, C_{dyn}); AHR: Mch, Citric Acid
Sheep	Natural, A. suum	A. suum, aerosol	IAR (R_L); LAR (R_L); AHR: Crbl
Horse	Natural, barn environment	Environmental exposure	IAR (R_L, C_{dyn}); AHR: Hist, Mch, Citric Acid
Monkey			
Squirrel	Natural, A. suum	A. suum, aerosol	IAR (R_L, C_{dyn}); LAR (R_L, C_{dyn}); AHR: IgE
Rhesus	Natural, A. suum	A. suum, aerosol	IAR (f, R_L, C_{dyn}); AHR: Hist, Crbl, $PGF_{2\alpha}$

Note: B. pertus., *Bordetella pertussis*; IAR, immediate-onset response; R_L, pulmonary resistance; OA, ovalbumin; f, frequency; t, temperature; v, volume; vt, tidal volume; A. tenius, *Alternaria tenius*; C_{dyn}, dynamic compliance; Ach, acetylcholine; Mch, methacholine; Hist, histamine; A. suum, *Ascaris suum*; Crbl, carbachol.

From Karol, M. H., *Toxicology of the Lung*, Raven Press, New York, 1993, 417. With permission.

B. *IN VIVO* ANIMAL MODELS

The need to predict the sensitization capability of chemicals before they enter large-scale production, together with currently incomplete mechanistic information regarding chemical sensitization, necessitates the use of animal systems to assess the sensitization potential of chemicals.

Many animal models of pulmonary hypersensitivity have been developed and recently critiqued.[25] Considerable differences exist among models in the following aspects: the animal species utilized, the route of antigen administration, protocol for both induction and elicitation of responses, type of response measured, and evaluation of significant reactions. A listing of various types of models is provided in Table 6-3.[26]

It should be noted that most of the models listed in Table 6-3 were developed using HMW allergens, notably proteins. Few have been tested with LMW allergens.

One recent test which was designed for LMW allergens is the Mouse IgE Test.[27] The test is intended to evaluate the potential of LMW chemicals to cause respiratory sensitization. The method is based on the assumption that LMW chemicals with respiratory sensitizing ability induce the production of IgE antibody following topical administration in mice, whereas contact sensitizing LMW chemicals do not. The test is performed by applying the chemical, in vehicle, to the shaved flank of the mouse. Seven days later, mice are challenged on the ear with the chemical. Serum is taken on days 14 to 21 for assessment of total IgE.

This test holds promise because of its ease of operation and its low cost. However, to date, a standard method has not been published and only a limited number of chemicals have been assessed. Further development is needed with regard to definition of IgE titer and "elevated" levels of IgE, and doses appropriate for testing. Additionally, testing of recognized chemical allergens, including contact and respiratory allergens such as those listed in Tables 6-1 and 6-2, and of nonallergens is needed for further evaluation of this methodology.

A guinea pig model which was developed for assessment of the sensitizing capability of LMW chemicals has been successfully utilized to distinguish contact from respiratory LMW sensitizers.[28] The model utilizes inhalation as the route of exposure for both the sensitization phase, and the elicitation phase of the response. It also has the capacity to assess IAR, as well as LAR. The latter is made possible by the use of minimally restrained animals, housed in an exposure chamber supplied with a dynamic air supply in which the allergen can be freshly generated, and continuous 24-hr monitoring of respiratory function of animals. The latter feature makes the model particularly appropriate for evaluation of LMW chemicals where LAR is frequent.

The advantages of using the guinea pig model are listed in Table 6-4. These include: use of inhalation as the relevant route of exposure; generation of atmospheres of reactive chemicals; measurement of physiologic responses including IAR, LAR, fever, and hyperreactive airways; measurement of specific antibody production; and histopathologic evaluation of pulmonary tissue. Disadvantages of the model are the cost, and the employment of guinea pigs. The latter is a species in which the major class of hypersensitivity antibody is IgG_1, rather than IgE (the latter is the predominant class in humans).

The model has the ability to detect IAR and LAR irrespective of their times of onset. Using the model, differences were readily apparent between sensitization to HMW vs. LMW allergens. Responses to ovalbumin (OA) and bacterial subtilisin (HMW allergens) consisted of severe IAR (in 90 to 100% of animals) and less frequent occurrence of LAR (50% of animals).[29] By contrast, sensitization to MDI (LMW allergen) consisted predominantly of LAR.[30] This

TABLE 6-4
The Guinea Pig Model of Respiratory Hypersensitivity

Feature	Benefit
Animals are nonsedated	Avoids effects of anesthetics on breathing control and patterns
Animals are nonrestrained	Allows measurements of unaltered breathing patterns for at least 24 hr
Inhalation route for sensitization	Utilizes natural exposure route, permits access to appropriate antigen-presenting cells
Inhalation route for challenge	Direct assessment of chemical's ability to cause response in respiratory tract
Monitor development of IAR and LAR	Since mechanisms of LAR unknown, allows detection of both types of responses, continuous monitoring of respiratory pattern allows recognition of irritant effect of chemical exposure
Detects hyperreactive airways (HAR)	HAR associated with LAR and is a cardinal feature of asthma
Continuous passive monitoring for mild febrile reactions	Fever has been associated with inflammation and LAR, passive procedure does not affect measurements of breathing patterns
Assessment of specific and nonspecific antibody formation	Determine effects of chemical exposure on the humoral immune system
Assessment of dermal sensitivity	Determine effects on humoral and cellular elements of immune system, compare potency of chemical toward skin vs. respiratory system
Histologic evaluation of pulmonary tissue	Assessment of chemical damage to lungs resulting from sensitization or challenge exposures, composition of inflammatory cells, and cytokine response essential to mechanistic interpretation of response

finding reflects the human experience with LAR being a frequently observed response in humans to LMW allergens.

Febrile reactions have been detected in the animal model by passive monitoring using temperature-sensitive radio transmitters.[31] This response has been found to accompany LAR in the guinea pig airways and likely reflects the inflammation characteristic of LAR.

Airway hyperreactivity, a cardinal feature of clinical asthma, is also measured in the animal model.[22] The degree of hyperreactivity is assessed by increased pulmonary responsiveness of animals to inhaled histamine.

The model has been validated in two ways. Firstly, it has been reproduced in several laboratories.[18,24,32-34] In each, responses of animals to inhaled TDI have been verified. This confirms the robustness of the model. Secondly, the model has been found to accurately distinguish pulmonary sensitizers from dermal sensitizers and nonsensitizers. For example, inhalation exposure of animals to TDI and MDI resulted in pulmonary sensitization, whereas similar exposure to formaldehyde and hydrogenated MDI, two recognized contact sensitizers, resulted in dermal sensitivity of guinea pigs.[30,35] Validation achieved with the animal model engenders confidence that mechanistic studies will yield information applicable to human disease.

V. CONCLUSIONS

Numerous methods have been proposed to predict the respiratory sensitizing capability of chemicals. Each method has distinct advantages as well as disadvantages. Few of the methods have been validated to assess their abilities to distinguish positive from negative agents.

Major progress in predictive methodologies is anticipated from current mechanistic studies exploring cytokine production and activities, as well as from studies seeking to identify *in vivo* reactions of chemical allergens. Until such progress is made, combined utilization of *in vitro* and *in vivo* methodologies can provide meaningful information for regulatory purposes.

ACKNOWLEDGMENT

This article was supported by grant #ES05651 from the National Institute of Environmental Health Sciences.

REFERENCES

1. **Karol, M. H.,** Occupational asthma and allergic reactions to inhaled compounds, in *Principles and Practice of Immunotoxicology,* Miller, K., Turk, J. and Nicklin, S., Eds., Blackwell Scientific, Oxford, 1992, 228.
2. **Chan-Yeung, M.,** A clinician's approach to determine the diagnosis, prognosis, and therapy of occupational asthma, *Med. Clinics N. Amer.,* 74, 811, 1990.
3. **Emmett, E. A.,** Allergic contact dermatitis in polyurethane plastic moulders, *J. Occup. Med.,* 18, 802, 1976.
4. **Israeli, R., Smirnov, V., and Sculsky, M.,** Vergiftungsercheinungen bei Dicyclohexyl-Methan-4-4'-Diisocyanat-Exposition, *Int. Arch. Occup. Environ. Health,* 48, 179, 1981.
5. **Chang, K. C. and Karol, M. H.,** Diphenylmethane diisocyanate (MDI)-induced asthma: evaluation of the IgE response and application of an animal model for isocyanate sensitivity, *Clin. Allergy,* 14, 329, 1984.
6. **Stadler, J. and Karol, M. H.,** Experimental delayed hypersensitivity following inhalation of dicyclohexylmethane-4,4'diisocyanate: a concentration–response relationship, *Toxicol. Appl. Pharmacol.,* 74, 244, 1984.
7. **Karol, M. H. and Magreni, C. M.,** Extensive skin sensitization with minimal antibody production in guinea pigs as a result of exposure to dicyclohexylmethane-4-4'diisocyanate, *Toxicol. Appl. Pharmacol.,* 65, 291, 1982.
8. **Chan-Yeung, M. and Lam, S.,** Occupational asthma, *Am. Rev. Resp. Dis.,* 133, 686, 1986.
9. **Spiegelberg, H. L.,** Biological role of different antibody classes, *Int. Arch. Allergy Appl. Immunol.,* 90, 22, 1989.

10. **Watson, J. W., Conklyn, M., and Showell, H. J.,** IgG$_1$-mediated acute pulmonary hypersensitivity response in the guinea pig. Involvement of specific lipid mediators, *Am. Rev. Respir. Dis.,* 142, 1093, 1990.

11. **Griffiths-Johnson, D., Jin, R., and Karol, M. H.,** The role of purified IgG$_1$ in pulmonary hypersensitivity responses of the guinea pig, *J. Toxicol. Environ. Health,* 40, 117, 1993.

12. **Karol, M. H., Jin, R., and Rubanoff, B.,** Clinical and experimental evaluation of isocyanate lung injury, *Comments in Toxicol.,* 3, 117, 1989.

13. **Karol, M. H., Tollerud, D., Campbell, T. P., Fabbri, L., Maestrelli, P., Saetta, M., and Mapp, C. E.,** The predictive value of airways hyperresponsiveness and circulating IgE for identifying types of responses to toluene diisocyanate inhalation challenge, *Am. J. Respir. Crit. Care Med.,* 149, 611, 1994.

14. **Jin, R. and Karol, M. H.,** Diisocyanate antigens that detect specific antibodies in exposed workers and guinea pigs, *Chem. Res. in Toxicol.,* 1, 288, 1988.

15. **Venables, K. M. and Newman-Taylor, A. J.,** Exposure–response relationships in asthma caused by tetrachlorophthalic anhydride, *J. Allergy Clin. Immunol.,* 85, 55, 1990.

16. **DeKruff, R. H., Fang, Y., and Umetsu, D. T.,** IL-4 synthesis by in vivo primed keyhole limpet hemocyanin-specific CD4+ T cells, *J. Immunol.,* 149, 3468, 1992.

17. **Garssen, J., Nijkamp, F. P., Wagenaar, S. S., Zwart, A., Askenase, P. W., and Van Loveren, H.,** Regulation of delayed-type hypersensitivity-like responses in the mouse lung, determined with histological procedures: serotonin, T suppressor inducer factor and high antigen dose tolerance regulate the magnitude of T cell dependent inflammatory reactions, *J. Immunol.,* 68, 51, 1989.

18. **Sarlo, K. and Karol, M. H.,** Guinea pig predictive tests for respiratory allergy, in *Immunotoxicology and Immunopharmacology,* 2nd ed., Dean, J., Luster, M., Munson, A. and Kimber, I., Eds., Raven Press, New York, 1994, 703.

19. **Karol, M. H.,** Concentration-dependent immunologic response to toluene diisocyanate (TDI) following inhalation exposure, *Toxicol. Appl. Pharmacol.,* 68, 229, 1983.

20. **Ptak, W., Geba, G. P., and Askenase, P. W.,** Initiation of delayed-type hypersensitivity by low doses of monoclonal IgE antibody, *J. Immunol,* 146, 3929, 1991.

21. **Chatelain, R., Varkila, K., and Coffman, R. L.,** IL-4 induces a Th2 response in *Leishmani major*-infected mice, *J. Immunol.,* 148, 1182, 1992.

22. **Griffiths-Johnson, D. and Karol, M.,** Validation of a non-invasive technique to assess development of airway hyperreactivity in an animal model of immunologic pulmonary hypersensitivity, *Toxicology,* 65, 283, 1991.

23. **Wass, U. and Belin, L.,** An in vitro method for predicting sensitizing properties of inhaled chemicals, *Scand. J. Work Environ. Health,* 16, 208, 1990.

24. **Sarlo, K. and Clark, E. D.,** A tier approach for evaluating the respiratory allergenicity of low molecular weight chemicals, *Fund. Appl. Toxicol.,* 18, 107, 1992.

25. **Karol, M. H.,** Animal models of occupational asthma, *Eur. Respir. J.,* 7, 555, 1994.

26. **Karol, M. H., Griffiths-Johnson, D. A., and Skoner, D. P.,** Chemically induced pulmonary hypersensitivity, airway hyperreactivity and asthma, in *Toxicology of the Lung,* 2nd ed., Gardner, D. E., Crapo, J. D., and Massaro, E. J., Eds., Raven Press, New York, 1993, 417.

27. **Dearman, R. J. and Kimber, I.,** Differential stimulation of immune function by respiratory and contact allergens, *Immunology,* 72, 563, 1991.

28. **Karol, M. H.,** Assays to evaluate pulmonary hypersensitivity, in *Modern Methods in Immunotoxicology,* Vol. 2, Burleson, G., Dean, J., and Munson, A., Eds., Wiley-Liss Publishers, New York, 1995, 401.

29. **Thorne, P. S. and Karol, M. H.,** Association of fever with late-onset pulmonary hypersensitivity responses in the guinea pig, *Toxicol. Appl. Pharmacol.,* 100, 247, 1989.

30. **Karol, M. H. and Thorne, P. S.,** Pulmonary hypersensitivity and hyperreactivity: implications for assessing allergic responses, in *Toxicology of the Lung,* Gardner, D. E., Crapo, J. D., and Massaro, E. J., Eds., Raven Press, New York, 1988, 427.

31. **Thorne, P. S., Yeske, C. P., and Karol, M. H.,** Monitoring guinea pig core temperature by telemetry during inhalation exposures, *Fund. Appl. Toxicol.,* 9, 398, 1987.
32. **Pauluhn, J. and Eben, A.,** Validation of a non-invasive technique to assess immediate or delayed onset of airway hypersensitivity in guinea-pigs, *J. Appl. Toxicol.,* 11, 423, 1991.
33. **Shiotsuka, R. N., Sangha, G. K., and Lyon, J. P.,** Comparison of respiratory sensitization responses to 2,4 and 2,6 toluene diisocyanate, *Toxicologist,* 13, 41, 1993.
34. **Stadler, J. C. and Loveless, S. E.,** Guinea pigs exhibit extended latency period for the development of sensitivity to an amine, *Toxicologist,* 12, 44, 1992.
35. **Lee, H. K., Alarie, Y., and Karol, M. H.,** Induction of formaldehyde sensitivity in guinea pigs, *Toxicol. Appl. Pharmacol.,* 75, 147, 1984.

Chapter 7

CHEMICAL RESPIRATORY ALLERGY: EXPOSURE AND DOSE–RESPONSE RELATIONSHIPS

Ian Kimber

CONTENTS

I. INTRODUCTION

Analysis of toxic phenomena and approaches to preventing adverse effects caused by chemicals requires an understanding of the route, extent, and duration of exposure, and the relationship between chemical disposition and dose and the appearance of symptoms.

Allergic hypersensitivity is, by definition, an immunological disease, dependent upon induction by the inducing allergen of a specific immune response in the susceptible individual. For sensitization to be achieved following encounter with allergen, an immune response of sufficient magnitude and of the appropriate type or quality must be provoked. If the sensitized individual is exposed subsequently to the same or a structurally similar allergen, then an accelerated and more aggressive response is elicited which causes local tissue changes, leading to the adverse effects recognized clinically as allergic hypersensitivity. With respect to chemical respiratory allergy, such symptoms may have an immediate or delayed onset and may range from rhinitis to frank asthma. In the context of defining safe occupational exposure levels and threshold-limit values for potential allergens, it is apparent that consideration must be given to the biphasic nature of allergic disease. There is the concentration of allergen which is necessary to induce sensitization in the immunologically naive, but susceptible, individual. Secondly, there is the concentration required to elicit hypersensitivity reactions in a previously sensitized individual. The available evidence points to the dose required to provoke the symptoms of hypersensitivity being considerably lower than that necessary for initial sensitization.

TABLE 7-1

Examples of Occupational Chemical Respiratory Allergens

Allergen	Ref.
Anhydrides	
Trimellitic anhydride	4–6
Phthalic anhydride	4, 6–8
Tetrachlorophthalic anhydride	6, 9, 10
Maleic anhydride	6
Hexahydrophthalic anhydride	11
Isocyanates	
Diphenylmethane diisocyanate	12–14
Toluene diisocyanate	15, 16
Hexamethylene diisocyanate	17
Reactive dyes	18–20
Platinum salts	21–23
Chloramine-T	24, 25
Plicatic acid	26

It is possible to identify a number of factors which may influence either or both the induction and elicitation phases of respiratory hypersensitivity to chemicals. In addition to host-related factors such as genetic predisposition (and possibly atopy) and the pre-existence of other airway disease or dysfunction, these may include:

(a)　The inherent sensitizing potential of the chemical allergen
(b)　The route of exposure during the sensitization phase
(c)　The extent, duration, and frequency of exposure
(d)　The influence of external factors other than the availability of antigen itself.

II. FACTORS INFLUENCING CHEMICAL RESPIRATORY HYPERSENSITIVITY

A. CHEMICAL ALLERGEN

A variety of chemicals have been implicated as the cause of occupational respiratory allergy.[1-3] Some of those more commonly associated with hypersensitivity are listed in Table 7-1.

It is difficult at the present time to derive from clinical experience alone a clear indication of the relative sensitizing potential of these chemicals in man. Variation in the extent, duration, and nature of exposure and differences in individual susceptibility make such comparisons unrealistic. It may, however, be possible to obtain some information regarding comparative sensitizing activity based upon the use of appropriate animal models in which conditions of exposure can be controlled. If, operationally, potency is measured as a function of the lowest concentration of the test chemical which, under defined experimental conditions, provokes specific sensitization, then activity may be assessed either by the appearance of specific antibody or by the development of hypersensitivity as judged by pulmonary responses to subsequent inhalation challenge. Certainly in guinea pigs it has proven possible to define no-effect levels for respiratory sensitization with both toluene diisocyanate (TDI)[27] and with the proteolytic enzyme subtilizin.[28] Irrespective of the route of exposure, it is possible that the ability of a chemical to induce sensitization will be influenced also by the irritant properties of the material.

Studies of this type will, of course, give an indication only of intrinsic hazard. Although estimates of inherent sensitizing potential are unlikely to correlate directly with activity in man following occupational or environmental exposure, hazard evaluation does form an important component of the risk assessment process.

B. ROUTE OF EXPOSURE

It is assumed commonly that sensitization of the respiratory tract to chemical allergens is provoked almost exclusively following inhalation exposure. This is not necessarily the case, however. While it is likely that occupational respiratory hypersensitivity results most frequently from inhalation exposure to atmospheres containing the inducing allergen, there are reports of respiratory sensitization to chemicals induced apparently by dermal contact.[29,30] Moreover, it has been found in guinea pigs that intradermal[31-34] or topical[34,35] exposure to chemical respiratory allergens is able to induce sensitization. Karol et al.[35] reported that epidermal application of TDI to guinea pigs resulted in the appearance of homocytotropic antibody and, in a proportion of exposed animals, a substantial increase in respiratory rate following inhalation challenge with aerosols containing a TDI-protein conjugate. In addition, it is clear that topical exposure of mice to chemical respiratory allergens such as trimellitic anhydride (TMA) induces hapten-specific IgE antibody.[36]

It should not really come as any surprise that sensitization via routes other than inhalation can provoke the magnitude and quality of immune response necessary for effective sensitization of the respiratory tract. The implication is that dermal contact with high concentrations of chemical respiratory allergens, resulting probably from spillages, splashing, or other industrial accidents, may induce in susceptible individuals pulmonary hypersensitivity. Clearly, control of airborne concentrations of chemical allergens is necessary for the prevention of occupational asthma.[37] However, in addition to defining safe atmospheric concentrations, consideration should be given to the potential for skin exposure.

C. THE EXTENT, DURATION, AND FREQUENCY OF EXPOSURE

It is probable, but difficult to prove, that the amount of chemical required for the induction of sensitization is greater than that required for provocation of pulmonary hypersensitivity reactions in previously sensitized individuals. There is evidence for dose–response relationships and for the existence of threshold concentrations with regard to occupational respiratory sensitization. Such is reflected by the fact that a reduction in the levels of occupational exposure to TDI has been associated with a decrease in the frequency of reported cases of respiratory allergy.[2] The importance of exposure concentration in the induction of respiratory sensitization is suggested also by the observation that the incidence of asthma among isocyanate workers correlates positively with the number of spills experienced.[38] Similarly, an association between high-dose exposure and disease prevalence has been found among individuals working with Western red cedar, the causative allergen being plicatic acid. Workers employed in jobs associated with exposure to the highest levels of wood dust displayed the highest frequency of occupational asthma.[38]

The importance of exposure concentration for sensitization is apparent also from animal studies. Karol[27] examined in guinea pigs the influence of increasing concentrations of inhaled TDI on the induction of specific antibody responses and on changes in respiratory rate following bronchial provocation with a TDI-protein conjugate. Exposure of animals for 3 hr per day for 5 consecutive days to concentrations of 0.36 ppm TDI or greater caused significant challenge-induced respiratory reactions and the appearance of homocytotropic antibody in a proportion of the treated guinea pigs. Similar exposure to lower concentrations of TDI (0.12 ppm) failed to induce either anti-hapten antibody or respiratory hypersensitivity.[27]

Little information is available regarding the influence of the frequency of exposure per se on the induction of sensitization. What evidence there is suggests that exposure concentration rather than cumulative dose is of greater relevance. In the studies performed by Karol and quoted above,[27] it was found that exposure of guinea pigs for 3 hr per day for 5 consecutive days to 0.61 ppm TDI induced respiratory hypersensitivity in approximately 25% of the test animals. In contrast, exposure for 70 days (over a 4-month period) to 0.02 ppm TDI for 6 hr/day (to provide an equivalent cumulative dose) failed to cause sensitization. It might be, therefore, that a certain critical amount of chemical allergen must be available during a finite time frame to facilitate efficient priming for sensitization. Such, of course, is consistent with the requirement to stimulate a specific immune response.

D. EXTERNAL FACTORS

The importance of occupational asthma has been emphasized recently by the results of a scheme for the Surveillance of Work-related and Occupational Respiratory Disease (SWORD).[39] The SWORD project group found that the true incidence of acute occupational respiratory disease in the U.K. may be some three times greater than appreciated previously. During the period of this survey (1989) the most frequent diagnosis was of asthma (26.4% of all new cases). In the majority of instances the causative agent or allergen was identified or suspected, the most common diagnoses being associated with exposure to isocyanates (22% of occupational asthma), flour and grain dusts (8%), wood dusts (6%), solder flux (6%), laboratory animals or insects (4%), or hardening agents (3%). Among the other materials associated with occupational asthma were azodicarbonamide, platinum salts, antibiotics, and proteolytic enzymes.[39]

For effective sensitization of a susceptible individual, the availability of sufficient amounts of the provoking allergen in the external environment is clearly a critical determinant. The efficiency of sensitization may, however, be influenced also by a variety of other environmental factors.

There is a general agreement that the incidence of asthma is increasing in industrialized societies. Comparisons of the prevalence of asthma among the offspring of migrants to urban areas with that in children of those remaining in the community of origin have frequently endorsed the influence of environmental factors on the disease. The extent to which changing patterns in the prevalence of asthma result from an increased incidence of allergic sensitization of the respiratory tract is unclear, although there is evidence that allergic disease has increased in the U.K. since 1945.[40] Such changes cannot usually be reconciled solely on the basis of altered or increased exposure to allergen and have prompted consideration of the impact of environmental pollution on asthma.[41]

There are several reports which indicate that tobacco smoking is associated with increased IgE antibody production and/or elevated serum concentrations of IgE.[42-45] Moreover, it has been shown that rats exposed to tobacco smoke display enhanced IgE responses.[46] In the context of occupational respiratory sensitization, smoking is of greatest relevance, and exerts its greatest influence, during the period following introduction to a new allergenic material. There is evidence also for an increased incidence of asthma and elevated serum IgE levels among the children of women who smoke.[43] Whether parental smoking translates directly into enhanced allergic sensitization in children,[47] and whether the changes observed are secondary to effects *in utero* or result from passive inhalation of tobacco smoke during infancy remains unclear. Nevertheless, the suggestion is that smoking may represent an important and preventable risk factor, particularly during the period of first exposure to a new occupational or environmental allergen.

Common environmental pollutants may also influence IgE production and respiratory sensitization. It has been shown that the coadministration to mice of ovalbumin mixed with diesel exhaust particulates, by either the intraperitoneal or intranasal routes, results in increased production of IgE antibody specific for the protein allergen.[48,49] In addition, exposure of guinea pigs to high concentrations of sulfur dioxide followed by sensitization with ovalbumin was shown to result in both elevated levels of specific IgE antibody and increased airway responses following inhalation challenge with the protein.[50]

Finally, it is considered that respiratory viral infections may play a role in the induction or exacerbation of asthma.[51]

Taken together, the available evidence suggests that the presence of pollutants in the external environment, smoking, and concurrent infection of the respiratory tract may individually or in concert have the potential to act as adjuvants for IgE antibody production and/or influence the development of allergic sensitization. The corollary is that environmental conditions, smoking, and the health status of the individual may all affect the concentration of allergen required to induce effective sensitization or the extent to which sensitization will result from allergen exposure.

III. CONCLUDING COMMENTS

It will be clear that risk assessment of hypersensitivity resulting from exposure to potential chemical respiratory allergens is still an inexact science. Although in recent years progress has been made toward the development of methods suitable for the prospective testing of chemicals and hazard identification, effective risk management and the derivation of appropriate occupational exposure levels will require a more precise evaluation of the relative potency of sensitizing materials and a clearer understanding of the factors which influence the development of respiratory allergy.

REFERENCES

1. **Salvaggio, J. E., Butcher, B. T., and O'Neil, C. E.,** Occupational asthma due to chemical agents, *J. Allergy Clin. Immunol.,* 78, 1053, 1986.
2. **Karol, M. H.,** Occupational asthma and allergic reactions to inhaled compounds, in *Principles and Practice of Immunotoxicology,* Miller, K., Turk, J., and Nicklin, S., Eds., Blackwell Scientific, Oxford, 1992, 228.
3. **Grammer, L. C.,** Occupational immunologic lung disease, in *Allergic Diseases. Diagnosis and Management,* 4th ed., Patterson, R., Grammer, L. C., Greenberger, P. A. and Zeiss, C. R., Eds., Lippincott, Philadelphia, 1993, 745.
4. **Bernstein, D. I., Patterson, R., and Zeiss, C. R.,** Clinical and immunologic evaluation of trimellitic anhydride- and phthalic anhydride-exposed workers using a questionnaire and comparative analysis of enzyme linked immunosorbent and radioimmunoassay studies, *J. Allergy Clin. Immunol.,* 69, 311, 1982.
5. **Zeiss, C. R., Wolkonsky, P., Chacon, R., Tuntland, R. N., Levitz, D., Prunzansky, J. J., and Patterson, R.,** Syndromes in workers exposed to trimellitic anhydride. A longitudinal clinical and immunologic study, *Ann. Int. Med.,* 98, 8, 1983.
6. **Topping, M. D., Venables, K. M., Luczynska, C. M., Howe, W., and Newman Taylor, A. J.,** Specificity of the human IgE response to inhaled acid anhydrides, *J. Allergy Clin. Immunol.,* 77, 834, 1986.
7. **Kern, R. A.,** Asthma and allergic rhinitis due to sensitization to phthalic anhydride. Report of a case, *J. Allergy,* 10, 164, 1939.
8. **Maccia, C. A., Bernstein, I. L., Emmett, E. A., and Brooks, S. M.,** In vitro demonstration of specific IgE in phthalic anhydride hypersensitivity, *Am. Rev. Respir. Dis.,* 113, 701, 1976.

9. Howe, W., Venables, K. M., Topping, M. D., Dally, M. B., Hawkins, R., Law, J. S., and Taylor, A. T., Tetrachlorophthalic anhydride asthma: evidence for specific IgE antibody, *J. Allergy Clin. Immunol.*, 71, 5, 1983.

10. Schlueter, D. P., Banaszak, E. F., Fink, J. N., and Barboriak, J., Occupational asthma due to tetrachlorophthalic anhydride, *J. Occup. Med.*, 20, 183, 1978.

11. Moller, D. R., Gallagher, J. S., Bernstein, D. I., Wilcox, T. G., Burroughs, H. E., and Bernstein, I. L., Detection of IgE-mediated respiratory sensitization in workers exposed to hexahydrophthalic anhydride, *J. Allergy Clin. Immunol.*, 76, 663, 1985.

12. Tansar, A. R., Bourke, M. P., and Blandford, A. G., Isocyanate asthma: respiratory symptoms caused by diphenylmethane diisocyanate, *Thorax*, 28, 596, 1973.

13. Zeiss, C. R., Kannellakes, T. M., Bellone, J. D., Levitz, D., Pruzansky, J. J., and Patterson, R., Immunoglobulin E-mediated asthma and hypersensitivity pneumonitis with precipitating anti-hapten antibodies due to diphenylmethane diisocyanate (MDI) exposure, *J. Allergy Clin. Immunol.*, 65, 346, 1980.

14. Zammit-Tabona, M., Sherkin, M., Kijek, K., Chan, H., and Chan-Yeung, M., Asthma caused by diphenylmethane diisocyanate in foundry workers. Clinical, bronchial provocation and immunologic studies, *Am. Rev. Respir. Dis.*, 128, 226, 1983.

15. Danks, J. M., Cromwell, O., Buckingham, J. A., Newman Taylor, A. J., and Davies, R. J., Toluene diisocyanate induced asthma: evaluation of antibodies in the serum of affected workers against a tolyl monoisocyanate protein conjugate, *Clin. Allergy*, 11, 161, 1981.

16. O'Brien, I. M., Harries, M. G., Burge, P. S., and Pepys, J., Toluene di-isocyanate-induced asthma. 1. Reactions to TDI, MDI, HDI and histamine, *Clin. Allergy*, 9, 1, 1979.

17. Grammer, L. C., Eggum, P., Silverstein, M., Shaughnessy, M. A., Liotta, J. L., and Patterson, R., Prospective immunologic and clinical study of a population exposed to hexamethylene diisocyanate, *J. Allergy Clin. Immunol.*, 82, 627, 1988.

18. Alanko, K., Keskinen, H., Bjorksten, F., and Ojanen, S., Immediate-type hypersensitivity to reactive dyes, *Clin. Allergy*, 8, 25, 1978.

19. Docker, A., Wattie, J. M., Topping, M. D., Luczynska, C. M., Newman Taylor, A. J., Pickering, C. A. C., Thomas, P., and Gompertz, D., Clinical and immunological investigations of respiratory disease in workers using reactive dyes, *Br. J. Ind. Med.*, 44, 534, 1987.

20. Topping, M. D., Forster, H. W., Ide, C. W., Kennedy, F. M., Leach, A. M., and Sorkin, S., Respiratory allergy and specific immunoglobulin E and immunoglobulin G antibodies to reactive dyes in the wool industry, *J. Occup. Med.*, 31, 857, 1989.

21. Cromwell, O., Pepys, J., Parish, W. E., and Hughes, E. G., Specific IgE antibodies to platinum salts in sensitized workers, *Clin. Allergy*, 9, 109, 1979.

22. Biagini, R. E., Bernstein, I. L., Gallagher, J. S., Moorman, W. J., Brooks, S., and Gann, P. H., The diversity of reaginic immune responses to platinum and palladium salts, *J. Allergy Clin. Immunol.*, 76, 794, 1985.

23. Murdoch, R. D., Pepys, J., and Hughes, E. G., IgE antibody responses to platinum group metals: a large scale refinery survey, *Br. J. Ind. Med.*, 43, 37, 1986.

24. Dijkman, J. G., Vooren, P. H., and Kramps, J. A., Occupational asthma due to inhalation of chloramine-T. 1. Clinical observations and inhalation-provocation studies, *Int. Arch. Allergy Appl. Immunol.*, 64, 422, 1981.

25. Wass, U., Belin, L., and Eriksson, N. E., Immunological specificity of chloramine-T-induced IgE antibodies in serum from a sensitized worker, *Clin. Allergy*, 19, 463, 1989.

26. Chan-Yeung, M., Barton, G., MacLean, L., and Grzybowski, S., Occupational asthma and rhinitis due to western red cedar (Thuja plicata), *Am. Rev. Respir. Dis.*, 108, 1094, 1973.

27. Karol, M. H., Concentration-dependent immunologic response to toluene diisocyanate (TDI) following inhalation exposure, *Toxicol. Appl. Pharmacol.*, 68, 229, 1983.

28. **Thorne, P. S., Hillebrand, J., Magreni, C., Riley, E. J., and Karol, M. H.,** Experimental sensitization to subtilisin. I. Production of immediate- and late-onset pulmonary reactions, *Toxicol. Appl. Pharmacol.*, 86, 112, 1986.
29. **Karol, M. H.,** Respiratory effects of inhaled isocyanates, *CRC Crit. Rev. Toxicol.*, 16, 349, 1986.
30. **Nemery, B. and Lenaerts, L.,** Exposure to methylene diphenyl diisocyanate in coal mines, *Lancet*, 341, 318, 1993.
31. **Botham, P. A., Rattray, N. J., Woodcock, D. R., Walsh, S. T., and Hext, P. M.,** The induction of respiratory allergy in guinea-pigs following intradermal injection of trimellitic anhydride: a comparison with the response to 2,4-dinitrochlorobenzene, *Toxicol. Lett.*, 47, 25, 1989.
32. **Pauluhn, J. and Eben, A.,** Validation of a non-invasive technique to assess immediate or delayed onset of airway hypersensitivity in guinea pigs, *J. Appl. Toxicol.*, 11, 423, 1991.
33. **Hayes, J. P., Daniel, R., Tee, R. D., Barnes, P. J., Chung, K. F., and Newman Taylor A. J.,** Specific immunological and bronchopulmonary responses following intradermal sensitization to free trimellitic anhydride in guinea pigs, *Clin. Exp. Allergy*, 22, 694, 1992.
34. **Rattray, N. J., Botham, P. A., Hext, P. M., Woodcock, D. R., Fielding, I., Dearman, R. J., and Kimber, I.,** Induction of respiratory hypersensitivity to diphenylmethane-4,4'-diisocyanate (MDI) in guinea pigs. Influence of route of exposure, *Toxicology*, 88, 15, 1994.
35. **Karol, M. H., Hauth, B. A., Riley, E. J., and Magreni, C. M.,** Dermal contact with toluene diisocyanate (TDI) produces respiratory tract hypersensitivity in guinea pigs, *Toxicol. Appl. Pharmacol.*, 58, 221, 1981.
36. **Dearman, R. J. and Kimber, I.,** Differential stimulation of immune function by respiratory and contact chemical allergens, *Immunology*, 72, 563, 1991.
37. **Venables, K. M.,** Preventing occupational asthma, *Br. J. Ind. Med.*, 49, 817, 1992.
38. **Brooks, S. M.,** The evaluation of occupational airways disease in the laboratory and workplace, *J. Allergy Clin. Immunol.*, 70, 56, 1982.
39. **Meredith, S. K., Taylor, V. M., and McDonald, J. C.,** Occupational respiratory disease in the United Kingdom 1989: a report to the British Thoracic Society and the Society of Occupational Medicine by the SWORD project group, *Br. J. Ind. Med.*, 48, 292, 1991.
40. **Fleming, D. M.,** Prevalence of asthma and hayfever in England and Wales, *Br. Med. J.*, 296, 279, 1987.
41. **Wardlaw, A. J.,** The role of air pollution in asthma, *Clin. Exp. Allergy*, 23, 81, 1993.
42. **Zetterstrom, O., Osterman, K., Machado, L., and Johansson, S. G.,** Another smoking hazard: raised serum IgE concentration and increased risk of occupational allergy, *Br. Med. J.*, 283, 1215, 1981.
43. **Kjellman, N.-I. M.,** Effect of parental smoking on IgE levels in children, *Lancet*, 1, 993, 1981.
44. **Burrows, B., Halonen, M., Barbee, R. A., and Lebowitz, M. D.,** The relationship of serum immunoglobulin E to cigarette smoking, *Am. Rev. Respir. Dis.*, 124, 523, 1981.
45. **Venables, K. M., Topping, M. D., Howe, W., Luczynska, C. M., and Hawkins, R.,** Interaction of smoking and atopy in producing specific IgE antibody against a hapten protein conjugate, *Br. Med. J.*, 290, 201, 1985.
46. **Zetterstrom, O., Nordvall, S. L., Bjorksten, B., Ahlstedt, S., and Stelander, M.,** Increased IgE antibody responses in rats exposed to tobacco smoke, *J. Allergy Clin. Immunol.*, 75, 594, 1985.
47. **Ownby, D. R. and McCullough, J.,** Passive exposure to cigarette smoke does not increase allergic sensitization in children, *J. Allergy Clin. Immunol.*, 82, 634, 1988.
48. **Muranaka, M., Suzuki, S., Koizumi, K., Takafuji, S., Miyamoto, T., Ikemori, R., and Tokiwa, H.,** Adjuvant activity of diesel-exhaust particulates for the production of IgE antibody in mice, *J. Allergy Clin. Immunol.*, 77, 616, 1986.

49. **Takafuji, S., Suzuki, S., Koizumi, K., Tadokoro, K., Miyamoto, T., Ikemori, R., and Muranaka, M.,** Diesel-exhaust particulates inoculated by the intranasal route have an adjuvant activity for IgE production in mice, *J. Allergy Clin. Immunol.*, 79, 639, 1987.
50. **Riedel, F., Kramer, M., Scherbenbugen, C., and Rieger, C. H. L.,** Effects of SO_2 exposure on allergic sensitization in the guinea pig, *J. Allergy Clin. Immunol.*, 82, 527, 1988.
51. **Frick, O. L., German, D. F., and Mills, J.,** Development of allergy in children. I. Association with virus infections, *J. Allergy Clin. Immunol.*, 63, 228, 1979.

Chapter 8

PREVENTION OF OCCUPATIONAL RESPIRATORY ALLERGY

Katherine M. Venables

CONTENTS

I. INTRODUCTION

A. CONCEPTS

For practical purposes, respiratory hypersensitivity induced by chemicals is an occupational, rather than an environmental, problem. More is known about occupational asthma than about rhinitis, and about hypersensitivity mediated by IgE or a similar mechanism than through the other mechanisms which are important in conditions such as extrinsic allergic alveolitis (hypersensitivity pneumonitis) or beryllium disease. This chapter therefore focuses on occupational asthma.

A belief in the importance of clarity in defining occupational asthma has guided the writing of this chapter. The incidence rate of asthma caused by workplace exposures will not fall unless those with responsibility for prevention have a clear idea of what they are trying to prevent. Much of the text preceding the outline of recommendations is therefore devoted to a description of the different pathways to asthma. The chapter focuses on the prevention of asthma caused by exposure to sensitizing agents in the workplace.

The chapter does not include working definitions for specific purposes of public health practice, clinical work, or epidemiological research. Such definitions grow from an understanding of conceptual definitions of respiratory hypersensitivity and take their final form depending upon several factors, including the context of the investigation (for example clinical practice as opposed to workplace screening), availability of resources, and potential for bias.

It would be wrong to assume that asthma induced or exacerbated by irritant exposures to chemicals in the workplace is of no importance. On the contrary, there is no information on the relative frequency of different types of asthma in the workplace; types related to irritant exposures may be the most frequently-occurring types in some industries.

B. APPROACH TO PREVENTION

The importance of primary prevention underpins the writing of this chapter. Control of the exposures causing asthma is the most direct way of reducing the number of incident cases. In order to achieve control, these exposures must be named; several hundred sensitizing agents are currently known to cause asthma; the relative frequency of occurrence of asthma caused by different agents can only be known by national surveillance; those responsible for prevention can only know where to take action if a list of sensitizing agents is published. Secondly, the relation between exposure and response must be quantified by longitudinal epidemiological studies and by inference from animal experimentation. Thirdly, the effectiveness of control measures in reducing the incidence of asthma must be measured.

No control measure can be assumed to eliminate asthma short of elimination of the causal agent. Therefore, cases of asthma will develop and screening programs are needed to detect them. Detection is useful as secondary prevention, because there is evidence that patients with occupational asthma who cease exposure promptly to the causal agent have a better long-term prognosis than those who do not. Detection is also useful as a way of monitoring primary prevention measures to control exposure, though it would be wrong for monitoring to rely on the detection of cases rather than on assessment of the workplace.

C. METHODS OF PREVENTION

Because the chapter envisages direct prevention as taking place in the workplace, another underlying assumption behind its writing is that the people with primary responsibility for prevention are in the workplace: managers and supervisors, worker representatives, and health and safety professionals working within industry. It does not assume that regulatory bodies, physicians outside industry, academic institutions, and others concerned with occupational health and safety have no influence. It is clear that they have considerable influence. However, in general, such agencies and individuals have no direct authority to control the work process and, where their influence is felt, they act indirectly through those with such authority. Those with responsibility are usually less knowledgeable than advisors and regulators about asthma and its relation to work, but more knowledgeable about practical ways in which control measures can be implemented. Successful prevention depends on co-operation between employers, workers and their representatives, regulators, and medical and nonmedical specialist advisors.

D. SUMMARY OF RECOMMENDATIONS

This chapter recommends that regulatory or advisory bodies with responsibility for occupational asthma publish a guidance document on occupational asthma explaining their current concepts of occupational asthma. Surveillance activities give information on how common asthma is relative to other occupational lung diseases and on the relative frequency with which different agents cause asthma. Publication of a list of sensitizing agents would aid those with responsibility for control of exposure in the workplace. Epidemiological research on exposure–response relations is necessary as a background to prevention. This chapter recommends such studies. Immunotoxicological research also has a role in testing hypotheses that cannot be tested in human subjects. Some standardization of screening programs in industry is desirable. A short, symptoms questionnaire is economical and acceptable to workers, but there are other approaches. Finally, evaluative research on preventive measures gives information on their effectiveness and efficiency.

II. ASTHMA

Asthma is common, found in up to 10% of adults (Gregg 1977). Scadding's definition is widely quoted: "Variable dyspnoea due to widespread narrowing of peripheral airways in the lungs, and varying in severity over short periods of time either spontaneously or as a result of treatment" (Scadding 1963).

A. ASTHMA AND BRONCHIAL HYPERRESPONSIVENESS

Bronchial hyperresponsiveness is generally accepted as the defining physiological characteristic, though it is not the only characteristic of asthma nor is it found exclusively in asthma. Its severity is variable and it may not be detectable in every patient with asthma on each occasion of testing. Bronchial hyperresponsiveness is normally expressed in terms of response to inhaled histamine or acetylcholine, or its analogs (Dautrebande and Phillipot 1941, Curry 1946). This response relates well to indices of the severity of asthma: the minimum treatment needed to control symptoms (Juniper et al. 1981), response to bronchodilator, diurnal variation in lung function (Ryan et al. 1982), and airway narrowing on exercise (Chatham et al. 1982), after inhalation of cold air (O'Byrne et al. 1982) or of dusts and aerosols (Dubois and Dautrebande 1958).

B. OCCUPATION AND ASTHMA

Any occupation where dusts, vapors or gases are inhaled is a potential cause of lung disease, or may exacerbate pre-existing lung disease. There are several ways in which occupational exposures cause or provoke asthma.

1. Exacerbation of Asthma

Firstly, asthma may be exacerbated by work involving exercise or exposure to cold, dust, or irritants to which hyperresponsive airways react. For example, people with asthma, but not normals, react with airway narrowing to sulfur dioxide at concentrations lower than recommended occupational standards (Sheppard et al. 1980). There is a prior history of symptoms, no latent interval of symptom-free exposure, and symptoms are improved by drug treatment and by avoiding heavy exposure to the provoking agent. The term "irritants" as used here does not imply a specific mechanism such as stimulation of irritant receptors and reflex bronchoconstriction.

2. Pharmacologically Active Agents

Pharmacologically active substances may be used at work and have been invoked as explanations for work-related asthma. For example, hexamethylene diisocyanate has *in vitro* anti-cholinesterase activity though it has no measurable *in vivo* cholinesterase inhibition in guinea pigs (Karol et al. 1984). It is not clear how commonly this occurs. It is not clear if provocation of airway narrowing by pharmacological means should be regarded as a form of "irritant response" as described above.

3. Heavy Exposures and Irritant-Induced Asthma

Heavy exposure to irritants in industrial accidents causes pulmonary edema and also airway mucosal damage (Charan et al. 1979). Survivors may have bronchial hyperresponsiveness which persists for months, perhaps longer (Gandevia 1970). This irritant-induced asthma has been termed "reactive airways dysfunction syndrome" (Brooks et al. 1985). It is not clear if this represents the severe end of the spectrum of "irritant responses." Less extreme exposures of volunteers give a temporary increase in bronchial responsiveness (Golden et al. 1978). The inflammatory response to viral respiratory tract infection (Empey et al. 1976) or influenza immunization (Ouellette and Reed 1965) is accompanied by temporarily increased bronchial responsiveness.

4. Sensitization

Lastly, exposure at work causes sensitization which, in a proportion of those affected, is associated with asthma. Although asthma caused by chemicals is sometimes discussed as a separate category from asthma caused by biologically-derived agents, there is no evidence that the asthma is qualitatively different. Allergic asthma caused by work is therefore discussed here regardless of the nature of the causal agent.

The term "allergy" was first used by von Pirquet (1906) for any altered response to foreign material, either immunity or hypersensitivity, but its use for immunity has lapsed. There is no evidence that allergic asthma caused by an occupational exposure differs from that caused by common allergens such as pollens or house dust. The term "allergy" is often used to imply an IgE-mediated mechanism. Asthma caused by some occupational exposures is IgE-mediated, for example to rat urine (Newman Taylor et al. 1977), though it is not always possible to detect specific IgE antibodies in every case at the time of testing. With some other causes of occupational asthma, for example Western red cedar wood, it is possible that other immunological mechanisms mediate the asthma in a high proportion of patients. The symptoms and physiological findings are indistinguishable from those in IgE-mediated asthma. The term "sensitization" is used here, rather than "allergy," to avoid implying that IgE is the only immunological mechanism in occupational asthma.

Clinically, there is a latent interval of weeks to years and patients may have no prior history of asthma, unless they are already hypersensitive to some other allergen. After sensitization, there may be responsiveness at extremely low levels of exposure. Potentially fatal attacks of asthma can occur after exposures which are orders of magnitude below those which irritate. Often the only way symptoms can be minimized is by leaving the job. The clinical picture is thus different from irritant-induced asthma where, although there is nonspecific airway hyperresponsiveness to a range of airway irritants, patients do not have extreme and specific sensitivity to low concentrations of the causal agent.

The pattern of response is not specific to a particular agent. For example, the acid anhydrides, a group of low molecular weight agents, can cause IgE-mediated asthma, stimulate specific IgG and other antibody production, exacerbate pre-existing asthma, and cause direct toxicity to the airways and lung parenchyma if inhaled in sufficient concentration (Venables 1989).

III. SENSITIZER-INDUCED ASTHMA

A. MECHANISM

In the airways, inhaled antigen links to antibody which starts a chain of mediator release leading to smooth muscle contraction, mucosal edema, and mucus hypersecretion. The physiological evidence of these processes is airway narrowing which may be immediate, late, or dual (Pepys and Hutchcroft 1975).

B. LATE ASTHMATIC RESPONSE

It is thought that the late response is the response component which most closely resembles spontaneously occurring asthma. The late response is maximal at 4 to 8 hr after exposure. Increases in nonspecific bronchial responsiveness occur in parallel with the late response and were first reported by Cockcroft et al. (1977). It may be weeks before baseline values are regained, and during this time the patient experiences a heightened response to irritants and exercise with increased diurnal variation in airway caliber. This phenomenon resembles the increase in bronchial responsiveness after infection or irritant exposure described above, and the common factor is likely to be airway mucosal inflammation which Hogg (1982) has suggested increases mucosal permeability, allowing chemical mediators to reach smooth muscle and irritant receptors.

C. TERMINOLOGY

The term "occupational asthma" is often used loosely. In this chapter, two types of occupationally-induced asthma are recognized: sensitizer-induced asthma (or allergic asthma) and irritant-induced asthma (reactive airways dysfunction syndrome). In both pathways to asthma, a work exposure has caused a fundamental change in the worker's airway response to his work and home environment. This is qualitatively different from work-exacerbated asthma where work exposure is only one of many triggers which provoke asthma attacks. The complexity of the interactions between occupational and nonoccupational factors, and between asthma and other airway diseases, has been discussed in more detail by Hendrick (1983). He makes the important points that defining one's terms "is not simply a matter of semantics, because the attendant problems of recognition, management, prevention, and compensation may differ profoundly" and that an unduly broad definition "assumes a license that few would consider helpful."

Harber (1992) has recently delineated 12 "occupation-asthma interactions." As well as categories of no relationship and uncertain relationship, he includes the two causal pathways noted in this chapter: induction by a sensitizer or by a high-level toxic exposure. He identifies permanent and temporary exacerbation of asthma symptoms as separate categories. He comments that workplace factors may indirectly worsen asthma by, for example, precluding the use of medication. He states that there may be factors at work which improve the severity of asthma. He proposes that asymptomatic bronchial hyperresponsiveness may become symptomatic with a work exposure, and includes the development of symptoms earlier than otherwise would have occurred as a separate interaction. Finally, he comments that asthma may limit the worker's ability to carry out his work, or may increase his future risk at work.

D. IMPORTANCE: FREQUENCY ESTIMATES

Blanc (1987) has published from the U.S. 1978 Social Security Administration Survey of Disability and Work, a survey of over 6,000 persons weighted toward benefit applicants and those with activity limitation. A self-reported diagnosis of asthma was noted in 7.7% of respondents and 15.4% of these reported that it was "caused by bad working conditions."

The best national data for occupational asthma appear to be from Finland (Keskinen et al. 1978). There were 156 cases in a working population of about 2.2 million in 1981 (Keskinen 1983), a crude annual incidence rate of 71 per million. In Finland, the condition is assumed to be mediated through a sensitization mechanism. Workplace or laboratory challenge testing is normally required to establish the diagnosis. The U.K. currently has a voluntary scheme for Surveillance of Work-related and Occupational Respiratory Disease (SWORD), which noted 554 cases in a working population of about 25 million in 1989 (Meredith et al. 1991), a crude annual incidence rate of 22 per million. The cases are of physician-diagnosed "occupational or work-related" asthma and reports come from most of the country's specialist chest physicians and an unknown proportion of specialist occupational physicians. Asthma is the most common disease in the SWORD data. Some rates for comparison are 13 per million for malignant mesothelioma, 13 per million for all pneumoconioses, 5 for allergic alveolitis (hypersensitivity pneumonitis), and 1 for byssinosis.

1. Underestimation of Frequency

The U.K. estimate for asthma in 1989 is assumed to be an underestimate, with the true rate closer to that reported from Finland, as regional rates varied from 8 to 63 per million and a large part of the variation is thought to be explained by ascertainment. 1989 was the first year of the U.K. scheme, whereas the Finnish Occupational Disease Register dates from 1964, with improvements in 1975. There has been an increase in the number of cases of occupational asthma in Finland from less than 10 per year in the 1960s to 80 in 1976 (Keskinen et al. 1978), presumably reflecting increased ascertainment.

2. Denominators for Frequency Estimates

In these analyses from Europe, the population denominator for each estimate is the working population. Keskinen has estimated that less than one-quarter of the Finnish working population is exposed to the more common causes of occupational asthma (Keskinen et al. 1978). SWORD data show considerable variation in incidence by occupational group, from less than 10 to 114 per million, with much higher rates in some subgroups, for example 639 per million in automobile and spray painters. Much higher rates have been noted in individual workplaces or working groups, so that the unit of incidence (or, more commonly, prevalence) is usually given as a percentage rather than as per million (Venables 1987, Chan-Yeung 1990). For example, about one-quarter of a cohort of platinum refinery workers developed both a positive skin prick test to platinum salts and also respiratory symptoms in up to 4 years of follow-up, most within the first year (Venables et al. 1989a). Examples of cross-sectional studies are more numerous; an early example in the platinum refining industry estimated that 46% of workers had occupational asthma (Hunter et al. 1945).

E. IMPORTANCE: MORBIDITY

Occupational asthma is an important cause of morbidity in the general community because the onset of allergic asthma appears to represent a fundamental switch in the individual's way of responding to his environment. Once switched on, asthma does not, in general, go away. Several follow-up studies show that asthma, although improved, does persist after elimination or reduction of exposure to Western red cedar (Chan-Yeung et al. 1982), solder fume (Burge 1982b), toluene diisocyanate (Paggiaro et al. 1984), tetrachlorophthalic anhydride (Venables et al. 1987), snow crab (Malo et al. 1988), and various agents (Hudson et al. 1985, Venables et al. 1989b). These studies report persistent symptoms in up to 100% of patients; most also report persistent bronchial hyperresponsiveness; two were able to note persistent specific immunological responsiveness to the causal agent (Venables et al. 1987, Malo et al. 1988); one study suggested a plateau of improvement at about 2 years after cessation of the causal exposure (Malo et al. 1988). Therefore, sensitization in the workplace is a source of new cases of adult-onset asthma in the community.

F. IMPORTANCE: SEVERITY

Lastly, occupational asthma is important because the asthma may be severe. Death is possible in an acute attack of asthma (Fabbri et al. 1988). Severe attacks requiring hospitalization are not uncommon. In the U.S. program to eradicate screwworm fly by aerial distribution of irradiated screwworms, 70% of pilots and dispersers developed allergic symptoms in about 2 weeks to 6 months after joining the program and there were several emergency landings performed because of severe asthmatic symptoms (Gibbons et al. 1965).

The chronic asthma remaining after cessation of exposure is usually of mild to moderate severity. Early diagnosis and early removal from exposure are good prognostic factors (Chan-Yeung et al. 1982, Hudson et al. 1985). Patients who remain in exposure may remain stable with medication, respiratory protective equipment, and modifications to work, or may deteriorate despite such measures (Paggiaro et al. 1984). Bronchodilator treatment at the same time as allergen exposure is known to suppress the immediate asthmatic response, allow exposure to a greater dose of allergen, and increase the size of the late asthmatic response (Lai et al. 1989).

IV. PREVENTION OF SENSITIZER-INDUCED ASTHMA

Preventive measures are conventionally grouped as primary, secondary, and tertiary prevention (Last 1987). Primary preventive action prevents occupational exposure, and, if appropriate, any other determinants of occupational asthma. Secondary prevention aims to detect asthma early and take appropriate and timely action to minimize its duration and severity. Tertiary prevention is applicable only to patients with established asthma as the aim is prevention of deterioration and complications by means of appropriate health care. Primary and secondary prevention take place in the workplace and are addressed further here. Tertiary prevention in occupational asthma, although sometimes carried out by physicians with specialist expertise in occupational medicine, is essentially the same as effective management of any type of asthma (NIH 1991, BTS 1993).

A. ATTITUDES TOWARD PREVENTION
1. Adverse Influence of Concepts of Allergy

It used to be assumed that there was little to be done to prevent sensitizer-induced asthma because the cause was "constitutional," that is, inherent to the individual. There has been a suggestion that laboratory animal allergy could be associated with HLA B15 and DR4 (Löw et al. 1988). In occupational medicine, this way of thinking has often led to a focus on making the workers fit for the hazards of work, rather than making the work environment safe for all, or most, workers. Some workplace programs consist almost entirely of systems to monitor the health of exposed individuals with little attention to the workplace. Managers with responsibility for health and safety often believe that asthma cannot be controlled and is the province of doctors, rather than of occupational hygienists and safety engineers.

2. Confusion Between Induction and Provocation of Asthma

This view has been encouraged by confusion between the levels of allergen exposure which will elicit symptoms in sensitized individuals and the levels of exposure sufficient to induce the sensitized state in persons who are not sensitized. Sensitized patients may develop attacks of asthma after very low exposures. For example, in inhalation challenge tests with tetrachlorophthalic anhydride, patients with occupational asthma caused by this chemical showed late and dual asthmatic responses after 30-min exposures to airborne levels orders of magnitude lower than those recommended as occupational standards (Venables et al. 1990). Patients with occupational asthma caused by colophony in solder responded to only a few breaths of solder fume, whereas unaffected individuals carry out hand soldering operations over a normal work shift (Burge et al. 1978).

B. EXPOSURE–RESPONSE RELATIONS

There is, however, no evidence to suggest that the exposure–response relations for asthma induction are any different from those for the induction of other diseases. In a cross-sectional study of British bakers, for example, indices associated with occupational asthma were related to either duration of exposure or intensity of exposure measured as an 8-hour time-weighted average of inhalable dust (Musk et al. 1989). In a study in the U.S., the prevalence of laboratory animal allergy increased with number of hours per week of exposure and with number of species handled (Bland et al. 1986). Much of the epidemiological data on occupational asthma is derived from surveys of its prevalence in population cross-sections, rather than of its incidence in cohorts. Because asthma often leads to severe respiratory symptoms, it is common for affected workers to leave exposure by leaving the employer or transferring to alternative work with no or low exposure. This healthy worker survival bias is the likely explanation for a lack of exposure–response relations observed in some populations. For example, in a group of pharmaceutical research workers, symptoms consistent with occupational asthma caused by laboratory animals showed no relationship with job type and showed an inverse relationship with increasing duration of exposure (Venables et al. 1988).

1. Effect of Short-Term High Exposures

The assumption that intensity and duration of exposure are equivalent is questionable for many occupational diseases. In asthma, there is some evidence that short periods of high exposure could be more important than the equivalent dose accumulated at a lower exposure over a longer time. In one of the few longitudinal studies, Weill et al. (1981) followed workers exposed to toluene diisocyanate (TDI). Of 12 who developed asthma in 5 years, half had been exposed to spills of TDI. Animal models of asthma caused by subtilizin and by isocyanates have shown an increase in percentage of guinea pigs sensitized by inhalation with increasing concentration of sensitizing agent (Karol et al. 1985). The animal model of asthma caused by subtilizin suggests that long-term exposure to low levels of enzyme had less effect than an equivalent dose administered over a short time (Thorne et al. 1986).

2. Exposure Windows

It is possible that exposure during a time "window" shortly after first exposure is the major determinant of occupational asthma. The latent interval between the first exposure to a sensitizing agent and the development of asthma appears to be short for most agents and most cases that will develop appear to do so in the first year or two of exposure (Venables et al. 1989a, Meredith and McDonald 1994). Exposure accumulated later may not increase the risk further.

3. Potency

Some occupational allergens appear to be exceptionally potent sensitizing agents; the effect occurs at low levels of exposure. For example, the hessian sacks used to transport castor beans are recycled for transport of green coffee bean. The sacks contain enough castor bean antigen to sensitize dock workers unloading coffee (De Zotti et al. 1988). The sacks are also recycled to make upholstery felt and this has been responsible for sensitization of furniture workers (Topping et al. 1981).

4. Methods of Measurement

Another source of confusion is that there have been few ways of measuring high molecular weight allergens until recently, and these allergens represent a major class of agents that cause occupational asthma (Chan-Yeung 1990). The field has developed in recent years and techniques of immunoassay are proliferating (Newman-Taylor and Tee 1990, Reed 1990).

C. PRIMARY PREVENTION

Elimination of the sensitizing agent is the most secure way of preventing exposure. If this is not practicable, it means reduction in exposure, combined or not with measures to limit the numbers of people exposed. It also means control of any other known determinants, where they are susceptible to control.

In occupational asthma, atopy and smoking have been proposed as additional determinants. Corn (1983) has provided a general description of the assessment and control of environmental exposure, with special reference to allergenic agents. He lists 20 control methods. Fifteen relate directly to primary prevention: elimination, substitution, isolation, enclosure, ventilation, process change, product change, housekeeping, dust suppression, maintenance, sanitation, work practices, personal protective devices, waste disposal practices, and administrative controls. Corn includes "medical controls" as a sixteenth. Four are indirectly related to primary prevention in that they are needed in order to carry out the remaining sixteen: education, labeling and warning systems, environmental monitoring, and management programs.

1. Prevention by Elimination

There are several examples of successful elimination of exposures causing asthma. Venables et al. (1985a) investigated an outbreak of asthma in a steel coating plant, where at least 21 cases of occupational asthma developed in a workforce of around 200. Simple outbreak investigation techniques suggested that toluene diisocyanate was the cause. This was confirmed by inhalation challenge testing of the two index cases. The isocyanate had been introduced into the process by a supplier. When it was replaced by a different chemical, no new cases developed and a repeat survey showed that the existing cases had improved.

2. Prevention by Reduction of Exposure

Good examples of prevention by exposure reduction come from enzyme detergent production (Juniper et al. 1977) and platinum refining (Hughes 1980). After the epidemics of occupational asthma caused by *Bacillus subtilis* enzyme detergents in the 1960s and early 1970s, enzyme was handled as a liquid slurry rather than a dust; it was encapsulated; some plants were enclosed; exhaust ventilation was increased; protective clothing was modified; education and administrative systems were set up; and job applicants deemed at increased risk of allergy were excluded from employment. These measures led to a reduction in measured enzyme exposure and also in occupational asthma caused by enzyme. A similar package of multiple measures, with particular reliance on enclosure of processes, was used to reduce exposure to platinum salts. It has been noted previously that when multiple measures are used, it is impossible to separate the effects of one component of a package from another (Venables 1987). Formal evaluation of the effectiveness or efficiency of control measures is still unusual in any area of preventive medicine.

3. Restriction of Employment

In the examples of asthma caused by enzyme detergents and platinum salts quoted above, the criterion for increased risk of allergy was atopy, defined as a history of allergic illness, or a positive skin prick test to common environmental allergens. Depending on how a positive skin prick test is defined, this approach may mean denying employment to around a third of job applicants, only a proportion of whom would develop occupational asthma (Venables et al. 1988). It does not eliminate the problem because the association with atopy is not absolute. Other pre-employment screening criteria besides atopy are used (Lutsky et al. 1983). Newill and her colleagues (1986) have made theoretical estimates of the efficacy of these various pre-employment screening criteria for preventing laboratory animal allergy (Lutsky et al. 1983). The study suggests that screening procedures are introduced without consideration of their likely effects or evaluation of their actual effects and that the use of screening as a method of preventing laboratory animal allergy was premature. Some pre-employment tests are carried out quite appropriately for reasons other than the prevention of occupational asthma. Pre-exposure clinical data are useful as a baseline for results obtained during later health monitoring. It also seems wise to identify the rare individual with important respiratory impairment, as he should not be exposed to further risk of impairment.

4. Control of Smoking, and of Respiratory Irritants

Smoking has been shown to be a risk factor for the development of specific IgE antibody against occupational agents (Zetterström et al. 1981, Cartier et al. 1984, Venables et al. 1985b and 1989a) though not necessarily for asthma (Meredith and McDonald 1994). Smoking (Zetterström et al. 1985) potentiates experimental sensitization of animals. This adjuvant effect is shared with irritant gases such as ozone and sulfur dioxide (Matsumura 1970, Osebold et al. 1980, Biagini et al. 1986). This raises the interesting possibilities that, firstly, programs to reduce smoking may have the additional benefit of preventing IgE-mediated sensitization. Secondly, attention to control of respiratory irritants may have the same effect. Control of respiratory irritants would also control their direct toxic effects on the airways, which were noted above.

5. Respiratory Protection

A helmet respirator (Racal) has been shown to be partially effective in reducing the consequences of exposure in patients with occupational asthma due to laboratory animals (Slovak et al. 1985) and aluminium potroom emissions (Kongerud and Rambjör 1991). This is a form of incomplete tertiary prevention, but it raises the possibility that respiratory protective equipment may have some role in preventing the induction of sensitization and asthma. Some workers exposed to respiratory sensitizers use respiratory protection for that reason, by individual preference or in following an employer's policy.

D. SECONDARY PREVENTION
1. Methods

Secondary prevention includes the periodic screening tests that are used in industry. The proportion of employers with an asthma risk who offer screening to their workforce is unknown. A wide variety of questionnaires are in use.

Some employers use spirometric tests, but often without any attempt to time the tests so as to have a high probability of detecting asthma. A small number use skin prick tests with specific antigens, for example enzyme extract (Juniper et al. 1977) and complex platinum salts (Hughes 1980). It is the impression of this author that most employers would welcome guidance on the choice of appropriate screening tests and on quality control procedures for their conduct.

2. Symptoms Questionnaires

The best known respiratory symptoms questionnaire is that published by the Medical Research Council in the U.K. (Medical Research Council 1960). It has given rise to others, such as that published by the American Thoracic Society. The roots of the Medical Research Council questionnaire and of its progeny lie in the bronchitis research conducted in the 1950s and 1960s. Asthma was of interest only as a potential modifying variable. The questions are not suitable for detecting asthma. Some fresh approaches to questionnaire design for asthma have been made in recent years. Mortagy and colleagues (1986) published a questionnaire on "bronchial irritability" symptoms. A similar questionnaire was published under the sponsorship of the International Union against Tuberculosis and Lung Diseases (Burney et al. 1989). Another was developed by Venables and colleagues (1993). All three questionnaires aim to note variable symptoms provoked by known triggers of asthma, such as cold air. All show that "bronchial irritability" symptoms are associated with bronchial hyperresponsiveness. Questions about improvement in symptoms during weekends and on vacation are widely used but have not been standardized or validated. A short questionnaire would have many advantages as a first-line screening tool for exposed workers.

3. Pulmonary Function Tests

Burge (1982a) has observed that spirometry is an insensitive technique for detecting asthma, even when two measurements are made, before and after the workshift. Peak flow records over several weeks are both sensitive and specific for occupational asthma. They are feasible tests in surveys and interpretation of the resultant graphs by an experienced observer is reproducible (Venables et al. 1984). However, this is a highly labor-intensive technique and unlikely to gain acceptance as a first-line screening tool. It would be better applied in a second phase of screening.

4. Immunological Tests

It is not clear if evidence of sensitization precedes the development of asthma. Sensitization can be detected, for example, by means of skin prick tests or testing for serum specific IgE antibody. One study made use of routinely collected occupational health screening data in order to assemble and follow a cohort of platinum refinery workers (Venables et al. 1989a). This showed that symptoms were as likely to be noted after as before detection of a positive skin prick test to platinum salts. The symptoms variable studied was any lower respiratory tract symptoms, so it is possible that use of a tailored questionnaire would have shown different results. It is, however, also possible that it is difficult to achieve fine detail about the timing of onset of symptoms in other than a research setting. More information is needed before recommending immunological tests as suitable for the early detection of asthma.

5. Procedures

It is as important to have a practicable procedure for investigating workers who have positive screening tests. If a worker has occupational asthma, his prognosis is improved by early detection and prompt removal from exposure (Chan-Yeung et al. 1982, Hudson et al. 1985). From a public health viewpoint, one worker with occupational asthma is evidence that primary prevention has failed. His working conditions should be reviewed in case some failure of control has occurred. Workers with similar exposures should be tested in case others are affected. A proactive approach and systematization of the procedures following detection of a case of occupational asthma should ensure action is taken promptly.

E. TERTIARY PREVENTION

This chapter does not address the care of the affected worker. Once he has developed asthma, his situation is entirely different. He will experience attacks of asthma after exceedingly low exposures. These attacks may be severe. He is at risk of death in an acute attack (Fabbri et al. 1988). The appropriate treatment is prompt removal from exposure with preservation of income. In many countries, there is often a conflict between removal from exposure and preservation of income. This conflict can be resolved politically (by bargaining within a firm or nationally through the party political process) and is not a strictly medical or scientific issue. It is, however, important that employers and employee representatives are fully aware of what constitutes good medical practice in the management of patients with sensitizer-induced asthma.

V. RECOMMENDATIONS

There is a great deal that could be done to prevent occupational asthma. The following strategy is implementable with existing knowledge and would go some way towards prevention:

1. National regulatory or advisory bodies, or other agencies with responsibility for prevention, should publish a guidance document on the prevention of occupational asthma. This should establish general principles, such as terminology, rather than enumerate details.
2. Existing surveillance should be extended or new surveillance established, to collect information on clinical diagnoses of occupational asthma.

This could be combined with collection of data on other occupational lung diseases. It could also be combined with estimation of the relative frequency of sensitizer-induced asthma, irritant-induced asthma, and work-exacerbated asthma.

3. Existing lists of sensitizers that cause asthma should be collated in order to publish a list that could form the basis for standards in product labeling and exposure control strategies. Such a list would require regular updating.

4. Prospective epidemiological studies of exposure–response relations should be carried out. Such studies also allow scope for studying the effect of modifying variables, such as family history of asthma, or personal smoking. They also can act as a framework for additional studies, for example on the time course of sensitization.

5. Agreement should be reached on a short questionnaire which could form the basis for screening tests in industry, or on other forms of screening.

6. Pilot prevention projects in industry should be encouraged coupled with evaluative research to assess their effectiveness and efficiency.

A. GUIDANCE DOCUMENT

Agreement on a guidance document would clarify an agency's view of the definition of occupational asthma, its causes, mechanisms, natural history, and consequences. This document can form the basis of a guidance statement with a target audience consisting of those with a professional interest in occupational asthma. It is suggested, in addition, that a short summary document is composed for distribution as necessary to a wider audience.

B. SURVEILLANCE

Few countries have national case-finding for occupational lung disease. The frequency with which asthma occurs in relation to the pneumoconioses and other occupational lung diseases is information which aids in setting priorities for preventive activities and for research. The relative frequency of asthma caused by different sensitizers is, similarly, useful information.

The SWORD project (Meredith et al. 1991) in the U.K. collects simple data from specialist physicians. SENSOR (Sentinel Event Notification System for Occupational Risks) in the U.S. is described by Baker (1989) and Matte et al. (1990). Unlike SWORD, SENSOR does not aim to include all specialists who might make a diagnosis of occupational asthma; it includes diagnostic criteria for occupational asthma, it is not anonymous, and it is accompanied by intervention in the workplace.

C. LIST OF SENSITIZERS
1. Purpose of List

Several hundred agents are known to cause occupational asthma by inducing specific sensitization. These should be drawn together into a list. The list is intended for use in controlling exposure by indicating materials and processes in industry which are a hazard because they may cause asthma. It thus indicates areas and processes where exposure controls should be applied and identifies occupational groups at risk of asthma who should be offered health screening programs. Such a "controlled exposure" list could form the basis for regulations, for example, on safe handling, permissible exposure, and product labeling.

Lists of occupational sensitizing agents can be used for other purposes. For example, physicians who see patients with possible occupational asthma need a list of agents as a focus for taking occupational histories. Such a list should include agents forming the "controlled exposure" list above and additionally should include agents which are suspected to cause asthma.

2. Criteria for List

Chan-Yeung's list (Chan-Yeung 1990, Bernstein et al. 1993) is, perhaps, the one consulted most frequently by English-speaking physicians with an interest in occupational asthma. The Health and Safety Executive (HSE) of the U.K. government has published guidance on respiratory sensitizers to enable compliance with the regulations on Control of Substances Hazardous to Health (COSHH); this document contains a list (Venables 1992, HSE 1994). There may be other reputable lists.

This chapter recommends to responsible agencies that existing lists be, firstly, collated. Clinical practice has changed in this century and in many centers today there is considerable reliance on tests such as immunological tests or bronchial provocation tests which were not available to physicians 20 or 30 years ago. Therefore, this chapter recommends, secondly, that explicit criteria for inclusion be determined retrospectively by inspection of the relevant publications about these agents. This process will have the benefits of codifying and formalizing the process of professional judgment where this was not made clear by an author, of identifying gaps in the medical literature where an agent is accepted in professional practice to cause asthma but where the published explicit evidence is slight, and of providing a rationale for updating the list as new sensitizers are introduced to industry and cause asthma.

3. Generating the List

There are several approaches to generating such a list. This chapter recommends that one person, rather than a group, is responsible for the initial draft. It further recommends that inclusion criteria should be applicable by nonspecialists with experience in regulation and also be practicable for at least 5 years from publication. To allow for debate and disagreement, a consultation period is recommended between initial and final publication for active solicitation of comments from practicing physicians, academic institutions, regulatory bodies, and trade and union representatives.

The final list should be widely disseminated and then should be updated regularly, perhaps annually. This could be accomplished by means of an annual literature search supplemented by active solicitation of new candidates for the list. Those meeting the inclusion criteria can then be published.

D. DESCRIBING EXPOSURE–RESPONSE RELATIONS

Primary prevention measures to control exposure to agents which cause asthma must be informed by knowledge of exposure–response relations and of the factors which modify these relations. This chapter suggests that these should be studied by means of prospective longitudinal studies with measurement of allergen exposure. Suitable populations are those at high risk of sensitization and asthma and with good markers of response to exposure.

Studies in animals can test hypotheses about sensitization that have been generated in epidemiological studies and also may be the only way to obtain information on mechanisms.

E. SCREENING

This chapter recommends use of a validated, reproducible questionnaire on work-related respiratory symptoms for use in screening programs for detecting cases of occupational asthma and also in epidemiological research. Such a questionnaire would normally be used in conjunction with other tests but may, in some settings, be used alone. The design of the questionnaire and details of its validation will vary according to the context of its use.

F. EVALUATION OF PILOT PREVENTION PROJECTS

The need is for studies that will give clearcut answers which are generalizable to other populations with comparable exposures and, possibly, populations with different exposures. Topical general questions include: does a measure to reduce exposure reduce the subsequent incidence rate of sensitization and of asthma, and does a screening program to detect cases of asthma at an early stage reduce the subsequent functional impairment and consequent disability?

ACKNOWLEDGMENTS

The work for this chapter was made possible by a Fellowship in Environmental Health and Public Policy at Harvard School of Public Health and by leave of absence from the National Heart and Lung Institute. Much of its text was included in July 1992 in an internal report commissioned by the Division of Respiratory Disease Studies of the U.S. National Institute for Occupational Safety and Health and entitled "Prevention of Occupational Asthma."

REFERENCES

Baker, E. L., Sentinel event notification system for occupational risks (SENSOR): the concept, *Am. J. Pub. Health,* 1989;79(Suppl.):18–20.

Bernstein, I. L., Chan-Yeung, M., Malo, J.-L., and Bernstein, D. I., *Asthma in the Workplace,* Marcel Dekker, New York, 1993.

Biagini, R. E., Moorman, W. J., Lewis, T. R., and Bernstein, I. L., Ozone enhancement of platinum asthma in a primate model, *Am. Rev. Respir. Dis.,* 1986;134:719–725.

Blanc, P., Occupational asthma in a national disability survey, *Chest,* 1987;92:613–617.

Bland, S. M., Levine, S. M., Wilson, P. D., Fox, N. L., and Rivera, J. C., Occupational allergy to laboratory animals: an epidemiologic study, *J. Occup. Med.,* 1986;28:1151–1157.

British Thoracic Society (and others), Guidelines on the management of asthma, *Thorax,* 1993;48(Suppl.):S1–S24.

Brooks, S. M., Weiss, M. A., and Bernstein, I. L., Reactive airways dysfunction syndrome (RADS): persistent asthma syndrome after high level irritant exposures, *Chest,* 1985;88:376–384.

Burge, P. S., Single and serial measurements of lung function in the diagnosis of occupational asthma, *Eur. J. Respir. Dis.,* 1982a;63(Suppl. 123):47–59.

Burge, P. S., Occupational asthma in electronics workers: follow-up of affected workers, *Thorax,* 1982b;37:348–353.

Burge, P. S., Harries, M. G., O'Brien, I. M., and Pepys, J., Respiratory disease in workers exposed to solder flux fumes containing colophony (pine resin), *Clin. Allergy,* 1978;8:1–14.

Burney, P. J. G., Chinn, S., Britton, J. R., Tattersfield, A. E., and Papacosta, A. O., What symptoms predict the bronchial response to histamine? Evaluation in a community survey of the bronchial symptoms questionnaire (1984) of the International Union against Tuberculosis and Lung Disease, *Int. J. Epidemiol.,* 1989;18:165–173.

Cartier, A., Malo, J.-L., Forest, F., Lafrance, M., and Pineau, L., St-Aubin, J.-J., and Dubois, J.-Y., Occupational asthma in snow crab-processing workers, *J. Allergy Clin. Immunol.,* 1984;74:261–269.

Chan-Yeung, M., Occupational asthma, *Chest,* 1990;98(Suppl. 5):148–161.

Chan-Yeung, M., Lam, S., and Koener, S., Clinical features and natural history of occupational asthma due to western red cedar (*Thuja plicata*), *Am. J. Med.,* 1982;72:411–415.

Charan, N. B., Myers, C. G., Lakshminarayan, S., and Spencer, T. M., Pulmonary injuries associated with acute sulfur dioxide inhalation, *Am. Rev. Respir. Dis.,* 1979;119:555–560.

Chatham, M., Bleecker, E. R., Smith, P. L., Rosenthal, R. R., Mason, P., and Norman, P. S., A comparison of histamine, methacholine, and exercise airway reactivity in normal and asthmatic subjects, *Am. Rev. Respir. Dis.,* 1982;126:235–240.

Cockcroft, D. W., Ruffin, R. E., Dolovich, J., and Hargreave, F. E., Allergen-induced increase in non-allergic bronchial reactivity, *Clin. Allergy,* 1977;7:503–513.

Corn, M., Assessment and control of environmental exposure, *J. Allergy Clin. Immunol.,* 1983;72:231–241.

Curry, J. J., The action of histamine on the respiratory tract in normal and asthmatic subjects, *J. Clin. Invest.,* 1946;25:785–791.

Dautrebande, L. and Phillipot, E., Crise d'asthme expérimental par aérosols de carbaminoylcholine chez l'homme traitée par dispersat de phénylaminopropane: étude de l'action sur la respiration de ces substances par la determination du volume respiratoire utile, *Presse Méd.,* 1941;49:942–946.

De Zotti, R., Patussi, V., Fiorito, A., and Larese, F., Sensitization to green coffee bean (GCB) and castor bean (CB) allergens among dock workers, *Int. Arch. Occup. Environ. Health,* 1988;61:7–12.

Dubois, A. B. and Dautrebande, L., Acute effects of breathing inert dust particles and of carbachol aerosol on the mechanical characteristics of the lungs in man. Changes in response after inhaling sympathomimetic aerosols, *J. Clin. Invest.,* 1958;37:1746–1755.

Empey, D. W., Laitenen, L. A., Jacobs, L., Gold, W. M., and Nadel, J. A., Mechanisms of bronchial hyperreactivity in normal subjects after upper respiratory tract infection, *Am. Rev. Respir. Dis.,* 1976;113:131–139.

Fabbri, L. M., Danieli, D., Crescioli, S., Bevilacqua, P., Meli, S., Saetta, M., and Mapp, C. E., Fatal asthma in a toluene diisocyanate sensitized subject, *Am. Rev. Respir. Dis.,* 1988;137:1494–1498.

Gandevia, B., Occupational asthma, *Med. J. Aust.,* 1970;ii:332–335, 372–376.

Gibbons, H. L., Dille, J. R., and Cowley, R. G., Inhalant allergy to the screwworm fly, *Arch. Environ. Health,* 1965;10:424–430.

Golden, J. A., Nadel, J. A., and Boushey, H. A., Bronchial hyperirritability in healthy subjects after exposure to ozone, *Am. Rev. Respir. Dis.,* 1978;118:287–294.

Gregg, I., Epidemiology, in *Asthma,* Clark, T. J. H. and Godfrey, S., Eds., Chapman and Hall, London, 1977, 214–240.

Harber, P., Assessing occupational disability from asthma, *J. Occup. Med.,* 1992;34:120–128.

The Health and Safety Executive, Prevention of asthma at work: how to control respiratory sensitisers, HSE Books, Sudbury, U.K., 1994.

Hendrick, D. J., Occupational asthma — problems of definition, *J. Occup. Med.,* 1983;25:488–489.

Hogg, J. C., The pathophysiology of asthma, *Chest,* 1982;(Suppl 1.):8–12.

Hudson, P., Cartier, A., Pineau, L., Lafrance, M., St-Aubin, J.-J., Dubois, J.-Y., and Malo, J.-L., Follow-up of occupational asthma caused by crab and various agents, *J. Allergy Clin. Immunol.,* 1985;76:682–688.

Hunter, D., Milton, R., and Perry, K. M. A., Asthma caused by the complex salts of platinum, *Br. J. Ind. Med.,* 1945;2:92–98.

Hughes, E. G., Medical surveillance of platinum refinery workers, *J. Soc. Occup. Med.,* 1980;30:27–30.

Juniper, C. P., How, M. J., Goodwin, B. F. J., and Kinshott, A. K., *Bacillus subtilis* enzymes: a 7-year clinical, epidemiological and immunological study of an industrial allergen, *J. Soc. Occup. Med.,* 1977;27:3–12.

Juniper, E. F., Frith, P. A., and Hargreave, F. E., Airway responsiveness to histamine and methacholine: relationship to minimum treatment to control symptoms of asthma, *Thorax,* 1981;36:575–579.

Karol, M. H., Hansen, G. A., and Brown, W. E., Effects of inhaled hexamethylene diisocyanate (HDI) on guinea pig cholinesterases, *Fund. Appl. Toxicol.,* 1984;4:284–287.

Karol, M. H., Stadler, J., and Magreni, C., Immunotoxicologic evaluation of the respiratory system: animal models for immediate- and delayed-onset pulmonary hypersensitivity, *Fund. Appl. Toxicol.,* 1985;5:459–472.

Keskinen, H., Epidemiology of occupational lung disease: asthma and allergic alveolitis, in Proc. XI Int. Cong. Allergology Clin. Immunol., London 1982, Kerr, J. W., and Ganderton, M. A., Eds., Macmillan, London, 1983, 403–407.

Keskinen, H., Alanko, K., and Saarinen, L., Occupational asthma in Finland, *Clin. Allergy,* 1978;8:569–579.

Kongerud, J. and Rambjör O., The influence of the helmet respirator on peak flow rate in aluminum potroom, *Am. Ind. Hyg. Assoc. J.,* 1991;52:243–248.

Lai, C. K. W., Twentyman, O. P., and Holgate, S. T., The effect of an increase in inhaled allergen dose after rimiterol hydrobromide on the occurrence and magnitude of the late asthmatic response and the associated change in non-specific bronchial responsiveness, *Am. Rev. Respir. Dis.,* 1989;140:917–923.

Last, J. M., *Public Health and Human Ecology,* Appleton and Lange, East Norwalk, CT, 1987.

Löw, B., Sjöstedt, L., and Willers, S., Laboratory animal allergy — possible association with HLA B15 and DR4, *Tissue Antigens,* 1988;31:224–226.

Lutsky, I., Kalbfleisch, J. H., and Fink, J. N., Occupational allergy to laboratory animals: employer practices, *J. Occup. Med.,* 1983;25:372–376.

Malo, J.-L., Cartier, A., Ghezzo, H., Lafrance, M., McCants, M., and Lehrer, S. F. B., Patterns of improvement in spirometry, bronchial hyperresponsiveness, and specific IgE antibody levels after cessation of exposure in occupational asthma caused by snow-crab processing, *Am. Rev. Respir. Dis.,* 1988;138:807–812.

Matte, T. D., Hofman, R. E., Rosenman, K. D., and Stanbury, M., Surveillance of occupational asthma under the SENSOR model, *Chest,* 1990;98(Suppl.):173–178.

Matsumura, Y., The effects of ozone, nitrogen dioxide, and sulfur dioxide on the experimentally induced allergic respiratory disorder in guinea pigs. 1: The effect on sensitization with albumin through the airway, *Am. Rev. Respir. Dis.,* 1970;102:430–437.

Medical Research Council, Committee on the aetiology of chronic bronchitis, Standardized questionaries on respiratory symptoms, *Br. Med. J.,* 1960;ii:1665.

Meredith, S. K. and McDonald, J. C., Occupational asthma in chemical, pharmaceutical, and plastics processors and manufacturers in the United Kingdom, 1989–1990, *Ann. Occup. Hyg.,* 38:833–837, 1994.

Meredith, S. K., Taylor, V. M., and McDonald, J. C., Occupational respiratory disease in the United Kingdom 1989: a report to the British Thoracic Society and the Society of Occupational Medicine by the SWORD project group, *Br. J. Ind. Med.,* 1991;48:292–298.

Mortagy, A. K., Howell, J. B. L., and Waters, W. E., Respiratory symptoms and bronchial reactivity: identification of a syndrome and its relation to asthma, *Br. Med. J.,* 1986;293:525–529.

Musk, A. W., Venables, K. M., Crook, B., Nunn, A. J., Hawkins, R., Crook, G. D. W., Graneek, B. J., Tee, R. D., Farrer, N., Johnson, D. A., Gordon, D. J., Darbyshire, J. H., and Newman-Taylor, A. J., Respiratory symptoms, lung function, and sensitisation to flour in a British bakery, *Br. J. Ind. Med.,* 1989;46:636–642.

Newill, C. A., Evand, R., and Khoury, M. J., Preemployment screening for allergy to laboratory animals: epidemiologic evaluation of its potential usefulness, *J. Occup. Med.,* 1986;28:1158–1164.

Newman-Taylor, A. J. and Tee, R. D., Environmental and occupational asthma: exposure assessment, *Chest,* 1990;98(Suppl.):209–211.

Newman-Taylor, A. J., Longbottom, J. L., and Pepys, J., Respiratory allergy to urine proteins of rats and mice, *Lancet,* 1977;ii:847–849.

National Institutes of Health, Guidelines for the diagnosis and management of asthma, U.S. Department of Health and Human Services (NIH) Publ. No. 91-3042, U.S. Government Printing Office, 1991.

O'Byrne, P. M., Ryan, G., Morris, M., McCormack, D., Jones, N. L., Morse, J. L. C., and Hargreave, F. E., Asthma induced by cold air and its relation to nonspecific bronchial responsiveness to methacholine, *Am. Rev. Respir. Dis.,* 1982;125:281–285.

Osebold, J. W., Gershwin, L. J., and Zee, Y. C., Studies on the enhancement of allergic lung sensitization by inhalation of ozone and sulfuric acid aerosol, *J. Environ. Path. Toxicol.,* 1980;3:221–234.

Ouellette, J. J. and Reed, C. E., Increased response of asthmatic subjects to methacholine after influenza vaccine, *J. Allergy,* 1965;36:558–563.

Paggiaro, P. L., Loi, A. M., Rossi, O., Ferrante, B., Pardi, F., Roselli, M. G., and Baschieri, L., Follow-up study of patients with respiratory disease due to toluene diisocyanate (TDI), *Clin. Allergy,* 1984;14:463–469.

Pepys, J. and Hutchcroft, B. J., Bronchial provocation tests in etiologic diagnosis and analysis of asthma, *Am. Rev. Respir. Dis.,* 1975;112:829–859.

Reed, C. E., Clinical management when the environment can be changed, *Chest,* 1990;98(Suppl.):216–219.

Ryan, G., Latimer, K. M., Dolovich, J., and Hargreave, F. E., Bronchial responsiveness to histamine: relationship to diurnal variation of peak flow rate, improvement after bronchodilator, and airway calibre, *Thorax,* 1982;37:423–429.

Scadding, J. G., Meaning of diagnostic terms in broncho-pulmonary disease, *Br. Med. J.,* 1963;ii:1425–1430.

Sheppard, D., Wong, W. S., Uehara, C. F., Nadel, J. A., and Boushey, H. A., Lower threshold and greater bronchomotor responsiveness of asthmatic subjects to sulfur dioxide, *Am. Rev. Respir. Dis.,* 1980;122:873–878.

Slovak, A. J., Orr, R. G., and Teasdale, E. L., Efficacy of the helmet respirator in occupational asthma due to laboratory animal allergy (LAA), *Am. Ind. Hyg. Assoc. J.,* 1985;46:411–415.

Thorne, P. S., Hillebrand, J., Magreni, C., Riley, E. J., and Karol, M. H., Experimental sensitization to subtilisin I: production of immediate- and late-onset pulmonary reactions, *Toxicol. Appl. Pharmacol.,* 1986;86:112–123.

Topping, M. D., Tyrer, F. H., and Lowing, R. K., Castor bean allergy in the upholstery department of a furniture factory, *Br. J. Ind. Med.,* 1981;38:293–296.

Venables, K. M., Epidemiology and the prevention of occupational asthma, *Br. J. Ind. Med.,* 1987;44:73–75.

Venables, K. M., Low molecular weight chemicals, hypersensitivity, and direct toxicity: the acid anhydrides, *Br. J. Ind. Med.,* 1989;46:222–232.

Venables, K. M., Preventing occupational asthma, *Br. J. Ind. Med.,* 1992;49:817–819.

Venables, K. M., Burge, P. S., Davison, A. G., and Newman-Taylor, A. J., Peak flow rate records in surveys: reproducibility of observers' reports, *Thorax,* 1984;39:828–832.

Venables, K. M., Dally, M. B., Burge, P. S., Pickering, C. A. C., and Newman-Taylor, A. J., Occupational asthma in a steel coating plant, *Br. J. Ind. Med.,* 1985a;42:517–524.

Venables, K. M., Topping, M. D., Howe, W., Luczynska, C. M., Hawkins, R., and Newman-Taylor, A. J., Interaction of smoking and atopy in producing specific IgE antibody against a hapten protein conjugate, *Br. Med. J.,* 1985b;290:201–204.

Venables, K. M., Topping, M. D., Nunn, A. J., Howe, W., and Newman-Taylor, A. J., Immunologic and functional consequences of chemical (tetrachlorophthalic anhydride)-induced asthma after four years of avoidance of exposure, *J. Allergy Clin. Immunol.,* 1987;80:212–218.

Venables, K. M., Tee, R. D., Hawkins, E. R., Gordon, D. J., Wale, C. J., Farrer, N. M., Lam, T. H., Baxter, P. J., and Newman-Taylor, A. J., Laboratory animal allergy in a pharmaceutical company, *Br. J. Ind. Med.,* 1988;45:660–6.

Venables, K. M., Dally, M. B., Nunn, A. J., Stevens, J. F., Stephens, R., Farrer, N., Hunter, J. V., Stewart, M., Hughes, E. G., and Newman-Taylor, A. J., Smoking and occupational allergy in workers in a platinum refinery, *Br. Med. J.,* 1989a;299:939–942.

Venables, K. M., Davison, A. G., and Newman-Taylor, A. J., Consequences of occupational asthma, *Respir. Med.,* 1989b;83:437–440.

Venables, K. M. and Newman-Taylor, A. J., Exposure-response relationships in asthma caused by tetrachlorophthalic anhydride, *J. Allergy Clin. Immunol.,* 1990;85:55–58.

Venables, K. M., Farrer, N. M., Sharp, L., Graneek, B. J., and Newman-Taylor A. J., Respiratory symptoms questionnaire for asthma epidemiology: validity and reproducibility, *Thorax,* 1993;48:214–219.

von Pirquet, C., *Allergie,* Münch Med Wochenschr 1906;30:1457, Tr., Prausnitz, C., in *Clinical Aspects of Immunology,* Gell, P. G. H., Coombs, R. R. A., Eds., Blackwell, Oxford, 1963. 805–807.

Weill, H., Butcher, B., Dharmarajan, V., Glindmeyer, H., Jones, R., Carr, J., O'Neil, C., and Salvaggio, J., Respiratory and immunologic evaluation of isocyanate exposure in a new manufacturing plant, NIOSH Technical Report No. 81-125, U.S. Government Printing Office, 1981.

Zetterström, O., Osterman, K., Machado, L., and Johansson, S. G. O., Another smoking hazard: raised serum IgE concentration and increased risk of occupational allergy, *Br. Med. J.*, 1981;283:1215–1217.

Zetterström, O., Nordvall, S. L., Björkstén, B., Ahlsted, S., and Stelander, M., Increased IgE antibody responses in rats exposed to tobacco smoke, *J. Allergy Clin. Immunol.*, 1985;75:594–598.

Chapter 9

THE ROLE OF INDOOR AND OUTDOOR AIR
POLLUTION IN ALLERGIC DISEASES

Heidrun Behrendt, Karl-Heinz Friedrichs, Ursula Krämer,
Bettina Hitzfeld, Wolf-M. Becker, and Johannes Ring

CONTENTS

ABSTRACT

The role of indoor and outdoor air pollution on induction, elicitation, and
maintenance of allergic inflammatory reactions ("allergotoxicology") is
demonstrated by means of environmental epidemiology and experimental
studies including human *in vitro* cell systems as well as pollen samples.
We can distinguish two types of air pollution: type I air pollution is
present predominantly in eastern Europe and is associated with infectious
and chronic inflammatory airway reactions. Type II air pollution is present
in the industrialized western countries in both the indoor and outdoor
environment and is composed of mainly NO/NO_2, O_3, volatile organic
compounds (VOCs), and fine particles. Type II air pollution is associated
with allergic diseases and allergic sensitization. The risk of becoming
sensitized is two to three times higher when living in a household near
roads with heavy traffic or when unvented gas is used for cooking/
heating. Substances adsorbed to airborne particles which had been collected in west German regions are able to induce the release of mediators
from human basophils and neutrophils and exhibit priming effects on
them. Airborne pollutants also interact with the pollen grains, as is shown
by morphology as well as by immunochemistry. It is concluded that air
pollutants interfere with parameters of allergy at the level of sensitization,
elicitation of symptoms, and chronification of disease.

I. NATURE, SOURCE, AND CLASSIFICATION OF AIR POLLUTANTS

Environmental pollution has been recognized to be harmful to human health, especially with respect to cytotoxicity, to mutagenicity, and to cancerogenicity. As far as allergic reactions are concerned, air pollutants are proposed to be at least one important factor among those contributing to the increase of allergies worldwide. This suggestion is confirmed by experimental work showing that a variety of substances participating in air pollution, i.e., SO_2, NO_2, O_3, and particles,[1,2] are able to either exhibit adjuvant activity for allergen-specific IgE production in experimental animals[3-8] or to modulate mediator release,[9-12] and to have irritant effects on effector organs in the allergic response.[13]

Air pollution is not restricted to single substances or to single contaminated spots. It is a worldwide problem related to densely populated urban areas and to heavily industrialized regions. A great number of substances of either gaseous or particulate origin are emitted into the outdoor and indoor environment, or into both (Tables 9-1 and 9-2). Some contaminants are even found to occur at far higher concentrations indoors than outdoors (Table 9-2). Outdoor air pollutants neglect national boundaries and can effect regions far distant from the emission sources. Their concentration mostly depends on various factors, for example, emission pattern, removal processes, climate, and weather conditions.[1]

According to Spengler[1] air pollutants are classified as primary, being emitted as such into the atmosphere (SO_2, NO, NO_2, volatile organic compounds, large particles, CO), and as secondary, being formed within the atmosphere by chemical or physical processes (ozone, fine particles, <1 μm). Both groups of pollutants have been shown to exhibit effects on the respiratory tract of asthmatics and nonasthmatics,[14,15] and may even enhance and aggravate symptoms during emission of peak concentrations of these pollutants.[13,16] Outdoor air pollution — as summarized in Table 9-1 — is represented by gaseous substances, by VOCs, by particles, and by metals which are usually adsorbed to the particles. All of these compounds are emitted by manmade sources: industrial processes including petrol refineries, metallurgy, and chemical plants. SO_2 and NO_x as well as particles are emitted by stationary combustion processes. Another source is waste incinerating plants emitting a great variety of substances including CO, NO_x, hydrocarbons, solvents, VOCs, aldehydes, toxic metals, and smoke. The major manmade source of outdoor air pollution today is, however, transportation, especially the automobile. Emissions from this source include gaseous compounds (CO, NO/NO_2, hydrocarbons) and particulates deriving from unburned fuel, fuel additives, or carbonaceous soot. Exposure to these pollutants may be measured by either detecting immission concentrations in the polluted region or by determination of internal body load using blood, urine, or teeth as a vehicle.[17] Table 9-2 gives an impression of the great variety of indoor air pollutants. It indicates that indoor contaminants may

TABLE 9-1
Main Outdoor Air Pollutants

Gaseous Agents	Volatile Organic Chemicals (VOCs)	Particulates	Metals
SO_2	Benzene	TSP	As, Cd, Cr,
CO	Toluene	PM_{10}	CU, Pb, Hg,
NO/NO_2	Xylene	$PM_{2.5}$	Ni
O_3	Methylene chloride		

come from outdoors only, or may have both indoor and outdoor sources. In addition, more than 300 contaminants are known to occur only in the indoor environment, some of them being highly carcinogenic. The concentration of indoor air pollutants varies and largely depends on the rate of ventilation, on additional sources of indoor contaminants (e.g., heating/cooking with unvented gas), smoking, and on human activity.

As far as air pollution in general is concerned, two types are to be distinguished at the moment (Figure 9-1). Type I air pollution is characterized by the existence of the primary pollutants SO_2, particles (total suspended particles,

TABLE 9-2
Main Indoor Pollutants

Source and Type	Indoor Concentration	Indoor/Outdoor Ratio
Pollutants from outdoors		
Sulfur oxides	$0-15 \ \mu g/m^3$	<1
Ozone	0-10 ppb	≪1
Pollutants from indoors and outdoors		
Nitrogen oxides	$10-700 \ \mu g/m^3$	≫1
Carbon monoxide	5-50 ppm	≫1
Carbon dioxide	2000-3000 ppm	≫1
Particulate matter	$10-1000 \ \mu g/m^3$	1
Pollutants from indoors		
Radon	0.01-4 pCi/L	≫1
Formaldehyde	0.01-0.5 ppm	>1
Synthetic fibers	0-1 fiber mL	1
Organic substances		>1
Polycyclic hydrocarbons		
Aerosols		>1
Microorganisms		>1
Allergens		>1

Modified from Spengler, J. D., *Principles and Practice of Environmental Medicine*, Tarcher, A. B., Ed., Medical Book Co., New York, 1992, 21.

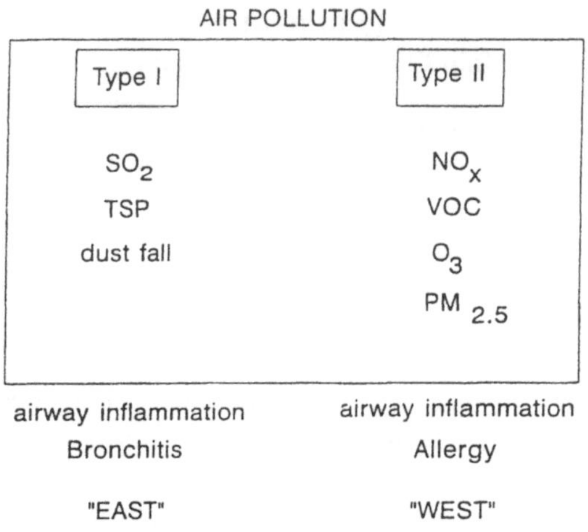

AIR POLLUTION

```
┌─────────────────────────────────────────────────────┐
│   ┌────────┐              ┌─────────┐                │
│   │ Type I │              │ Type II │                │
│   └────────┘              └─────────┘                │
│                                                       │
│      SO₂                      NOₓ                    │
│      TSP                      VOC                    │
│    dust fall                   O₃                    │
│                              PM ₂.₅                  │
│                                                       │
└─────────────────────────────────────────────────────┘
```

airway inflammation   airway inflammation
      Bronchitis   Allergy

        "EAST"   "WEST"

FIGURE 9-1. Classification of air pollution.

TSP), and dust fall, and is emitted predominantly from outdoor sources. Today this type is still present in the eastern part of Germany and in other countries of eastern Europe. It is associated with adverse health effects on the airways, i.e., viral and bacterial airway inflammation and infectious diseases.[18] Type II air pollution is characterized by the presence of primary and secondary pollutants, each of them being emitted from outdoor as well as from indoor sources. This type represents the highly populated, industrialized urban areas in the Western world. And it is this type II air pollution which is predominantly associated with allergic sensitization in terms of prevalence rates. One has to keep in mind that the biological effects of environmental pollutants are generally induced by the repeated uptake of low doses of multiple agents which coexist in the atmosphere and to which the individual is exposed either simultaneously or consecutively. Therefore, it still remains a problem to directly correlate ambient levels of toxic substances with chronic adverse health effects. As far as allergic reactions are concerned, the situation is further complicated by the fact that the interaction between allergens and pollutants does not only occur within the exposed individuals but also at the level of the allergen carrier itself. The pollen being released into a polluted atmosphere will itself be exposed to the contaminants.

II. EPIDEMIOLOGICAL EVIDENCE FOR AN ASSOCIATION BETWEEN TYPE II AIR POLLUTION AND ALLERGY

The aims of our epidemiological cross-sectional studies were (a) to monitor the effects of air pollution on the health of children as a basis of control, and (b) to establish dose–response relationships by studies in areas with highly different types and intensities of air pollution. For this purpose more than 3600 preschool children, 5 to 6 years old, were investigated in the reunited Germany during the spring of 1991 in two areas with different types of pollution (Type I/East vs. Type II/West) and with different intensities of pollution within these areas, including one control region for each type. Both study areas were comparable in geographic latitude, weather, and climate conditions as well as in the genetic background of the population, but differed with respect to type of intensity of air pollution as well as in living conditions.[18,19] The results obtained show that in west Germany an association between emittants of type II air pollution and parameters of allergic sensitization can be observed. This is true for the indoor source variable "cooking/heating with unvented gas" as well as for the outdoor source variables "living more than 2 years within 50 m of a road with heavy traffic" and "staying/moving more than 1 hr per day on a road with heavy traffic."[20] When unvented gas is used in the household, the prevalence rates for children with total serum IgE elevation above 180 kU/L is 17% (control 8%), for children with specific IgE antibodies against grass or mite allergens it is 21% vs. 6.5% and 24% vs. 8.2%. The adjusted-odds ratios indicating a positive and significant association between these two variables are given in Table 9-3.

A similar positive association between exposure and sensitization to at least one inhalable allergen or lifetime prevalence of allergic diseases has been found for outdoor sources of type II air pollution, automobile exhaust (Table 9-4). Therefore, in our studies the exposure to emission sources of type II air pollutants seem to be a risk factor for atopic sensitization after controlling for other covariates.

If this is true, what do we then learn from comparative studies in east and west German regions exhibiting different types and different intensities of air pollution? There is evidence from our own studies[18-21] and from others[22] that the prevalence of viral and bacterial infections, tonsillitis, pneumonia, and bronchitis is higher in the east as compared to the west, and within the east, the prevalence is higher, the more the area is polluted with SO_2 and dust. In contrast, there were lower rates of doctor's diagnosed hay fever in the east. In skin prick tests, the lowest rates of sensitization were found against birch pollen,[18] being even lower than in the control area in west Germany, although the pollen counts did not differ between these areas (Behrendt, unpublished observation). The most striking difference between east and west is, however, the level of total serum IgE, the mean being about three times higher in east German children, a fact that was not reflected by specific IgE values against seven common allergens in the RAST.[19] The risk of having high serum IgE levels is significantly higher when living in east Germany, independent of the intensity of air pollution, even after controlling for all other relevant covariates, including parasitic infestation. Therefore, elevated levels of total serum IgE do not necessarily reflect allergic sensitization, and the reason for this has still to be elucidated and is not known at the moment.

TABLE 9-3
Indoor Source of Air Pollution: Cooking/Heating
with Unvented Gas

	OR (95% CI)	n (year)
Serum IgE > 180 kU/L	2.90 (1.60, 4.20)	488 (1988)
	2.94 (1.47, 5.87)	593 (1991)
RAST grass	3.36 (1.52, 7.44)	590 (1991)
RAST mite	2.97 (1.45, 6.10)	590 (1991)
SPT mite	2.22 (1.29, 5.37)	1052 (1991)

Note: Results from two studies in Germany among preschool children
in 1988 (n = 488) and 1991 (n = 1052). Results are adjusted for
age, sex, passive smoking, family history of allergy, social
status.

III. EXPERIMENTAL EVIDENCE FOR AIRBORNE PARTICLES AS PRIMING FACTORS

Airborne particles seem to play a crucial role in provoking adverse health effects due to air pollution. They contain a large number of chemical substances that can induce cytotoxic, mutagenic, and carcinogenic effects on mammalian cells. Substances adsorbed to airborne particles which had been collected in a highly polluted region in west Germany are also able to release histamine from enriched human basophils in nontoxic dose-dependent fashion.[23] They also show a priming effect on anti-IgE-induced release of histamine and LTC_4. The increase in mediator release is significantly higher in basophils from allergic donors than from nonallergic persons.[10] This indicates that cells from atopic individuals are more sensitive to the particles associated with type II air pollution, thus maybe describing a "risk population." Similar findings have been detected in purified human neutrophils after incubation with extracts of airborne particles: the cells are morphologically activated.[10,23] They show enhanced production of oxygen radicals in chemiluminescence, as is true for the generation and release of LTB_4, LTC_4, PGE_2, and interleukin-8.[10,12] There

TABLE 9-4
Outdoor Source of Air Pollution: Automobile Exhaust

Lifetime Prevalence	OR (95% CI)	p
Bronchial asthma	1.7 (1.0, 2.9)	<0.05
Hay fever	0.9 (0.5, 1.4)	
Atopic eczema	1.4 (1.1, 1.8)	<0.05
Allergy	1.2 (1.0, 1.5)	<0.1
Sensitization (RAST max)	3.7 (1.4, 9.2)	<0.01

Note: 1052 preschool children, Germany 1991. Adjusted for covariates.

was an interesting difference, however, in mediator release between allergic and nonallergic donors, namely increased LTB_4 and decreased PGE_2 secretion in allergic donors as compared to normals.

From these *in vitro* experiments we conclude that substances adsorbed to airborne particles are able to interfere with cells involved in inflammatory processes and exhibit priming effects on them. Therefore, the exposure of individuals to adequate concentrations of particles may provoke enhancement of symptoms of allergic diseases and continuation of allergic inflammation.

IV. EVIDENCE FOR THE INTERACTION BETWEEN POLLEN AND AIR POLLUTANTS

Pollen grains which are the major source of outdoor aeroallergens also incorporate pollutants when released into a polluted atmosphere. They are able to accumulate heavy metals, e.g., lead, cadmium, and mercury as well as sulfur, and the amount of sulfur per pollen dry weight seems to be a bioindicator for the burden of the atmosphere with sulfuric aerosols.[24] We investigated the behavior of pollen grains which had been collected from four differently polluted regions in Northrhine-Westphalia/west Germany by means of Burkhard traps and/or high volume samplers.[24,25] The results obtained show that there is an overall higher concentration of pollen grains in polluted urban areas than in rural regions, independent of weather conditions. In addition, emission peaks of SO_2, NO/NO_2, or atmospheric fine dust, but not of O_3, usually precede peaks of high pollen concentrations. Pollen grains which had been collected from industrial regions polluted with high amounts of organic substances are agglomerated with airborne particles (Figure 9.2). The same holds true for hazel pollen which had been collected from trees near a road with heavy traffic, but not for pollen from park trees.[25] The particles agglomerated to the pollen surface are heteromorphic, smaller than 5 μm in diameter, and may be aggregated. A substantial amount of them belongs to the submicronic range. Agglomeration of particles to pollen surfaces in areas with emission of organic compounds deriving from the fuel industry is supported by semiquantitative data confirming both a high degree of agglomeration as well as a high number of defective and destroyed pollen grains.[24,26] Based on these observations we asked whether pollen–particle interactions may affect allergen formation and release at the level of the carrier itself. Results obtained from *in vitro* experiments with native *Dactylis glomerata* pollen and extracts of airborne particulate matter show that substances which are present in the aqueous phase of the particles induce the release of proteins from pollen grains and give rise to the formation of submicronic particles.[24] Using SDS-polyacrylamide gel electrophoresis and immunoblotting, an alteration of allergenic protein bands, i.e., shifting of binding pattern intensity of IgE reactive bands to the acidic side, has been demonstrated.[25] The altered allergens, however, do not prove the altered allergenicity per se.

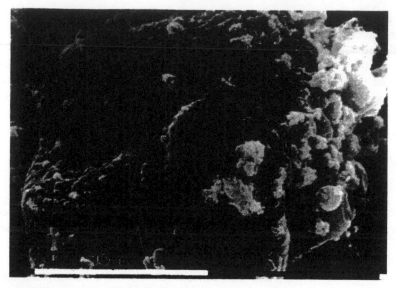

FIGURE 9-2. Agglomeration of airborne particles onto the surface of a grain of birch pollen collected in a highly polluted region. Scanning electron micrograph, bar 10 μm.

The conclusion from these experiments is that organic substances adsorbed to airborne particles mediate agglomeration of particles onto pollen surfaces followed by local preactivation of the coated pollen. Under appropriate conditions aqueous compounds may then induce local allergen release, resulting in either allergenic extrusions followed by generation of allergenic aerosols or in adsorption of pollen-derived proteins to airborne particles. In regions with high

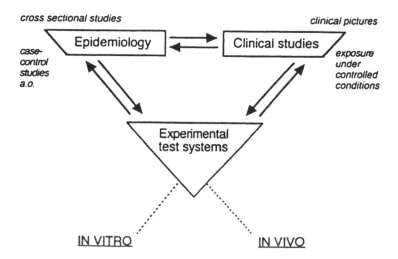

FIGURE 9-3. Course of investigation in allergotoxicology.

air pollution, particles may therefore not only carry pollutants but also allergens, and pollen hosts not only allergens but also pollutants. Therefore, pollen counts may not necessarily reflect the actual load of the atmosphere with outdoor allergens within polluted regions, and methods have to be developed to directly measure allergens in the outdoor environment.

In order to further elucidate the role of indoor and outdoor air pollution in induction, elicitation, and maintenance of allergic inflammatory reactions ("allergotoxicology")[27] a huge interdisciplinary effort has to be undertaken including environmental epidemiology, *in vitro* experiments, and clinical investigations (Figure 9-3).

ACKNOWLEDGMENTS

This work is supported by a grant of the BMFT (07ALL04) and by a grant of MURL/NRW.

REFERENCES

1. **Spengler, J. D.**, Outdoor and indoor air pollution, in *Principles and Practice of Environmental Medicine*, Tarcher, A. B., Ed., Medical Book Co., New York, 1992, 21.
2. **Quackenboss, J. J., Lebowitz M. D., and Crutchfield, C. D.**, Indoor-outdoor relationships for particulate matter: exposure classifications and health effects, *Environ. Int.*, 1989; 15:353.
3. **Riedel, F., Krämer, M., Scheibenbogen, C., and Reiger, C. H. L.**, Effects of SO_2 exposure on allergic sensitization in the guinea pig, *JACI*, 1988;82:527.
4. **Osebold, J. W., Zee, Y. C., and Gershwin, L. J.**, Enhancement of allergic lung sensitization in mice by ozone inhalation, *Proc. Soc. Exp. Biol. Med.*, 1988;188:259.
5. **Muranaka, M., Suzuki, S., Koizumi, K., Takafuji, S., Miyamoto, T., Ikemori, R., and Tokiwa, H.**, Adjuvant activity of diesel-exhaust particulates for the production of IgE antibody in mice, *JACI*, 1986;77:616.
6. **Takafuji, S., Suzuki, S., Koizumi, K., Tadokoro, K., Miyamoto, T., Ikemori, R., and Muranaka, M.**, Diesel-exhaust particulates inoculated by the nasal route have an adjuvant activity for IgE production in mice, *JACI*, 1987;79:639.
7. **Takafuji, S., Suzuki, S., Koizumi, K., Tadokoro, K., Ohashi, H., Muranaka, M., and Miyamoto, T.**, Enhancing effect of suspended particulate matter on the IgE antibody production in mice, *Int. Arch. Allergy Appl. Immunol.*, 1989;90:1.
8. **Yanai, M., Ohuri, T., Aikawa, T., Okayama, H., Sekizawa, K., Maeyama, K., Sasaki, H., and Takashima, T.**, Ozone increases susceptibility to antigen inhalation in allergic dogs, *J. Appl. Physiol.*, 1990;68:2267.
9. **Behrendt, H., Weiczorek, M., Wellner, S., and Winzer, A.**, Effect of some metal ions (Cd, Pb, Mn) on mediator release from mast cells in vivo and in vitro, in *Environmental Hygiene*, Seemayer, N. H. and Hadnagy, W., Eds., Springer, Berlin-Heidelberg, 1988, 105.
10. **Hitzfeld, B., Friedrichs, K. H., and Behrendt, H.**, In vitro interaction between human basophils and polymorphonuclear granulocytes: effect of airborne particulate matter, *Int. Arch. Allergy Immunol.*, 1992;99:390.
11. **Raulf, M. and König, W.**, In vitro effects of polychlorinated biphenyls on human platelets, *Immunology*, 1991;72:287.

12. **Hitzfeld, B., Friedrichs, K. H., Tomingas, R., and Behrendt, H.,** Organic atmospheric dust extracts and their effects on functional parameters of human polymorphonuclear leukocytes (PMN), *J. Aerosol. Sci.,* 1992;23(Suppl. 1):S532.

13. **Pierson, W. E. and Koenig, J. Q.,** Respiratory effects of air pollution on allergic disease, *JACI,* 1992;90:557.

14. **Molfino, N. A., Wright, S. C., Katz, I., Tarlo, S., Silverman, F., McClean, P. A., Szalai, J. P., Raizenne, M., Slutsky, A. S., and Samel, N.,** Effect of low concentrations of ozone on inhaled allergen responses in asthmatic subjects, *Lancet,* 1991;338:199.

15. **Magnussen, H., Jörres, R., Wagner, H. M., and von Nieding, G.,** Relationship between the airway response and inhaled sulphur dioxide, isocapnic hyperventilation and histamine in asthmatic subjects, *Int. Arch. Occup. Environ. Health,* 1990;62:485.

16. **Ruszak, C., Devalia, J. L., and Davies, R. J.,** The impact of pollution on allergic disease, *Allergy,* 1994;49:21.

17. **Jermann, E., Hajimiragha, H., Brockhaus, A., Freier, I., Ewers, U., and Roscovanu, A.,** Exposure to benzene, toluene, lead, and carbon monoxide of children living in a central urban area with high traffic density, *Zbl. Hyg.,* 1989;189:50.

18. **Schlipköter, H. W., Krämer, U., Behrendt, H., Dolgner, R., Stiller-Winkler, R., Ring, J., and Willer, H. J.,** Impact of air pollution on children's health — results from Saxony-Anhalt and Saxony as compared to Northrhine Westphalia, in *Health and Ecological Effects. Critical Issues in the Global Environment,* Vol 5, Air & Waste Management Association, Pittsburgh, PA, 1992, IU-A 2103.

19. **Behrendt, H., Krämer, U., Dolgner, R., Hinrichs, J., Willer, H., Hagenbeck, H., and Schlipköter, H. W.,** Elevated levels of total serum IgE in East German children: atopy, parasites, or pollutants? *Allero J.,* 1993;2:31.

20. **Krämer, U., Behrendt, H., Dolgner, R., Kainka-Sänicke, E., Oberbarnscheidt, J., Sidaoui, H., and Schlipköter, H. W.,** Auswirkung der Umweltbelastung auf allergologische Parameter bei 6jährigen Kindern, in *Epidemiologie allergischer Erkrankungen,* Ring, J., Ed., MMV, München, 1991, 165.

21. **Ring, Johannes,** Clinical and epidemiological aspects of hypersensitivity reactions of the skin, in *Allergic Hypersensitivities Induced by Chemicals: Recommendations for Prevention,* Vos, J. G., Younes, M., and Smith, E., Eds., World Health Organizaiton, Geneva/CRC Press, Boca Raton, 1995, p. 203.

22. **von Mutius, E., Fritzsch, C., Weiland, S. K., Röll, G., and Magnussen, H.,** Prevalence of asthma and allergic disorders among children in united Germany: a descriptive comparison, *Br. Med. J.,* 1992;305:1395.

23. **Behrendt, H., Friedrichs, K. H., Fischer, I., and Tomingas R.,** Airborne particulate matter induces histamine release from and degranulation of enriched human basophils, *ACI News,* 1991;(Suppl. 1):218.

24. **Behrendt, H., Friedrichs, K. H., Kainka-Sänicke, E., Darsow, U., Becker, W. M., and Tomingas, R.,** Allergens and pollutants in the air — a complex interaction, in *New Trends in Allergy III,* Ring, J. and Przybilla, B., Eds., Springer, Berlin-Heidelberg, 1991, 467.

25. **Behrendt, H., Becker, W. M., Friedrichs, K. H., Darsow, U., and Tomingas R.,** Interaction between aeroallergens and airborne particulate matter, *Int. Arch. Allergy Immunol.,* 1992;99:425.

26. **Kainka-Stänicke, E., Behrendt, H., Friedrichs, K. H., and Tomingas, R.,** Morphological alterations of pollen and spores induced by airborne pollutants: observations from two differently polluted areas in West Germany, *Allergy,* 1988;43(Suppl. 7):57.

27. **Behrendt, H.,** Allergotoxikologie: Ein Forschungskonzept zur Untersuchung des Einflusses von Umweltschadstoffen auf die Allergieentstehung, in *Allergieforschung: Probleme, Strategien und klinische Relevanz,* Ring, J., Ed., MMV, München, 1992, 123.

Skin Allergy

Chapter 10

MECHANISMS OF ALLERGIC CONTACT DERMATITIS TO CHEMICALS*

Rik J. Scheper and B. Mary E. von Blomberg

CONTENTS

I. SKIN REACTIONS DUE TO EXPOSURE TO CHEMICALS

The skin provides an efficient mechanical barrier towards external noxious agents, including microorganisms and chemicals. This, however, does not mean that skin exposure to toxic agents is without limitations. In contrast, such exposure may rapidly induce skin inflammatory reactions along different pathways. These reflect evolutionary defense mechanisms to escape potentially life-threatening insults, such as sepsis and toxicemia. Cutaneous reac-

*This chapter has been adapted from *Cellular Mechanisms in Allergic Contact Dermatitis*, by R. J. Scheper and B.M.E. von Blomberg, which appeared in *Textbook of Contact Dermatitis*, Rycroft, R. J. G., Menné, T., Frosch, P. J., and Benezra, C., Eds., Springer-Verlag, Berlin, 1992. Copyright Springer-Verlag, 1992. With permission.

tions to chemical exposure may be defined as nonimmunological or immunological inflammatory reactions. The first are regularly classified as irritant reactions, whereas the latter involve both IgE-mediated, immediate contact reactions and T-cell-mediated, allergic contact dermatitis.

Acute irritant reactions are primarily due to chemicals which directly damage skin cells. These may trigger keratinocytes to release pro-inflammatory cytokines, such as IL-1, IL-6, and TNF-α, mast cells to release histamine and associated mediators, and nerve cells to generate itching and pain. Agents that may cause acute irritant reactions include a vast array of widely applied detergents, acids, alkalis, and oils. Prolonged exposure to these agents may cause chronic irritant skin reactions, which increasingly depend on further genetic, climatic, and mechanical factors. No state of allergic hypersensitivity is, however, induced by these agents. This means that: (1) skin reactions are limited to the site(s) of contact, and (2) no immunological memory is induced, that is, essentially similar dosages of the same chemical are required each time to induce the irritant reactions.

In sharp contrast, subgroups of chemicals may, upon skin contact, induce immunological hypersensitivities. These involve either the development of chemical agent-specific homocytotropic (notably IgE) antibodies, resulting in immediate contact reactions, or T-lymphocytes, resulting in allergic contact hypersensitivity. In sensitized individuals, immediate contact reactions appear within minutes to an hour after the responsible agents, such as distinct cosmetics, drugs, and fruit- and vegetable-derived molecules, have been in contact with the skin, and they regularly disappear within one day. Symptoms include itching, wheal and flare, and erythema. Foodstuffs are the most common primary cause of immediate allergic contact reactions. The orolaryngeal area is a site where immediate reactions are frequently provoked by food allergens, most frequently among atopic individuals. Since skin hyperreactivity of this type is often secondary to mucosal contacts with the eliciting agents, the mechanisms of IgE-mediated hypersensitivities receive primary attention in the chapters on respiratory allergy (Section II) and gastrointestinal allergy (Section IV). Here, we will focus on the mechanisms of allergic contact dermatitis, caused by T-lymphocytes specific for distinct chemical agents.

II. CELL-MEDIATED HYPERSENSITIVITY OF THE SKIN

A. CELL-MEDIATED IMMUNITY

Evolutionarily, cell-mediated immunity has fully developed in vertebrates, for their benefit, by facilitating effective eradication of microorganisms and toxins. Elicitation of allergic contact hypersensitivity by very low, nontoxic doses of small molecular allergens indicates that often the T-cell repertoire is broader than one might wish. This hypersensitivity thus represents an untoward side effect of a well-functioning immune system.

Subtle differences can be noted in macroscopic appearance, time course, and histopathology of allergic contact reactions in various vertebrates, including rodents and man. Nevertheless, essentially all basic features are shared. Since both mouse and guinea pig models have strongly contributed to our present knowledge of allergic contact dermatitis, in this chapter data from animal studies have been taken together with results obtained from studies on T-cell-mediated reactions in man. Allergic contact dermatitis can be regarded as a prototype of delayed hypersensitivity, as classified by Turk[3] and Gell and Coombs[4] (type IV hypersensitivity).

B. DEVELOPMENT OF ALLERGIC CONTACT HYPERSENSITIVITY

It is convenient to distinguish afferent and efferent limbs in the development of allergic contact hypersensitivity.[3,5,6] The afferent limb includes the events following a first contact with the chemical (allergen) and is complete when the individual is sensitized and capable of giving a positive elicitation reaction. The efferent limb is activated during the elicitation (challenge) phase. Typically, elicitation of allergic skin reactions may be obtained with 10- to 1000-fold lower dosages of the allergen than required for primary skin inflammatory reactions. The entire process of the afferent limb requires from at least 3 days to several weeks, whereas full development of the elicitation phase only requires 24 to 48 hr. The main afferent (a–d) and efferent (e) events (Figure 10-1) are:

a. *Binding of allergen to skin components.* Contact allergens penetrating the skin associate with skin constituents, notably class II molecules. These molecules, coded in humans by HLA-D region genes, are abundantly present in the epidermis on dendritic cells, the Langerhans' cells (LCs). This association may be due to a covalent bond, as in the case of poison oak/ivy allergens, or to a coordination bond, as with nickel cations. Allergen-carrying LC travel via the afferent lymphatics to the regional lymph nodes, where they settle in the T-cell (paracortical) areas.

b. *Recognition of allergen-modified LCs by specific T-cells.* In the paracortical areas, conditions are optimal for allergen-carrying LCs to encounter T-cells that specifically recognize the allergen–class II molecule complexes. Notably, in nonsensitized individuals the frequency of T-cells with corresponding specificities is far below one per thousand. The dendritic morphology of the LC strongly facilitates multiple cell contacts, ultimately leading to binding and activation of sufficient numbers of allergen-specific T-cells.

c. *Proliferation of specific T-cells in draining lymph nodes.* Supported by interleukin-1, a mediator released by the allergen-presenting cells, activated T-cells start producing several other mediators called cytokines, including interleukin-2 (IL-2). A partly autocrine cascade follows in

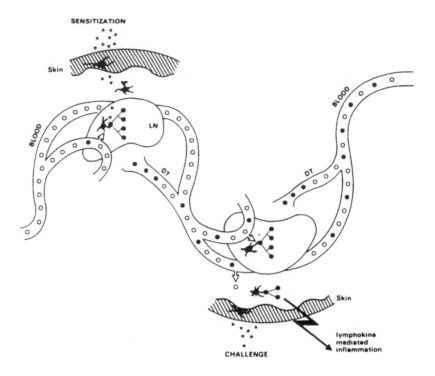

FIGURE 10-1. Pathogenesis of allergic contact hypersensitivity: sensitization (top, left) and challenge (bottom, right). Allergen-bearing Langerhans' cells (★) travel to the draining lymph nodes (LN) where they present the allergen (▲) to specific T-lymphocytes (●; T-lymphocytes with irrelevant specificities: ○). Their progeny reaches the blood through the thoracic duct (DT). Subsequent challenge of the skin with allergen results in a local cytokine-mediated inflammation and further increases the frequency of allergen-specific cells in the circulation.

which the density of IL-2 receptors on activated T-cells increases transiently, leading to full reception of the IL-2 signal, resulting in vigorous blast formation and proliferation within a few days.

d. *Propagation (dissemination) of specific T-cell progeny over the body.* The expanded progeny (effector cells) is subsequently released from the lymph nodes by the efferent lymphatics, reaches the circulation, and may again enter the peripheral tissues, including the skin.

e. *Efferent action.* Allergic contact dermatitis is based on the increased frequency of T-cells with a given specificity throughout the body of a sensitized individual, as well as on the enhanced capacity of effector cells to enter peripheral tissues. At sites of primary allergenic skin contact, too few specific T-cells are available locally to allow for macroscopically detectable skin reactivity. When challenging a sensitized individual, however, a local inflammatory reaction will follow. Allergen-presenting cells and specific T-cells meet locally in the skin, causing rapid release of cytokines, notably IFN-γ. Moreover, contact allergens may be particularly active in triggering keratinocytes to produce a broad repertoire of proinflammatory cytokines, including IL-1α, TNF-α, and chemotactic cytokines.[41,42] These cause the arrival of more T-cells, thus further amplifying local mediator release. Gradually an eczematous reaction develops which reaches a maximum after 18 to 48 hr. Cytokines that play a major role in the development of contact hypersensitivity are those with stimulatory effects on other lymphocytes (IL-2, interferon-gamma), on mononuclear phagocytes (chemotactic factor, migration inhibition factor, interferon-gamma), and on mast cells and vasculature (skin reactive factor, interferon-gamma). In the following paragraphs we will focus on the roles of two pivotal cell types in the above-described cascade of events: the Langerhans' cells and T-lymphocytes.

C. ALLERGEN-PRESENTING CELLS

1. Langerhans' Cells

In healthy skin, the epidermal dendritic Langerhans' cell, bearing high numbers of class II molecules on its cell membrane, is the primary allergen-presenting cell (APC). LC stem from the bone marrow, but their continuous presence in the epidermis is at least partly maintained by local LC proliferation.[7-9] In mice, the sensitization rate was found to be determined by the density of LCs at the site of allergen application. Moreover, absence or functional inactivation of LC, e.g., by UV-treatment, was associated with tolerance induction[10,11] (see below).

Fortunately, the risk of severe skin inflammatory reactions at the first encounter of chemical allergens is low. This is important in view of the extreme amplification power of the cytokine cascade, as illustrated by experimental studies in which measurable edema could be triggered by only one specific T-cell.[12] Besides rigid activation requirements and limited entry of naive (unprimed) T-cells into the skin (see Section D), the necessity of LCs to further mature before full presenting power is obtained also appears important.[1,13-15] Such maturation may occur when LCs travel from the skin and settle within the paracortical areas of the lymph nodes, or farther away within the peri-arteriolar lymph sheaths in the spleen. So-called "dendritic cells" isolated from these organs display rapid and efficient allergen presentation toward naive T-cells. Similar stimulatory capacity with LCs, freshly isolated from skin blister roofs, was observed only after 1 to 2 days of preculturing. Notably, even without delay, LCs are excellent in stimulating primed, memory T-cells with contact allergens.[2,6,15-17]

2. Allergen Presentation

Most contact allergens are small, chemically-reactive molecules. Upon penetration through the epidermis, they readily bind to LC surface-exposed MHC class II molecules (mainly HLA-DR, DP, and DQ), besides binding to the plethora of other skin constituents. The ability to associate with MHC class II molecules, either directly or via binding to the antigenic peptides in the MHC class II groove, is a prerequisite for contact sensitization. Whereas most allergens bind spontaneously, some need enzyme- or photo-induced activation before they bind. The latter allergens are called contact photoallergens.[74] Class II modification may also result from allergen binding to nearby cell-surface molecules or even free proteins. Allergen-binding proteins, however, require degradation by allergen-presenting cells, or nearby macrophages, to small fragments, before being presented at the cell membrane. Close association with class II molecules, essential for T-cell triggering, is facilitated by peptide-binding sites present on class II molecules.[18,19]

It should be recalled that effective presentation of contact allergens to T-lymphocytes depends not just simply on class II molecule expression by LC, but also on their capacity to create and maintain close cell contacts with T-lymphocytes (clustering). To this end, the intricate structure of paracortical areas in the lymph nodes and the characteristic membrane ruffling of LC/dendritic cells provide optimal conditions for T-cell triggering. Intimate cellular contacts are further facilitated by sets of specialized interaction molecules.

D. LYMPHOCYTES
1. Effector T-Cells
The nature of the cells that mediate contact sensitivity has intrigued dermatologists and immunologists for decades. The earliest relevant information was obtained in experimental studies by transferring immune cells to naive recipients. Both in guinea pigs and mice, contact hypersensitivity could be transferred with thymus-derived T-, and not with B-lymphocytes.[5] Then monoclonals became available, discriminating between primarily cytokine-producing (CD4+, "helper"), and cytotoxic (CD8+, "cytotoxic/suppressor") T-lymphocytes. Although cytotoxic T-cells can mediate distinct skin damage in allergic contact dermatitis, the helper subset is held to be primarily responsible for mediating delayed hypersensitivity/allergic contact dermatitis.[2,6,20,26]

Within the helper T-cell population, different subsets can now be distinguished. The ability to mount delayed type hypersensitivity reactions resides in the Th1 subset, able to produce IL-2, IFN-γ, and TNF-β.[67,68] The Th2 subset produces IL-4, IL-5, and IL-10, and may account for both the persistent production of antibodies (notably IgE) and eosinophilia observed in helminthic infections and immediate type I hypersensitivity reactions. Both subsets produce IL-3, TNF-α, and granulocyte–macrophage colony-stimulating factor (GM-CSF). The phenomena of distinct cytokine profiles for T-cells may result from an intrinsic, mutually exclusive regulation of Th1 and Th2 cytokine genes. Type 2 cytokines such as IL-4 shift T-cell differentiation away from the production of type 1 cytokines, whereas the type 1 cytokine IFN-γ is very potent in preventing the development of Th2 cells.[59-61] Bacteria, viruses, and most contact allergens may preferentially induce Th1 responses through stimu-

TABLE 10-1

Some Characteristics of CD4⁺ Naive and Effector/Memory T-Cells

	Naive	Effector/Memory
Relative expression of surface molecules		
CD2 (LFA-3 receptor)[1]	+	++
CD3 (TCR-assoc.mol)	+	+
CD11a (LFA-1)	+	++
CD29 (VLA-beta chain)	+	++
CD45RA (LCA)	+++	+
CD45RO (truncated LCA)	+	++++
CD58 (LFA-3)	+	+++
Relative functional capacity		
Interleukin-2 production	+++ (late)	+++ (early)
Interferon-gamma production	+	+++
Allergen-driven prolif. in vitro	+	+++
Mitogen-driven prolif. in vitro	++++	++

[1] Cluster Designation (CD) number (recognized membrane component; some monoclonal antibodies) (References 21, 22, 61).

lation of interleukin-12 (IL-12) production by macrophages or other cells. IL-12 stimulates IFN-γ production by T-cells. Allergenic contacts and infections along the mucosal surfaces rather induce the differentiation of Th2 cells.[59,62,63] Here, IL-4 production by cells of the mast cell/basophil lineages, in the absence of local IFN-γ release, determines T-cell differentiation. The frequent induction of Th2 responses after antigen introduction along mucosal surfaces is probably further promoted by high local densities of B-cells as compared to the skin compartment: B-cells are excellent IL-10 producers, and antigen presentation by B-cells is known to favor Th2 responses.[57,58] Development of antigen-specific Th2 cells along mucosal surfaces bears the advantage that responsiveness focuses on rapid antibody (IgE/IgA)-mediated effector mechanisms, preventing microorganisms to enter the body, in the absence of potentially damaging tissue inflammatory reactions. In line with these recent insights, nickel-specific CD4⁺ effector T-cell clones were found to exhibit the Th1 cytokine profile, whereas human atopical allergen-specific CD4⁺ T-cell clones released typical Th2 cytokines. Thus, CD4⁺ Th1-type T-effector cells may be considered as central in causing allergic contact dermatitis.

Recently, a qualitative distinction between (difficult to stimulate/afferently acting) naive and (easy-to-stimulate/efferently acting) effector/memory T-cells was confirmed (see Table 10-1). A high molecular weight isoform of the leukocyte common antigen (CD45RA) characterizes naive T-cells, whereas only a truncated form of the molecule is expressed on effector/memory T-cells (CD45R0). Importantly, T-cells with the naive phenotype show excellent proliferative capacity, and good (albeit slow-onset) IL-2 production. On the other hand, effector/memory T-cells not only show additional IFN-γ production, but also full-scale IL-2 production as early as 24 hr after stimulation.[21,22] These results strongly suggest that the major effector cells in allergic contact dermatitis belong to the latter subset. These cells indeed dominate in the skin.[23] In contrast to naive cells, which really need LC/dendritic cells for allergen presentation, effector/memory lymphocytes can also be stimulated by other cell types presenting allergen-modified class II molecules, e.g., monocytes, endothelial cells, and B-cells.[24-26] Clearly, effector/memory cells display higher numbers of cellular adhesion molecules (CAMs), allowing for more promiscuous cellular interactions.

2. Cellular Adhesion Molecules

Thus, presentation of allergen by class II molecules does not provide sufficient binding strength. Additional sets of CAMs facilitate the strongest binding and full activation of T-cells. Three of these sets deserve mention here (see Figure 10-2). First, the CD4 (helper) molecule bears affinity to a constant region of the class II molecule, providing additional binding stability in T-cell/APC interactions. Second, LFA-1 (leukocyte-function associated antigen, a 180/95 kD two-chain cell-surface molecule) is abundantly present on T-cells, and at least one of its counterstructures (ICAM-1) is present on LC. For CD2 (a T-cell antigen formerly known for its binding to sheep red blood cells) the counter-structure is a 25 to 29 kD molecule (LFA-3), anchored in the plasma membrane of most nucleated cells. In particular, CD2/LFA-3 interaction may provide an activational signal synergizing with the T-cell-receptor/class II allergen-mediated signal.[27,28]

Importantly, triggering of sensitized (effector/memory) T-cells in contact hypersensitivity is facilitated by enhanced expression of various CAMs. Moreover, expression of the counterstructures on several (epi-) dermal cells is upregulated by T-cell cytokines, notably IFN-γ. This further promotes development of allergic contact dermatitis, since eventually T-cells may also be triggered by allergen bound to various skin cells.[29-32]

On the other hand, priming of T-cells may lead to the loss of a distinct set of CAMs facilitating interactions with high endothelial venules in peripheral lymph nodes.[33] After sensitization, T-cells are less capable of recirculating through the lymphoid organs, but gain in ability to migrate into the peripheral tissues, including the skin. Indeed, interactions with peripheral endothelia are facilitated by the enhanced expression of CAMs. Moreover, the recent progeny of allergen-activated T-cells shows a strongly increased random migratory capacity.[34-36]

In conclusion, available evidence indicates that allergic contact dermatitis is primarily mediated by CD4+ effector/memory, Th1-type cells. Apparently, primary skin contacts with allergens are particularly prone to stimulate allergen-specific Th1 cells, whereas mucosal allergen contacts lead to Th2-type responses.

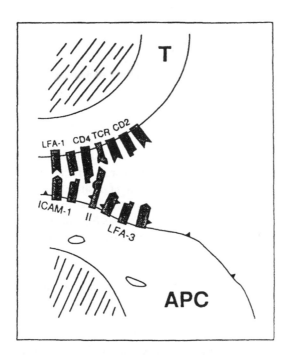

FIGURE 10-2. The cellular adhesion molecule-zipper connecting an allergen-presenting cell (APC) and a specific CD4+ T-cell (▲, allergen). The T-cell receptor (TCR) recognizes a modified MHC class II molecule, thereby allowing receptor-associated CD3-chains (not depicted) to transduce a signal for cellular activation. Additional binding and signals result from CD4/class II, CD2/LFA-3, and LFA-1/ICAM-1 interactions.

3. T-Cell Specificity: Cross-Reactions/Concomitant Sensitization

Allergen-specific T-cells display strong specificity: if for example nickel-specific T-cells are cultured *in vitro* with nickel (bivalent cations) presented by cells with different (incompatible) class II molecules, they will not react, even if all the other CAMs and counterstructures fit nicely. On the other hand, if the same cells are cultured with fully compatible APC, but with the physicochemically very similar metal allergen cobalt, they do not react either.[26,37] Such *in vitro* studies confirm the notion that simultaneously occurring contact allergies often derive from concomitant sensitization rather than from immunological cross-reactivity. Multiple sensitization may readily occur since allergens frequently go together (like nickel and cobalt in alloys, or drugs and preservatives in medicaments). Moreover, ongoing allergic reactions have strong immuno-potentiating power (e.g., from local IL-2 and IFN-γ release) and, thus, the risk of developing new allergies at these sites is high.[38] Also, chemical irritants may reduce the threshold of sensitization, possibly by augmented recruitment of inflammatory cells, including allergen-presenting cells and T-lymphocytes.[77]

Still, molecular mimicry may occur. The T-cell receptor (TCR) recognizes rather small molecular moieties, e.g., peptides with at least 7 to 8 amino acids.

Thus, it can be envisaged that to some degree, binding of truly different allergens to APC may generate similar class II epitopes, thus triggering T-cells unable to discriminate between these allergens. Evidence supporting this view was obtained in the above-mentioned studies: nickel-specific T-cell clones were either truly nickel-specific, or just as well stimulated by copper or palladium cations.[74] It is interesting that both these metals are immediately surrounding nickel in the periodic table. Thus, it may well be that development of nickel allergy in some patients inevitably also leads to palladium or copper allergy. Actually, the latter allergies have, to our knowledge, never been reported to occur in the absence of nickel allergy.[39] It can be concluded that skin testing alone does not reveal to what extent positive reactions to different allergens are based on the same T-cell clone(s), rather than on separate coexisting clones. Nevertheless, clinical consequences for patient and treatment are similar, whether allergies to different compounds stem from separate, overlapping, or identical sets of T-cell clones.

4. T-Cell/Allergen Retention in the Skin: Flare-up Reactions

Flare-up reactions are frequently observed at former skin reaction sites. From the basic mechanisms of contact hypersensitivity it can be inferred that allergen-specific flare-up reactions depend either on local allergen or T-cell retention at these skin sites. Flare-up reactions due to locally-persisting allergen can be readily observed in man when from about 1 week after primary sensitization, sufficient effector T-cells have entered the circulation to react with residual allergen at the sensitization site.[40] Another example is the inadvertent flare-up reaction that may be observed at a site of patch testing with a compound providing two different allergenic moieties: previously induced allergic reactivity and, thus, positive reactivity to one (e.g., penicillin) may potentiate primary sensitization to the other (e.g., formaldehyde), which then may cause a flare-up from about 1 week after skin testing. Local allergen retention is usually of short duration only. In experimental guinea pig studies using DNCB, chromium, and penicillin allergens, we never found local allergen retention in the skin to mediate flare-up reactions for periods exceeding 2 weeks (Scheper, unpublished results).

In contrast, allergen-specific T-cells may persist for much longer in the skin (up to several months). Thus, locally increased allergen-specific hyperreactivity, either detectable through accelerated "retest" reactivity (peaking at 6 to 8 hr after repeated allergenic contact at a former skin reaction site) or flare-up reactivity (similar, but after repeated allergen entry from the circulation, e.g., derived from food) may be observed for several months at former skin reaction sites.[71,72] Interestingly, histological examination of such previous skin reaction sites shows that only few residual T-cells remain present for such periods. Still, the remarkable flare-up reactivity at such sites can be understood by considering that only very few specific T-cells are required for macroscopic reactivity. Moreover, a high frequency of the residual T-cells may bear specificity to the

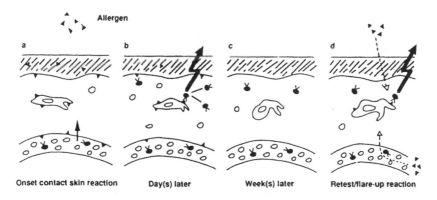

FIGURE 10-3. T-cell-mediated flare-up reactivity in allergic contact hypersensitivity. T-lymphocytes with irrelevant specificities (○), but in particular specific T-lymphocytes (●), enter a challenged skin site (a, b). Cytokine release and proliferation ensues. A small residual infiltrate, with a relatively high frequency of allergen-specific T-cells, persists (c) and may give rise to accelerated (specific cells already present) local reactivity upon renewed allergenic contacts, either via the skin (retest reactivity) or the circulation (flare-up reactivity) (d).

allergen: allergic skin reactions tend to selectively recruit allergen-specific T-cells from the circulation, whereas subsequent allergen-driven proliferation and retention of those cells may further increase the local frequency of allergen-specific cells (Figure 10-3).[34,43]

III. REGULATORY MECHANISMS IN ALLERGIC CONTACT DERMATITIS

A. TOLERANCE INDUCTION, SUPPRESSOR CELLS

Uncontrolled development and expression of T-cell-mediated immune function would be detrimental to the host. During evolution, several mechanisms developed to prevent lymph node "explosion" or excessive skin ulceration upon persisting antigen exposure. Epicutaneous allergenic contacts not only induce T-effector cells, but also lymphocytes that curtail further T-effector cell proliferation ("suppressor cells"). Sensitization, therefore, seems to be the result of a delicate balance between effector and suppressor mechanisms. Upon preferential stimulation of suppressor cells, a strong and stable allergen specific unresponsiveness may develop, known as immunological tolerance. This phenomenon can be readily demonstrated by administering allergen intravenously or orally to nonsensitized, immunologically naive individuals.[5,44-46] The cell type critical in determining whether exposure to an allergen leads to hypersensitivity or tolerance is the LC. By oral or intravenous administration, allergen presentation by LCs is bypassed. In support of this view, tolerance may be induced by applying allergens on skin sites where LCs have been functionally damaged, e.g., by UV-irradiation, or are naturally absent (e.g., in the tail skin of mice).[47,48]

The concept of suppressor cells controlling allergic contact dermatitis is based on two main facts. First, allergen-specific tolerance can be transferred from tolerant to normal (naive) animals by lymphoid cells. Second, contact sensitization can be enhanced, and tolerance reversed, by treatment with distinct cytostatic drugs, such as cyclophosphamide.[5,49] In mice, guinea pigs, and man, suppressor cells, or their precursors, have been demonstrated to be particularly sensitive to cytostatic drugs, thus allowing for exaggerated effector cell development. Still, suppressor cells are not a well-defined subpopulation of cells, and several different mechanisms of action have been proposed recently. These may involve a subset of allergen-specific T-cells shedding truncated T-cell receptor moieties which can block allergen presentation,[64] and Th2 cells releasing cytokines which interfere with effector T-cell functions (IL-4[65]; IL-10[66]).

Both clinical and experimental findings make clear that full and persistent tolerance can only be induced prior to any sensitizing allergen contacts. Experimentally, even sub-sensitizing doses of nickel applied to the skin prevented subsequent tolerance induction by feeding the metal allergen.[73] This may have contributed to incomplete tolerance induction in earlier clinical studies when feeding with poison ivy/oak-derived allergens.[50] Apparently, the progeny of naive allergen-specific cells, once on the stage, has escaped susceptibility to suppressor cell action. This explains why, to our knowledge, permanent reversal of existing allergic contact dermatitis has never been achieved clinically. Nevertheless, effector cells still seem to be susceptible to (transient) downregulation of allergen reactivity, as observed in desensitization procedures.[5,50]

B. SKIN HYPOREACTIVITY, DESENSITIZATION

For dermatologists, methods by which patients might be desensitized for an existing allergic contact dermatitis would be a welcome addition to the available therapies, and investigators have made a wide variety of attempts. A limited and transient degree of hyposensitization was obtained by Chase[51] when feeding DNCB-contact sensitized guinea pigs with the allergen, whereas for achieving persistent chromium-unresponsiveness in pre-sensitized animals, Polak and Turk[52] needed a harsh protocol involving up to lethal dosages of the allergen. Unfortunately, results from clinical studies on hyposensitization were in line with these experimental data. Therapeutic protocols involving ingestion of poison ivy allergen or penicillin were of only transient benefit to the patients.[50,53] The mechanism underlying specific desensitization in allergic contact dermatitis depends primarily on direct interference of allergen with effector T-cell function, by blocking or downregulating T-cell receptors.[54] No active suppressor cells are involved, as the onset of desensitization does not require time. Moreover, T-cells can be "desensitized" *in vitro* in the absence of putative suppressor cells.[55] Interestingly, the latter study also pointed towards class II bearing non-LCs (here, keratinocytes) as being most effective in rendering allergen-specific effector cells refractive to further effector function.

TABLE 10-2
Effects of Various Routes of Allergen Administration
on the Clinical Outcome[1]

Allergen Routes	Nonsensitized Individuals		Hypersensitive Individuals	
	Sensitization	Tolerance	ACD	Flare-ups/Desensitization
Skin	++[2]	—	++	(+)
Intravenous, oral[3]	—	++	—	++

[1] Clinical outcome may, depending on the immune status (nonsensitized vs. hypersensitive), vary from sensitization or tolerance induction, to allergic skin reactions, flare-ups, and/or desensitization.

[2] In this table, effects of different doses of allergen, and frequencies of allergenic contacts, have been omitted. Generally, higher doses and frequencies favor involvement of suppressor cell-mediated (tolerance) and flare-up/desensitization phenomena, also with the skin route.

[3] These tolerogenic routes may include skin deficient in, or depleted of, functionally active Langerhans' cells.

Thus, clonal anergy in desensitization may ensue when allergen is presented to T-cells in the absence of appropriate "second"or "costimulatory" signals that promote cell division. A major problem with desensitization *in vivo* may lie in the rapid replacement of peripherally inactivated effector cells from the relatively protected lymphoid organs.[54] Also, oral administration of allergens to individuals with existing allergic contact dermatitis often induces flare-up reactions of previously affected skin sites.

Upon repeated administration of allergens to sensitized individuals suppressor cells may, however, still develop and contribute to desensitization. It was found that local desensitization by repeatedly applying allergen at the same skin site did not result from local skin hardening or LC inactivation, as local reactivity to a nonrelated allergen at the site was unimpaired.[56] The role of suppressor cells in such local desensitization was further supported by the persistence of a cellular infiltrate at the site. Upon discontinuation of allergen exposure, local unresponsiveness rapidly dissolved (within 1 week). Collectively, these data illustrate the problems encountered in attempting to eradicate established effector cell function.

IV. SUMMARY AND CONCLUSIONS

Extensive research has recently led to the unravelling of the basic immunological mechanisms of ACH. Major cell types and mediators involved can be identified. How T-cells specifically recognize distinct allergens, and how these and other inflammatory cells interact to generate inflammation, has begun to be understood. The contrasting effects of various routes of allergen administration on the clinical outcome (see Table 10-2) are being explained. This rapid progress sharply contrasts with the slow motion in unravelling the regulatory mechanisms in cell-mediated immunity including ACH. Putative suppressor cell actions are still heavily disputed. The poor understanding of regulatory mechanisms in ACH hampers further therapeutic progress. So far, no methods of permanent desensitization have been devised.

Nevertheless, recently defined cellular interaction molecules and mediators provide promising targets for anti-inflammatory drugs, some of which have already entered clinical trials. Obviously, drugs found to be effective in preventing severe T-cell-mediated conditions, e.g., rejection of a vital organ graft, should be very safe before their use in ACH would seem appropriate. To date, prudence favors any measure to prevent ACH, be it through legal actions to outlaw the use of certain materials, or through avoiding personal contact with these materials. In the meantime, for difficult-to-avoid allergens, further studies on the potential value of tolerogenic treatment prior to possible sensitization seem warranted.

REFERENCES

1. **Wayne Streilein, J., Grammer, S. F., Yoshikawa, T., Demidem, A., and Vermeer, M.,** Functional dichotomy between Langerhans cells that present antigen to naive and to memory/effector T lymphocytes, *Immunol. Rev.,* 117, 159, 1990.
2. **Bergstresser, P. R.,** Sensitization and elicitation of inflammation in contact dermatitis, in *Immune Mechanisms in Cutaneous Disease*, Norris, D. A., Ed., Marcel Dekker, New York, 1989, 219.
3. **Turk, J. L.,** *Delayed Hypersensitivity*, 2nd ed., North-Holland Publishing Company, Amsterdam, 1975.
4. **Gell, P. D. H., Coombs, R. R. A., and Lachman, R.,** *Clinical Aspects of Immunology*, 3rd ed., Blackwell, London, 1975.
5. **Polak, L.,** Immunological aspects of contact sensitivity. An experimental study, *Monogr. Allergy*, 15, 1980.
6. **von Blomberg, B. M. E. Bruynzeel, D. P., and Scheper, R. J.,** Advances in mechanisms of allergic contact dermatitis: in vivo and in vitro research, in *Dermatotoxicology*, 4th ed., Marzulli, F. N. and Maibach, H. I., Eds., Hemisphere Publishing, Washington, 1991, p. 255.
7. **Stingl, G., Katz, S. I., Clement, L., et al.,** Immunological functions of Ia-bearing epidermal Langerhans cells, *J. Immunol.*, 121, 2005, 1978.
8. **Czernielewski, J. M. and Demarchez, M.,** Further evidence for the self-reproducing capacity of Langerhans cells in human skin, *J. Invest. Dermatol.*, 88, 17, 1987.
9. **Breathnach, S. M.,** The Langerhans cell. Centenary review, *Br. J. Dermatol.*, 119, 463, 1988.
10. **Toews, G., Bergstresser, P., Streilein, J., et al.,** Epidermal Langerhans cell density determines whether contact hypersensitivity or unresponsiveness follows skin painting with DNCB, *J. Immunol.*, 124, 445, 1980.
11. **Halliday, G. M. and Muller, H. K.,** Induction of tolerance via skin depleted of Langerhans cells by a chemical carcinogen, *Cell Immunol.*, 99, 220, 1986.
12. **Marchal, G., Seman, M., Milon, M., et al.,** Local adoptive transfer of skin delayed type hypersensitivity initiated by a single T Lymphocyte, *J. Immunol.*, 129, 954, 1982.

13. **Schuler, G. and Steinman, R. M.,** Murine epidermal Langerhans cells mature into potent immuno-stimulatory dendritic cells in vitro, *J. Exp. Med.,* 161, 526, 1985.

14. **Shimada, S., Caughman, S. W., Sharrow, S. O., et al.,** Enhanced antigen-presenting capacity of cultured Langerhans cells is associated with markedly increased expression of Ia antigen, *J. Immunol.,* 139, 2551, 1987.

15. **Inaba, K. and Steinman, R. M.,** Accessory cell-T lymphocyte interactions. Antigen-dependent and -independent clustering, *J. Exp. Med.,* 163, 247, 1986.

16. **Braathen, L. R. and Thorsby, E.,** Human epidermal Langerhans cells are more potent than blood monocytes in inducing some antigen-specific T cell-responses, *Br. J. Dermatol.,* 108, 139, 1983.

17. **Res, P., Kapsenberg, M. L., Bos, J. D., et al.,** The crucial role of human dendritic antigen-presenting cell subsets in nickel-specific T cell proliferation, *J. Invest. Dermatol.,* 88, 550, 1987.

18. **Cresswell, P.,** Antigen recognition by T lymphocytes, *Immunol. Today,* 8, 67, 1987.

19. **Claverie, J. M., Prochnicka-Chalufour, A., and Bougueleret, L.,** Implications of a Fab-like structure for the T cell receptor, *Immunol. Today,* 10, 10, 1989.

20. **Shimada, S. and Katz, S. I.,** TNP-specific Lyt-2⁺ cytolytic T cell clones preferentially respond to TNP-conjugated epidermal cells, *J. Immunol.,* 135, 1558, 1985.

21. **Sanders, M. E., Makgoba, M. W., and Shaw, S.,** Human naive and memory T cells: reinterpretation of helper-inducer and suppressor-inducer subsets, *Immunol. Today,* 9, 195, 1988.

22. **Dohlsten, M., Hedlund, G., Sjogren, H., et al.,** Two subsets of human CD4⁺ T helper cells differing in kinetics and capacities to produce interleukin 2 and interferon-gamma can be defined by the Leu 18 and UCHL1 monoclonal antibodies, *Eur. J. Immunol.,* 18, 1173, 1988.

23. **Bos, J. D., Zonneveld, I., Das, P. K., et al.,** The skin immune system (SIS): distribution and immunophenotype of lymphocyte subpopulations in normal human skin, *J. Invest. Dermatol.,* 88, 569, 1987.

24. **von Blomberg, B. M. E., van der Burg, C. K. H., Pos, O., et al.,** In vitro studies in nickel allergy: diagnostic value of a dual parameter analysis, *J. Invest. Dermatol.,* 88, 362, 1987.

25. **Hirschberg, H., Braathen, L. R., and Thorsby, E.,** Antigen presentation by vascular endothelial cells and epidermal Langerhans cells: the role of HLA-DR, *Immunol. Rev.,* 65, 57, 1982.

26. **Sinigaglia, F., Scheidegger, D., Garotta, G., et al.,** Isolation and characterization of Ni-specific T cell clones from patients with Ni-contact dermatitis, *J. Immunol.,* 135, 3929, 1985.

27. **Breitmeyer, J. B.,** Lymphocyte activation. How T cells communicate, *Nature,* 329, 760, 1987.

28. **Bierer, B. E. and Burakoff, S. J.,** T cell adhesion molecules, *FASEB J.,* 2, 2584, 1988.

29. **Issekutz, T. B., Stoltz, J. M., and von der Meide, P.,** Lymphocyte recruitment in delayed hypersensitivity; the role of interferon-gamma, *J. Immunol.,* 140, 2989, 1988.

30. **Messadi, D. V., Pober, J. S., Fiers, W., et al.,** Induction af an activation antigen on postcapillary venular endothelium in human skin organ culture, *J. Immunol.,* 139, 1557, 1987.

31. **Haskard, D. O., Cavender, D., Fleck, R. M., et al.,** Human dermal microvascular endothelial cells behave like umbilical vein endothelial cells in T cell adhesion studies, *J. Invest. Dermatol.,* 88, 340 1987.

32. **Dustin, M. L., Rothlein, R., Bhan, A. K., et al.,** Induction by IL-1 and interferon-gamma: tissue distribution, biochemistry, and function of a natural adherence molucule (ICAM-1), *J. Immunol.,* 137, 245, 1986.

33. **Hamann, A., Jablonski-Westrich, D., Scholz, K. W., et al.,** Regulation of lymphocyte homing. I. Alterations in homing receptor expression and organ-specific high endothelial venule binding of lymphocytes upon activation, *J. Immunol.,* 140, 737, 1988.

34. **Scheper, R. J. and von Blomberg, B. M. E.,** Allergic contact dermatitis: T cell receptors and migration, in *Current Topics in Contact Dermatitis*, Frosch, P. J., Dooms-Goossens, A., Lachapelle, J. M., Rycroft, R. J. G., and Scheper, R. J., Eds., Springer Verlag, Heidelberg, 1989, 12.

35. **Pals, S. T., Horst, E., Scheper, R. J., et al.,** Mechanisms of human lymphocyte migration and their role in the pathogenesis of disease, *Immunol. Rev.*, 108, 111, 1989.

36. **Duijvesteijn, A. and Hamann, A.,** Mechanisms and regulation of lymphocyte migration, *Immunol. Today*, 10, 23, 1989.

37. **Silvennoinen-Kassinen, S., Jakkula, H., and Karvonen, J.,** Helper T cells carry the specificity of nickel sensitivity reaction in vitro in humans, *J. Invest. Dermatol.*, 86, 18, 1986.

38. **Scheper, R. J., von Blomberg, B. M. E., Velzen, D., et al.,** Effects of contact sensitization and delayed hyper-sensitivity reactions on immune responses to non-related antigens. Modulation of immune responses, *Int. Arch. Allergy Appl. Immunol.*, 50, 243, 1976.

39. **Camarasa, J. G., Serra-Baldrich, E., Lluch, M., et al.,** Recent unexplained patch test reactions to palladium, *Contact Dermatitis*, 20, 388, 1989.

40. **Skog, E.,** Spontaneous flare-up reactions induced by different amounts of 1,3-dinitro-4-chlorobenzene, *Acta Derm.-Venereol.*, 46, 386, 1966.

41. **Barker, J. N. W. N., Mitra, R. S., Griffiths, C. E. M., Dixit, V. M., and Nickoloff, B. J.,** Keratinocytes as initiators of inflammation, *Lancet*, 337, 211, 1991.

42. **Enk, A. H. and Katz, S. I.,** Early molecular events in the induction phase of contact sensitivity, *Proc. Natl. Acad. Sci. U.S.A.*, 1398, 1992.

43. **Scheper, R. J., van Dinther-Janssen, A. C. H. M., and Polak, L.,** Specific accumulation of hapten reactive T cells in contact sensitivity reaction sites, *J. Immunol.*, 134, 1333, 1985.

44. **Miller, S. D., Sy, M.-S., and Claman, H. N.,** The induction of hapten-specific T cell tolerance using hapten-modified lymphoid membranes. II. Relative roles of suppressor T cells and clone inhibition in the tolerant state, *Eur. J. Immunol.*, 7, 165, 1977.

45. **Mowat, A.,** The regulation of immune responses to dietary protein antigens, *Immunol. Today*, 8, 93, 1987.

46. **van Hoogstraten, I. M. W., Andersen, J. E., von Blomberg, B. M. E., et al.,** Preliminary results of a multicenter study on the incidence of of nickel allergy in relationship to previous oral and cutaneous contacts, in *Current Topics in Contact Dermatitis*, Frosch, P. J., Dooms-Goossens, A., Lachapelle, J. M., Rycroft, R. J. G., and Scheper, R. J., Eds., Springer Verlag, Heidelberg, 1989, 178.

47. **Semma, M. and Sagami, S.,** Induction of suppressor T cells to DNFB contact sensitivity by application of sensitizer through Langerhans cell-deficient skin, *Arch. Dermatol. Res.*, 271, 361, 1981.

48. **Elmets, C. A., Bergstresser, P. R., Tigelaar, R. E., et al.,** Analysis of the mechanism of unresponsiveness produced by haptens painted on skin exposed to ultraviolet radiation, *J. Exp. Med.*, 158, 781, 1983.

49. **Zembala, M. and Asherson, G. L.,** Depression of T cell phenomenon of contact sensitivity by T cells from unresponsive mice, *Nature*, 244, 227, 1973.

50. **Epstein, W. L.,** The poison ivy picker of Pennypack Park: the continuing saga of poison ivy, *J. Invest. Dermatol.*, 88, 7, 1987.

51. **Chase, M. W.,** Inhibition of experimental drug allergy by prior feeding of the sensitizing agent, *Proc. Soc. Exp. Biol. Med.*, 61, 257, 1946.

52. **Polak, L. and Turk, J. L.,** Studies on the effect of systemic administration of sensitizers in guinea pigs with contact sensitivity to inorganic metal compounds. I. The induction of immunological unresponsiveness in already sensitized animals, *Clin. Exp. Immunol.*, 3, 245, 1968.

53. **Wendel, G. D., Stark, B. J., Jamison, R. B., et al.,** Penicillin allergy and desensitization in serious infections during pregnancy, *N. Engl. J. Med.*, 312, 1229, 1985.

54. **Polak, L. and Rinck, C.,** Mechanism of desensitization in DNCB-contact sensitive guinea pigs, *J. Invest. Dermatol.*, 70, 98, 1978.

55. **Gaspari, A. A., Jenkins, M. K., and Katz, S. I.,** Class II MCH-bearing keratinocytes induce antigen specific unresponsiveness in hapten-specific TH1 clones, *J. Immunol.*, 141, 2216, 1988.

56. **Boerrigter, G. H. and Scheper, R. J.,** Local and systemic desensitization induced by repeated epicutaneous hapten application, *J. Invest. Dermatol.*, 88, 3, 1987.

57. **Eynon, B. E. and Parker, D. C.,** Small B cells as antigen-presenting cells in the induction of tolerance to soluble protein antigens, *J. Exp. Med.*, 175, 131, 1992.

58. **Waal Malefyt, R. de, Yssel, H., Roncarolog, M. G., Spits, H., and Vries, J. E. de,** Interleukin-10, *Curr. Opinion Immunol.*, 4, 314, 1992.

59. **Romagnani, S.,** Induction of Th1 and Th2 responses: a key role for the 'natural' immune response? *Immun. Today*, 13(10) 379, 1992.

60. **Bloom, B. R., Salgame, P., and Diamons, B.,** Revisiting and revising suppressor T cells, *Immunol. Today*, 13(4) 131, 1992.

61. **Hsieh, C. S., Macatonia, S. E., Tripp, C. S., Wolf, S., O'Garra, A., and Murphy, K. M.,** Development of Th1 CD4+ T cells through IL-12 produced by Listeria-induced macrophages, *Science*, 260, 547, 1993.

62. **Lehner, T., Bergmeier, L.A., Panagiotidi, C., Tao, L., Brookes, R., Klavinskis, L. S., Walker, P., Ward, R. G., Hussain, L., Gearing, A. J. H., and Adams, S. E.,** Induction of mucosal and systemic immunity to a recombinant simian immunodeficiency viral protein, *Science*, 258, 1365, 1992.

63. **Scott, P.,** IL-12: Initiation cytokine for cell-mediated immunity, *Science*, 260, 496, 1993.

64. **Kuchroo, V. K., Byrne, M. C., Atsumi, Y., Greenfeld, E., Connol, J. B., Whitters, M. J., O'Hara, R. M., Collins, M., and Dorf, M. E.,** T cell receptor alpha chain plays a critical role in antigen-specific suppressor cell function, *Proc. Natl. Acad. Sci. U.S.A.*, 8700, 1991.

65. **Gautam, S. C., Chikkala, N. F., and Hamilton, T. A.,** Anti-inflammatory action of IL-4. Negative regulation of contact sensitivity to trinitrochlorobenzene, *J. Immunol.*, 148, 1411, 1992.

66. **Fiorentino, D. F., Zlotnik, A., Mosmann, T. R., Howard, M., and O'Garra, A.,** IL-10 inhibits cytokine production by activated macrophages, *J. Immunol.*, 147, 3815, 1991.

67. **Mosmann, T. R. and Coffmann, R. L.,** Th1 and Th2 cells: different patterns of lymphokine secretion lead to different functional properties, *Ann. Rev. Immunol.*, 7, 145, 1989.

68. **Cher, D. J. and Mossmann, T. R.,** Two types of murine helper T cell clone. Delayed type hypersensitivity is mediated by Th1 clones, *J. Immunol.*, 138, 3688, 1987.

69. **Hoogstraten, I. M. W. van, von Blomberg, B. M. E., Boden, D., Kraal, G., and Scheper R. J.,** Non-sensitizing epicutaneous skin contacts prevent subsequent induction of immune tolerance, *J. Invest. Dermatol.*, 102, 80, 1994.

70. **White, I. R.,** Phototoxic and photoallergic reactions, in *Textbook of Contact Dermatitis*, Rycroft, R. J. G., Menné, T., Frosch, P. J., and Benezra, C., Eds., Springer-Verlag, Berlin, 1992, 75.

71. **Scheper, R. J., von Blomberg, B. M. E., Boerrigter, G. H., et al.,** Induction of local memory in the skin. Role of local T cell retention, *Clin. Exp. Immunol.*, 51, 141, 1983.

72. **Yamashita, N., Natsuaki, M., and Sagami, S.,** Flare-up reaction on murine contact hypersensitivity. I. Description of an experimental model: rechallenge system, *Immunology*, 67, 365, 1989.

73. **McLelland, J., Shuster, S., and Matthews, J. N. S.,** 'Irritants' increase the response to an allergen in allergic contact dermatitis, *Arch. Dermatol.*, 127, 1016, 1991.

74. **Pistoor, F. H. M., Kapsenberg, M. L., Bos, J. D., Meinardi, M. M. H. M., von Blomberg B. M. E., Scheper, R. J.,** Cross-reactivity of human nickel-reactive T lymphocyte clones with copper and palladium, *J. Invest. Dermatol.*, 105, 92, 1995.

Chapter 11

CLINICAL AND EPIDEMIOLOGICAL ASPECTS OF HYPERSENSITIVITY REACTIONS OF THE SKIN

Johannes Ring

CONTENTS

I. INTRODUCTION

The skin as an organ represents the largest border area between the individual organism and the environment. Therefore, it is not astonishing that most — in fact almost all — known hypersensitivity reactions can also involve the skin and give rise to characteristic diseases.[1-4] These diseases differ according to the skin compartment involved, the eliciting agent, and the pathophysiologic mechanisms (Figure 11-1).

Consequently, they can be classified either according to the morphology (macroscopic or microscopic) (e.g., eczematous dermatitis, wheal and flare reaction, etc.), the eliciting agent (e.g., drug eruption, food allergy, etc.), the route of application (e.g., contact urticaria, systemically-induced contact dermatitis), or the pathomechanism involved (e.g., IgE-medicated reaction, immune–complex reaction, cellular hypersensitivity, etc.).

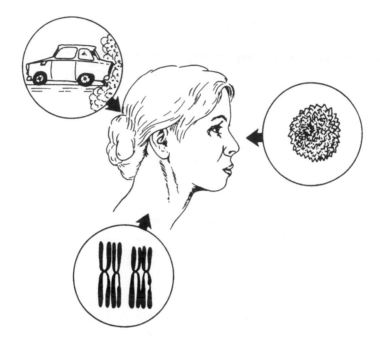

FIGURE 11-1. Three factors determining the development of allergic sensitization: genetic background, allergen exposure, and pollutants as adjuvants.

In this chapter, only the more frequent hypersensitivity diseases of the skin will be discussed with special focus on hypersensitivity reactions to chemicals. Furthermore, only hypersensitivity reactions against exogenous substances will be considered; this means the autoallergic mechanisms and autoimmune diseases are not part of this consideration.

There are several ways by which environmental substances can affect the organism, in this case, the skin, leading to disease. It is crucial to differentiate between pharmacologic/toxic effects of a given substance and hypersensitivity reactions of a certain individual (Table 11-1). This effect on the organism can cover a wide range between "impairment of feeling well," "irritation," and "actual disease."

"Hypersensitivity" is sometimes — not very precisely — equated with "allergy." However, there are a variety of hypersensitivity reactions that are pathophysiologically different from allergy (see the Glossary).

Besides allergy, there are other mechanisms of individual hypersensitivity, like pharmacological *intolerance*, when a well-known pharmacologic or toxic effect of a substance is observed at normal dose levels.

When hypersensitivity reactions occur without obvious relation to the pharmacologic effect and without involvement of the immune system, they are called *idiosyncrasy*. If the clinical symptoms mimic those of classic allergic diseases (e.g., urticaria) one also uses the term "pseudo-allergy."[5,6]

TABLE 11-1

Classification of Various Forms of Drug-Induced Adverse Reactions

Adverse Reaction

Toxicity	Hypersensitivity		
	Pharmacological intolerance	Idiosyncrasy/pseudo-allergy	Allergy
		Direct mediator release, Direct activation of biological systems, Enzyme defect Underlying disease	Immune reactions (type I–VI)
Pharmacological effect, Organ toxicity		(e.g., psychic) Allergy-like symptoms	Allergic disease

Among the allergic hypersensitivities, different types of immunopathologic reactions can be distinguished; the most commonly used classification was proposed by Coombs and Gell[2] with four types of reactions.

Type I: IgE-mediated reaction (e.g., anaphylaxis, urticaria, atopic eczema?)

Type II: Cytotoxic reaction (e.g., allergic thrombocytopenic purpura)

Type III: Immune complex reaction (e.g., allergic leukocytoclastic vasculitis, serum sickness)

Type IV: Cellular hypersensitivity (e.g., allergic contact dermatitis, exanthematous drug eruption?)

Additional types V and VI have been suggested:[4]

Type V: Granulomatous reaction (e.g., zirconium granuloma, collagen granuloma)

Type VI: Stimulating/neutralizing hypersensitivity (e.g., myasthenia gravis)

In the following section, the more frequent allergic skin diseases will be briefly discussed. However, it has to be kept in mind that the clinical morphology of all of these diseases can also be elicited by nonimmunological mechanisms in the sense of a "pseudo-allergic" or toxic reaction.

II. CLINICAL ASPECTS OF ALLERGIC SKIN DISEASE

A. URTICARIA AND ANGIOEDEMA

The primary lesion of urticaria is a wheal, defined as circumscribed white-to pink-colored compressible skin elevation produced by dermal edema. Urticaria means the exanthematic eruption of wheals surrounded by more or less pronounced erythematous flares. Characteristically, urticarial lesions are associated with intense pruritus which rarely leads to erosive scratching, but rather to superficial rubbing of the skin, leaving the epidermis intact.

If the edema formation involves deeper layers of the dermis or subcutaneous tissue, it is known as angioedema ("Quincke-edema," angioneurotic edema) which can occur either in combination with urticaria or alone.

In the etiopathophysiology, many different elicitors have to be considered. Allergy is only one mechanism. Toxic effects (e.g., nonimmunological contact urticaria) or pseudo-allergic reactions play a major role.[7,8]

A wide variety of physical stimuli can induce urticaria like cold, heat, pressure, etc. Most of the cases of allergic urticaria are mediated by IgE-antibodies against ubiquitous allergens in the food or in the air. However, urticaria can also be elicited by immune–complex reactions and represents the major symptom of serum sickness.

Furthermore, urticaria often is the earliest or most obvious symptom of anaphylaxis and anaphylactoid reactions.[3,9,10]

According to the clinical course, urticaria can be classified into an acute form (this disappears within a period of 6 weeks) or chronic urticaria (relapses of wheals exceeding a period of 6 weeks).[1,4] Among the forms of chronic urticaria, an intermittent type (longer symptom-free intervals) and a chronic, continuous type (daily eruptions of wheals) can be distinguished. Among the cases of angioedema, the hereditary angioedema has to be differentiated.[11] These patients lack the C1-inactivator, this nonallergic enzyme deficiency may be either structural (lack of the protein) or functional (loss of function) and does not respond to classical antiallergic therapy, but to infusion of concentrated C1-inactivator preparations.

B. ANAPHYLAXIS AND ANAPHYLACTOID REACTIONS

Anaphylaxis represents the maximal variant of an allergic immediate-type reaction involving the whole organism. Generally, the allergen contact takes place systemically (injection or ingestion); however, it also occurs after skin contact; cases of contact urticaria and contact anaphylaxis have been observed.[12,13]

Anaphylaxis is defined as an allergic reaction (i.e., an immunologically mediated reaction). However, in the daily routine, often the demonstration of an immunological sensitization is not readily possible, especially under emergency conditions. Therefore, the term "anaphylactoid reaction" has been used in order to describe all clinical reaction patterns mimicking anaphylaxis without describing pathophysiological mechanisms.[4,6,9] Some authors restrict the term "anaphylactoid" to nonimmunological, i.e., pseudo-allergic reactions.[5] Later on in the laboratory, anaphylactoid reactions can be classified into anaphylactic reactions (immunologically-mediated) or pseudo-allergic anaphylactoid reactions.

The clinical symptoms of anaphylaxis often start with skin symptoms (itch, flush, urticaria, angioedema) and can comprise respiratory, gastrointestinal,

TABLE 11-2
Grading of Anaphylactic/Anaphylactoid Reactions
According to Severity of Clinical Symptoms

	Symptoms			
Grade	Skin	Abdomen	Respiratory Tract	Cardiovascular System
I	Pruritus Flush Urticaria Angioedema	—	—	—
II	Pruritus Flush Urticaria Angioedema (not mandatory)	Nausea Cramping	Rhinorrhea Hoarseness Dyspnea	Tachycardia (Δ > 20 beats/min) RR change (Δ > 20 mmHg systolic) Arrhythmia
III	Pruritus Flush Urticaria Angioedema (not mandatory)	Vomiting Defecation Diarrhea	Laryngeal edema Bronchospasm Cyanosis	Shock
IV	Pruritus Flush Urticaria Angioedema (not mandatory)	Vomiting Defecation Diarrhea	Respiratory arrest	Cardiac arrest

Modified from Ring, J. and Meßmer, K., *Lancet*, 1, 466, 1977.

and cardiovascular symptoms; according to the intensity of the clinical symptomatology, a severity scale has been proposed (Table 11-2).[9]

The most frequent elicitors of anaphylactoid reactions are listed in Table 11-3.

C. ATOPIC DISEASES AND ATOPIC ECZEMA

Atopic eczema, together with allergic rhinoconjunctivitis (most commonly represented as hay fever) and allergic bronchial asthma, constitute the classical triad of atopic diseases.[14-17]

Genetic studies have shown clearly that these three atopic diseases are closely connected within families.[15,18]

The atopic skin disease is known under various synonyms (atopic dermatitis, neurodermitis constitutionalis, Prurigo Besnier, and endogenous eczema). The term "atopic eczema" seems to us the most modest, since it only describes eczema occurring in atopic individuals and does not imply mechanisms (e.g., neurological) or routes of elicitation (e.g., endogenous). In the U.S., the term "atopic dermatitis" is prevalent. This leaves open the debate regarding the

TABLE 11-3
Eliciting Agents of Anaphylactoid Reactions

Drugs
Additives in drugs and foods

Additives	Examples
Depot substances	In penicillins
Micelle formers	Cremophor EL in i.m. injections
Sulfites	Injections, sprays
Protein stabilizers	Caprylate in HSA
Benzyl alcohol	Injections, sterile H_2O or NaCl
Parabens	Local anesthetics
Colors	Tablets
Acetate	Dialysis

Occupational substances (e.g., latex)
Animal venoms
Aeroallergens
Contact urticariogens
Physical agents (cold, heat, UV irradiation)
C_1 inactivator deficiency
Exercise
Seminal fluid
Echinococcal cyst
"Summation-anaphylaxis"
Idiopathic(?)

difference between eczema and dermatitis, two terms not very well defined in the dermatological literature. However, dermatitis (meaning inflammation of the skin) is the broader term, while eczema describes a certain type of dermatitis.

Eczema can be defined as "noncontagious epidermodermitis with typical clinical (itch, erythema, papule, seropapule, vesicle, squames, crusts, lichenification) and dermatohistological findings (spongiosis, acanthosis, parakeratosis, lymphocytic infiltration) mostly due to a hypersensitivity."[1,4,15]

Atopic eczema often starts in childhood, sometimes in the third month of life and characteristically involves different areas in different age groups. In infants, the face, head, and extensor surfaces are predominantly involved; later on the large flexures, hands, and neck are involved. The eczema tends to become more and more dry in aspect and the skin shows lichenification; adult patients often develop excoriated nodules, the so-called pruriginous variant of atopic eczema. The atopic eczema goes along with so-called stigmata and minimal variants representing no or only a slight illness.

The diagnosis of atopic eczema can be made according to a number of criteria.[17] Table 11-4 shows the most important diagnostic criteria according to our experience. If four of these six criteria are positive, the diagnosis of atopic eczema can be made.

While the pathomechanisms of atopic respiratory diseases are rather well established, the role of allergic reactions in the elicitation and maintenance of atopic eczema is still under discussion. Several concepts try to explain the

TABLE 11-4
Atopic Eczema: Diagnostic Criteria

Morphology of eczematous skin lesions (age-related)
Itch
Localization of lesions (age-related)
Stigmata of atopy
Personal and/or family history of atopy
IgE-mediated sensitizations

complex etiopathophysiology of this disease. The strong genetic influence has been mentioned. Twin studies have shown a significantly elevated rate of concordance (70% in homozygous vs. 30% in heterozygous twins).

Increased IgE production is one of the hallmarks of atopic disease. Yet, the simple equation "atopic = IgE" is incorrect. Atopy is only one of many conditions leading to increased IgE production. On the other hand, atopy is more than IgE, since it also comprises altered nonspecific reactivity of skin and mucous membranes.[4,15]

According to the author, atopy can be defined as "familial tendency to develop certain diseases (allergic rhinoconjunctivitis, extrinsic bronchial asthma, atopic eczema) on the basis of hypersensitivity of the skin and mucous membranes against environmental substances, usually associated with increased IgE production and/or altered nonspecific reactivity."[4,15] Note that this definition is different from that given in the Glossary in that it also comprises the dimension of "altered reactivity!"

As in respiratory atopic diseases, not only IgE-mediated allergic reactions can elicit the clinical condition. The so-called dry or irritable skin is a major characteristic of atopic eczema. This implies not only increased transepidermal water loss, but also increased roughness, as well as a decrease in epidermal lipid content sometimes combined with decreased sebaceous activity.

The role of allergic reactions in atopic eczema has focused on IgE-mediated sensitization. The role of other types of allergic reactions (type II to VI) is less well studied. There is a vast amount of literature about a decreased tendency for patients with atopic eczema to develop contact allergy. However, most of these studies are retrospective and without adequate controls. We found no overall differences in frequency of contact sensitization between atopics and nonatopics.[19]

It also has been shown convincingly that food allergy plays a role in atopic eczema in oral provocation tests under double-blind conditions.[20] Chemical food additives also can provoke eczematous skin lesions in a placebo-controlled oral provocation test.[21,22]

The markedly elevated IgE-antibody production has been regarded by some authors as a mere epiphenomenon relevant for respiratory symptoms, but not for the skin lesions in former decades. Recently, there is increasing evidence that IgE-mediated sensitization also plays a decisive role in the pathogenesis of the eczematous skin lesions. This has become clear through investigations regarding the role of dendritic epidermal Langerhans cells which have been found to carry IgE, especially in patients with severe disease.[23,24] The nature of the IgE binding site on the Langerhans cells is the focus of current research. It is well known that Langerhans cells can be induced to express the low affinity IgE-receptor (CD23) (FCεR2) when incubated with interleukin 4 and interferon γ (in a synergistic manner).[24]

Recently, the high affinity IgE receptor (FCεR1) has been detected on human epidermal Langerhans cells.[25] This receptor had been regarded only to be present on mast cells and basophil leukocytes.

This unexpected new finding opens a new dimension of allergy research and possible new ways of understanding the pathophysiology of atopic eczema. Clinically, it is important that eczematous skin lesions can be provoked by epidermal application of allergens known to induce IgE responses (e.g., house dust mite, animal epithelia, etc.). This procedure has been called "atopy patch test."[26,27]

The clinical variant of so-called "protein dermatitis" was observed by Hjorth and Roed-Petersen 20 years ago and represents a subgroup of patients with atopic eczema developing the skin lesions after epidermal contact with protein allergens.[28]

It is well known that irritants (e.g., detergents, cigarette smoke, etc.) including chemicals, can provoke skin lesions in these patients.[4,15] As with other atopic diseases, one can classify atopic eczema, according to the relative role of immunological sensitization on one hand, and nonspecific altered reactivity on the other hand, into an "extrinsic" and an "intrinsic" form.[15,16]

D. ALLERGIC CONTACT DERMATITIS

Allergic contact dermatitis or allergic contact eczema is clinically characterized by the symptoms of eczema as defined above. In the literature, the terms dermatitis and eczema are sometimes used interchangeably; some authors prefer "dermatitis" for the more acute forms and reserve the term "eczema" for the more chronic form of this disease.

According to clinical appearance, specific subgroups can be described as so-called "dishydrotic eczema" (pompholyx) with characteristic blistering of palms and soles. Sometimes this form changes after years into hyperkeratotic-fissuring eczema. According to the preferred localization, one can classify allergic contact eczema of the hand, lower leg, etc. In the acute phase, the skin lesions are more wet and vesiculous; in the more chronic phase lichenification becomes prominent.

Allergic contact dermatitis usually starts at the sites of allergen contact; however, it characteristically spreads over these borders in the development of skin lesions. The lesions at the beginning are edematous, papulous, and later on, become vesicular-bullous. In some cases oozing can occur. Usually, allergic contact dermatitis goes along with intense pruritus. On the fingertips or

TABLE 11-5
Frequency of Positive Patch Test Reactions in Male (n = 4775) and Female (n = 7202) Patients of the Munich Dermatology Department

Male	Frequency	Female	Frequency
Fragrance mix	8.8%	Nickel sulfate	13.7%
Peru balsam	5.7%	Fragrance mix	9.0%
Potassium dichromate	5.1%	Peru balsam	6.8%
Cainemix	4.4%	Cobalt chloride	5.9%
p-Phenylendiamine	4.3%	Lanolin alcohol	4.7%
Lanolin alcohol	3.6%	Cainemix	4.1%
Benzocain	3.3%	p-Phenylendiamine	3.9%
Formaldehyde	3.1%	Formaldehyde	3.8%
Cobalt chloride	3.0%	Potassium dichromate	3.7%
Neomycin sulfate	3.0%	Neomycin sulfate	3.4%
Clioquinol	2.9%	Kolophonium	2.9%
Nickel sulfate	2.6%	Eucerin anhydr.	2.9%
Kolophonium	2.5%	Benzocain	2.8%
Mafenid	2.4%	Thiuram mix	2.7%
Thiuram mix	2.4%	Clioquinol	2.4%
Eucerin anhydr.	2.2%	Parabens	2.3%
PPD-Mix	1.9%	Mafenid	1.8%
Parabens	1.5%	Sublimate	1.4%
Gentamycin	1.4%	Gentamycin	1.4%
Epoxiresins	1.4%	PPD-Mix	0.9%
Terpentinperoxide	1.1%	Terpentinperoxide	0.9%
Sublimate	1.0%	Phenylmercuriborate	0.6%
Phenylmercuriborate	0.7%	Epoxiresins	0.6%
Mercaptomix	0.7%	Mercaptomix	0.5%
Ethylendiamine	0.6%	Ethylendiamine	0.4%
Naphthylmix	0.3%	Vaseline	0.3%
Vaseline	0.2%	Naphthylmix	0.2%

Adapted from Enders, F., Przybilla, B., Ring, J., Burg, G., and Braun-Falco, O., *Hautarzt,* 39, 779, 1988.

soles, sometimes hyperkeratosis and fissuring is the single symptom of allergic contact dermatitis, in this case accompanied by pain.

There are unusual variants of morphology of allergic contact dermatitis elicited by special allergens, for instance, erythema exsudativum muliforme-like lesions.

The pathophysiology of allergic contact dermatitis involves Langerhans cells[29,30] in the epidermis as antigen-presenting cells, and T-lymphocytes (predominantly Th1 cells) as effector cells[31,32] and is covered in Chapter 10.[33] Among the elicitors of allergic contact dermatitis, many substances of the daily life both from occupational and private exposure, have to be considered. Table 11-5 shows the frequency of positive patch test reactions against contact allergens in men and women from our own experience.[34]

TABLE 11-6
Classification of Photohypersensitivity Diseases

Photoallergic reactions
    Photoallergic contact dermatitis
    Photo-induced drug reactions
    Persistent light reaction
    (Actinic reticuloid)
Possibly photoallergic reactions
    Solar urticaria
    Polymorphous light eruption
Phototoxic reactions
    Acute and chronic light damage
    Phototoxic dermatitis
    Polymorphous light eruption (?)
    Photo-induced drug reaction
    Metabolic diseases (porphyria, Hartnup-syndrome, etc.)
    (questionable: Hydroa vacciniformia, Mallorca acne)
Light-provoked dermatoses (examples)
    Lupus erythematosus
    Dyskeratoisis follicularis (Darier)
    Porokeratosis, actinic type
    Lichen ruber
    Psoriasis
    Atopic eczema
    Herpes simplex

The selection of contact allergens used for standard tests is continuously changing according to the experiences of specialized national and international groups like the Deutsche Kontaktdermatitisgruppe, European Contact Dermatitis Research Group, North American Contact Dermatitis Research Group, and International Contact Dermatitis Research Group.[34-41] While in most cases allergic contact sensitization represents an individual reaction pattern of a minority of people exposed to an environmental substance, there are certain chemicals with a well-known strong sensitizing potential, i.e., they induce contact sensitization in more then 50% of individuals exposed. The classical example is poison ivy or dinitrochlorobenzene (DNCB). The question of how to measure the sensitizing potential of a certain chemical will be dealt with elsewhere in this book (see Chapter 13).

E. PHOTOSENSITIZATION

A subgroup of allergic contact dermatitis can be elicited by UV-irradiation, then called photoallergic reaction.[42] However, photosensitization is not limited to allergic contact dermatitis generally. There is also a phototoxic dermatitis as well as other phototoxic or photoallergic skin diseases. In Table 11-6 a list of possible UV-induced skin diseases both allergic or toxic or unknown in origin is given.

According to the opinion of many dermatologists, the prevalence of certain hypersensitivity reactions in connection with UV-irradiation, especially in the so-called polymorphous light eruption, seem to have increased in the last

TABLE 11-7
Photosensitizing Substances in the Natural and Cultural Environment

Plants
  Umbelliferæ
  (Heracleum, Angelica, Daucus carota [carrot], Ammi majus, Apium graveolens, Pastinaca)
  Rutaceæ
  Citrus bergamia (bergamot)
  Citrus sinensis (orange), Ruta graveolens, Dictamnus albus
  Moraceæ
  (Ficus carica)
  Leguminosæ
  Psoralea corylifolia
  Rosaceae
  Compositæ
Drugs and cosmetics
  Disinfectants (e.g., halogenated salicyl-anilin derivatives, hexachlorophen, chlorhexidin)
  Antimycotics (buclosamide)
  Chemotherapeutics (sulfonamides, tetracyclines, nalidixic acid)
  Sedatives (phenothiazine)
  Diuretics (hydrochlorothiazide)
  Non-steroidal anti-inflammatory drugs (e.g., benoxaprofen, carprofen)
  Antiarrhythmics (chinidin, amiodarone)
  Perfumes (musk-ambrette)
  Sun-protecting agents (e.g., paraaminobenzoic acid, benzophenone,
    isopropyldibenzoylmethan)
  Antidiabetics (sulfonylurea)

decade.[43] The most common photosensitizing agents are present in drugs or cosmetics (Table 11-7) or are of natural origin in plants, where they elicit the well-known dermatitis pratensis. Photosensitization is not uncommon against UV filter substances used in sunscreens.[44]

F. EXANTHEMATOUS ERUPTIONS

Adverse drug reactions frequently involve the skin.[46] An exanthematous drug eruption is the occurrence of disseminated skin lesions after systemic application of a drug. The clinical appearance of exanthematous drug eruptions can be quite variable (Table 11-8).

Classification of exanthematous drug eruptions according to eliciting agents and specific morphology has been attempted. This, however, is only possible in certain cases. We always have to keep in mind that any drug can elicit any kind of exanthematous eruption.[47-49]

The mechanism of exanthematous drug eruptions is not well established. Sensitized lymphocytes and positive lymphocyte transformation tests with eliciting drugs have been described. Therefore, they are commonly classified among type IV reactions.

A special and, for several reasons, very important form of exanthematous drug eruption is the drug-induced Lyell's syndrome or toxic epidermal necrolysis,

TABLE 11-8
Classification of Exanthematous Drug Eruptions
According to Morphology of Skin Lesions

Urticarial
Erythemato-vesicular (eczematous)
Maculo-papular
Exfoliative dermatitis
Purpuric, e.g.,
    Thrombocytopenia
    Vasculitis
    Purpura pigmentosa progressiva
Bullous e.g.,
    Erythema exsudativum multiforme
    Fixed drug eruption
    Lyell's syndrome (toxic epidermal necrolysis)
Lichenoid
Acneiforme
Psoriasiforme
Lymphohistiocytic infiltration

when the epidermis is shaded in big blisters as in a second degree burn. Sometimes the differentiation between severe types of erythema exsudativum multiforme with mucous membrane involvement (Stevens-Johnson syndrome) and toxic epidermal necrolysis is difficult. In spite of modern intensive therapy, the lethality of this condition is still around 30%! Among the drugs most commonly named as elicitors of drug-induced Lyell's syndrome (TEN) are antimicrobial drugs (e.g., sulfonamides), analgesics (e.g., pyrazolones), central nervous system acting drugs (e.g., barbiturates, phenytoin), allopurinol, and others.[50-53]

G. MISCELLANEOUS ALLERGIC SKIN DISEASES

Among other skin diseases possibly allergic in origin, several forms of purpura have to be mentioned.

Thrombocytopenic purpura can be induced by drug allergy of type II (cytotoxic reaction) and is a noninflammatory purpura.[2,4]
Purpura pigmentosa progressiva represents a variant of inflammatory purpura with lymphocytic infiltrates around superficial blood vessels and erythrocyte extravasation, and has been described after intake of food additives.[54]
Leukocytoclastic vasculitis is an immune–complex reaction and appears as hemorrhagic, necrotizing, or polymorphic-nodular variant.[55,56]

III. EPIDEMIOLOGY OF ALLERGIC SKIN DISEASES

There is no doubt that allergy is indeed an increasingly frequent diagnosis in clinical medicine. For a long time, however, it remained doubtful whether these observations really reflect an actual increase in prevalence of allergic diseases.

A critical evaluation has to take into account several factors:

The diagnosis "allergy" may be more fashionable than in former times.
Patients may be more informed about the possibility of allergies.
The improved diagnostic tools and better educated physicians are more ready to make the diagnosis of allergy.
Finally, the term "allergy" may be confused with any kind of other hypersensitivity or impairment of well being.

Therefore, an objective evaluation of a putative increase of allergic diseases can only be performed on the basis of repeated investigations of representative population samples using well-defined disease definitions and validated instruments.

A. EPIDEMIOLOGY OF ATOPY

So far, few studies have been performed where these criteria have been applied, mostly with regard to atopic diseases. These epidemiological studies showed an increase in prevalence of atopic diseases in the last decades.[57-60] This is particularly true for seasonal allergic rhinoconjunctivitis, where clear-cut increases in prevalence have been described in Switzerland, Japan, the U.K., and Sweden.[57-61] For a discussion on the prevalence of asthma and increased asthma mortality, see Chapters 4 and 5.

The situation is more difficult with regard to atopic eczema. However, there are a number of studies showing an increase in prevalence. The most impressive one was performed by Schulz-Larsen who examined a twin population in Denmark between 1960 and 1980 and found a rise in prevalence from 3% in 1960 to approximately 12% in 1980.[18] In an epidemiologic study performed between 1988 and 1991 in different parts of Bavaria (Germany), we found a prevalence of atopic eczema of approximately 10% in 5- to 6-year-old preschool children.[61]

In recent studies together with the Medical Institute for Environmental Hygiene in Düsseldorf, a comparison between different areas in east and west Germany was performed where we found rather high prevalence rates of atopic eczema both in east Germany and in several areas of Northrhine-Westphalia (ranging between 8 and 18%).[62,63]

B. EPIDEMIOLOGY OF URTICARIA AND ANGIOEDEMA

Epidemiological studies of urticaria focus mainly on prevalence rates among selected groups of patients in hospitals. There are few data about the prevalence of urticaria in the general population.[64]

The problem is further complicated by the fact that rarely is the diagnosis "urticaria" more precisely classified into etiopathophysiological categories or acute vs. chronic forms.

TABLE 11-9
Prevalence Rates of Urticaria in the General Population

Author	Year	Prevalence (%)	Country/Number
Lombolt	1963	0.05	Färöer Islands, 10,984
Hellgren	1983	0.1	Sweden, 35,343
Varonier			Switzerland (4 to 6 years old; 15 years old)
	1968	0.4; 0.7	4781; 2451
	1981	0.9; 0.5	3270; 3500
Freeman	1964	2.1	U.S., 2235 adolescents
Paul	1991	1.3	Estimated for part of Germany
Bakke	1990	9.0	Norway, 4992
Wang	1990	23.3	China, 10,144
Singh	1980	23.6	India, 424 newborns

Adapted from Schäfer, T. and Ring, J., in *Epidemiology of Clinical Allergy*, Karger, Basel, 1993, 49.

Still, it is obvious that urticaria belongs to the 20 most common skin diseases.[1] In the textbooks, estimates of cumulative prevalence rates reach 30%, meaning that about one third of the population experiences an urticarial episode at least once in their lifetime. In Table 11-9, data are given from studies dealing with prevalence rates in unselected populations in a general population, and Table 11-10 gives the prevalence rates in selected collectives of hospital patients.

It becomes clear that with a range between 0.05 and 23.6%, these studies are difficult to interpret. One is tempted to speculate that the studies showing prevalence rates of 23% and more have included acute urticaria, while the other studies focus more on the chronic forms.

Our own experiences in an epidemiological study in preschool children (including both acute and chronic forms) yielded a prevalence for urticaria of 2.8%.[61,64]

TABLE 11-10
Prevalence Rates of Urticaria: Selected Collectives

Author	Year	Prevalence (%)	Number/Source
Doeglas	1975	1.4	15,675, derm. out-patient clinic
Kleine-N.	1973	1.7	11,647, derm. clinic
Paul	1991	2.1	1500, general practice
Sarojini	1972	2.8	16,720, derm. out-patient clinic
Paul	1991	3.0	4000, derm. practice
Freeman	1964	8.8	545, allergic children
Myers	1961	16.0	216, allergic children

Adapted from Schäfer, T. and Ring, J., in *Epidemiology of Clinical Allergy*, Karger, Basel, 1993, 49.

TABLE 11-11
Prevalence of Hand or Contact Eczema in the General Population

Type of Eczema	Method	Population	Country	Prevalence	Author (year)
"Hand"	Q[a]	141,000	Sweden	2.3%	Agrup, 1965
"Manifest eczema"	E[b]	2,180	UK	9.0%	Rea et al., 1969
Manifest contact eczema	E	20,749	USA	5.4%	Johnson et al., 1971
Hand and forearm	E	3,140	Netherlands	6.2%	Coenraads, 1981
Hand	Q	14,667	Norway	9.0%	Kavli, Förde, 1979
Hand	Q + E	20,000 Q 1,238 E	Sweden	5.4%	Meding, 1984

[a] Q = questionnaire.
[b] E = examination.

Adapted from Burr, M. L., *Monographs in Allergy*, 31, Karger, Basel, 1993.

Urticaria seems to be more frequent in the female sex. There are different age-related prevalence rates: chronic urticaria is rare in children, where the acute form predominates. Chronic urticaria is much more common between the third and sixth decade.

Of special interest is the prevalence of physical urticarias, which is estimated between 12 and 57% within urticaria patients. Urticaria factitia is estimated to occur in 16 to 50%, cold urticaria in 5 to 33%, pressure urticaria in 0 to 21%, and cholinergic urticaria in 15 to 35%.[64]

C. EPIDEMIOLOGY OF ANAPHYLACTOID REACTIONS

In the above-mentioned study in preschool children, we found a prevalence of 3.1% for systemic insect sting reactions and 1.7% for angioedema.[61] For 7% of the children, the parents reported systemic untoward reactions (of urticarial, gastrointestinal, or cardiovascular symptomatology) after intake of foods possibly to be regarded as food allergy.

D. EPIDEMIOLOGY OF CONTACT ALLERGY

Most of the studies regarding the epidemiology of contact allergy deal with the experience of clinical and hospital populations where frequency rates of positive patch test reactions to tested allergens are given.[35,41,63]

There are some studies regarding the prevalence of contact eczema in the unselected populations (Table 11-11); however, some of these studies do not differentiate between allergic contact dermatitis and other forms of eczema, but only cover the term "hand eczema."[66-71]

Contrary to the general opinion that contact allergy is rare in childhood, we found in an epidemiologic study in preschool children a frequency of 15.8% positive patch test reactions to a standard battery of contact allergens. The most frequent positive allergen was merthiolate (8.8%), followed by nickel sulfate (7.1%), potassium dichromate (4.1%), and cobalt chloride (2%).[61]

There is a lot of information regarding the prevalence of contact sensitization in occupational settings covering the increased risk to develop eczema or contact sensitization for groups of workers in different occupations (see also Chapter 15).

IV. ENVIRONMENTAL POLLUTION AND ALLERGY

There is no doubt that high concentrations of pollutants such as suspended particles, automobile exhaust, ozone, sulfur dioxide, and nitrous oxides can be measured in rather high concentrations in the air over the countries where the above-mentioned studies[57-63] have been performed. However, some of these pollutants, like SO_2, have shown a decrease in concentration in the air during the last decades.

An epidemiologic study from Japan[72,73] reported an increasing prevalence of allergic rhinoconjunctivitis to Japanese cedar pollen during the last 30 years. In the same period, the SO_2 — and CO — concentrations, as well as the concentrations of photochemical oxidants in the air, have decreased in Japan. However, suspended particles emitted mostly from cars (e.g., diesel exhaust) have increased. In their study, the authors compared the prevalence of allergic rhinoconjunctivitis due to Japanese cedar pollen in five different living areas defined according to exposure to pollen allergens and/or traffic exhaust, respectively. While the lowest prevalence rates were in areas with few Japanese cedars or limited traffic, the highest incidences were found in people living along roads with Japanese cedars. These prevalence rates were significantly higher compared to living areas with equal pollen counts but less traffic.[73]

Another air pollutant which has been found to be associated with increased prevalence of atopic diseases is possibly derived from tobacco smoke. We found significantly increased risk ratios for the development of atopic diseases in children of mothers who smoked during pregnancy.[61]

There were striking differences in the prevalence of atopic diseases between different living areas which were not explained by the concentrations of SO_2, NO_X in the air, nor by genetic background. Krämer et al. found preliminary evidence for a possible role of exposure to tobacco smoke, unvented gas, and traffic-dependent pollutants.[74]

In the first controlled prospective trial comparing different living areas with various degrees of air pollution in west and east Germany in 1991, striking differences between different living areas both in east and west Germany were shown with rather decreased prevalence rates of allergic sensitization in heavily industrialized areas in the former GDR, but rather high rates of eczema.[75] Similar findings have been reported by von Mutius et al.[76]

The mechanism by which pollutants might influence the development of allergic diseases is by no means established. There is no doubt that high concentrations of pollutants can exert irritant effects on the skin and mucous membranes, thereby leading to an enhanced permeability of suspected allergens because of reduced barrier function.

Furthermore, interactions between air pollutants and allergic diseases have been studied on the level of IgE-mediated sensitization. Diesel exhaust particles, tobacco smoke, ozone, and SO_2 have been shown to act as adjuvants in the process of IgE formation in experimental animals.[77,78]

In the effector phase of an allergic response, pollutants may act as modulators of mediator release in sensitized individuals, as has been shown for metals such as cadmium[79] or pesticides such as dieldrine or polychlorinated biphenyls[80] in mast cells or basophils.

Recent investigations by Behrendt et al. (see this volume, Chapter 9) have shown that the influence of the air pollutants upon allergic sensitization is a much more complex one. Pollen collected from highly polluted areas were found to be loaded with chemical particles, consisting of inorganic and organic compounds adsorbed to the pollen surface. When pure pollen was incubated with dust particles and collected with high-volume samplers from polluted areas, it was activated to release more allergenic proteins. In electron microscopy studies, Behrendt et al. were able to show marked morphological changes in the pollen surface structure after contact with air pollutants.[81]

There is increasing evidence that air pollutants may play a role at several stages in the development of allergic diseases. It is necessary to further investigate these questions in epidemiologic studies under simultaneous consideration of allergen and pollutant exposure and in experimental research in order to elucidate the relevant substances and mechanisms.

REFERENCES

1. **Braun-Falco, O., Plewig, G., Wolf, H., and Winckelmann, R.,** *Dermatology*, Springer, Berlin, 1990.
2. **Coombs, R. R. A. and Gell, P. G. H.,** The classification of allergic reactions underlying disease, in *Clinical Aspects of Immunology*, Gell, P. G. H. and Coombs, R. R. A., Eds., Davis, Philadelphia, 1963, 317.
3. **Middleton, E., Reed, C. A., and Ellis, F. F.,** *Allergy. Principles and Practice*, 3rd ed., Mooby, St. Louis, 1988.
4. **Ring, J.,** *Angewandte Allergologie*, 2. Aufl., MMV Medizin, München, 1988.
5. **Dukor, P., Kallos, P., Schlumberger, H. D., and West, G. B., Eds.,** *Pseudo-Allergic Reactions I. Genetic Aspects and Anaphylactoid Reactions*, Karger, Basel, 1980.
6. **Ring, J.,** Pseudo-allergic drug reactions, in *Allergy. Theory and Practice*, Korenblat, P., Wedner, H. J., Eds., Saunders, Philadelphia, 1992, 243.
7. **Warin, R. P. and Champion, R. H.,** *Urticaria*, Saunders, London, 1974.
8. **Czarnetzki, B. M. and Grabbe, J., Eds.,** *Urtikaria*, Springer, Berlin, 1992.

9. **Ring, J. and Meßmer, K.,** Incidence and severity of anaphylactoid reactions to colloid volume substitutes, *Lancet*, 1, 466, 1977.
10. **Doeglas, H. M. G.,** Reactions to aspirin and food additives in patients with chronic urticaria, including the physical urticarias, *Brit. J. Derm.*, 93, 135, 1975.
11. **Rosen, F. S., et al.,** Hereditary angioneurotic edema: two genetic variants, *Science*, 148, 957, 1965.
12. **Lahti, A., von Krogh, G., and Maibach, H. I.,** Contact urticaria syndrome. An expanding phenomenon in *Dermatologic Immunology and Allergy*, Stone, J., Ed., Mosby, St. Louis, 1985, 379.
13. **Maibach, H. I. and Johnson, H. L.,** Contact urticaria syndrome. Contact urticaria to diethytoluamide, *Arch. Derm.*, 111, 726, 1975.
14. **Rajka, G.,** *Essentials in Atopic Dermatitis*, Springer, Berlin, 1990.
15. **Ruzicka, T., Ring, J., and Przybilla, B., Eds.,** *Handbook of Atopic Eczema*, Springer, Berlin, 1991.
16. **Wüthrich, B.,** *Zur Immunpathologie der Neurodermitis Constitutionalis*, Huber, Bern, 1975.
17. **Hanifin, J. B. and Rajka, G.,** Diagnostic features of atopic dermatitis, *Acta Derm. Venereol.*, Suppl. 1, 44, 1980.
18. **Schultz-Larsen, F.,** Atopic dermatitis: a genetic-epidemiologic study in a population-based twin sample, *J. Am. Acad. Dermatol.*, 28, 719, 1993.
19. **Enders, F., Przybilla, B., and Ring, J.,** Atopy and contact allergy, in *New Trends in Allergy III*, Ring, J. and Przybilla, B., Eds., Springer, Berlin, 1991, 409.
20. **Sampson, H. A. and Albergo, R.,** Comparison of results of skin, RAST and double-blind, placebo controlled challenge (DBFC) food tests in children with atopic dermatitis, *J. Allergy Clin. Immunol.*, 71, 161, 1983.
21. **van Bever, H. P., Docx, M., and Stevens, W. J.,** Food and food additives in severe atopic dermatitis, *Allergy*, 44, 588, 1989.
22. **Vieluf, D., Przybilla, B., Traenckner, I., and Ring, J.,** Oral provocation with food additives, *J. Allergy Clin. Immunol.*, 85, 206, 1990.
23. **Bruynzeel-Koomen, C., van Wichen, D. F., Toonstra, J., Berrens, L., and Bruynzeel, P. O. B.,** The presence of IgE molecules on epidermal Langerhans cells in patients with atopic dermatitis, *Arch. Derm. Res.*, 278, 199, 1986.
24. **Bieber, T., Ring, J., and Rieber, P.,** Comparison study of IgE bearing Langerhans cells and serum IgE level in atopic dermatitis, *J. Invest. Derm.*, 89, 313, 1987.
25. **Bieber, T., De la Salle, H., Wollenberg, A., Hakimi, J., Chizzonite, R., Ring, J., Hanau, D., and De la Salle, C.,** Human Langerhans cells express the high affinity receptor for IgE (FceRI), *J. Exp. Med.*, 175, 1285, 1992.
26. **Ring, J., Bieber, T., Vieluf, D., Kunz, B., and Przybilla, B.,** Atopic eczema, Langerhans cells and allergy, *Int. Arch. Allergy App. Immunol.*, 94, 194, 1991.
27. **Vieluf, D., Kunz, B., Bieber, T., Przybilla, B., and Ring, J.,** "Atopy Patch Test" with aeroallergens in patients with atopic eczema. *Allergy J.*, 2, 9, 1993.
28. **Hjorth, N. and Roed-Petersen, J.,** Occupational protein contact dermatitis in food handlers, *Contact Derm.*, 2, 28, 1976.
29. **Silberberg, I., Baer, R. I., and Rosenthal, S. A.,** The role of Langerhans cells in contact dermatitis in guinea pigs, *Acta Derm. Venereol.*, 54, 321, 1974.
30. **Stingl, G.,** New aspects of Langerhans cell function, *Int. J. Derm.*, 19, 189, 1980.
31. **Knop, J., Malorny, U., and Macher, E.,** Induction of T effector and T supressor lymphocytes in vitro by haptenized bone marrow-derived macrophages, *Cell Immunol.*, 88, 41, 1984.
32. **Braathen, L. R.,** T-cell subsets in patients with mild and severe atopic dermatitis, *Acta Derm. Venerol.*, Suppl. 114, 133, 1985.
33. **Scheper, R. J. and von Blomberg, B. M. E.,** Mechanisms of allergic contact dermatitis to chemicals, in *Allergic Hypersensitivities Induced by Chemicals: Recommendations for Prevention*, Vos, J. G., Younes, M., and Smith, E., Eds., CRC Press, Boca Raton, FL, 1995.

34. **Enders, F., Przybilla, B., Ring, J., Burg, G., and Braun-Falco, O.,** Epikutantest mit einer Standardreihe. Ergebnisse bei 12026 Patienten, *Hautarzt*, 39, 779, 1988.
35. **Frosch, P. J., Dooms-Goossens, A., Lachapelle, J. M., Rycroft, R. J. G., and Scheper, R. J., Eds.,** *Current Topics in Contact Dermatitis*, Springer, Berlin, 1989.
36. **Bandmann, H. J. and Dohn, W.,** *Die Epicutantestung*, 2nd ed., Bergmann, München, 1986.
37. **Fisher, A. A.,** *Contact Dermatitis*, 3rd ed., Lea and Febiger, Philadelphia, 1986.
38. **Cronin, E.,** *Contact Dermitilis*, Churchill Livingstone, London, 1980.
39. **Foussereau, J., Benezra, C., and Maibach, H. I.,** *Occupational Contact Dermatitis. Clinical and Chemical Aspects*, Munksgaard, Kopenhagen, 1982.
40. **Hausen, B. M., Brinkmann, J., and Dohn, W.,** *Lexikon der Kontaktallergene*, Ecomed, Landsberg, 1982.
41. **Schnuch, A. and Uter, W.,** Der Informationsverbund Dermatologischer Kliniken (IVDK) zur Dokumentation und wissenschaftlichen Auswertung der Kontaktallergien, *Allergologie*, 12, 471, 1989.
42. **Magnus, I. A.,** *Dermatological Photobiology. Clinical and Experimental Aspects*, Blackwell, Oxford, 1976.
43. **Plewig, G., Hölzle, E., Roser-Maaß, E., and Hofmann, C.,** Photoallergy, in *New Trends in Allergy*, Ring, J. and Burg, G., Eds., Springer, Berlin, 1985, 152.
44. **Schauder, S. and Ippen, H.,** Photoallergic and allergic contact dermatitis from dibenzylmethanes, *Photodermatology*, 3, 140, 1986.
45. **Arndt, K. A. and Jick, H.,** Rate of cutaneous reactions to drugs, *J. Am. Med. Assoc.*, 235, 918, 1976.
46. **Zürcher, K. and Krebs, A.,** *Cutaneous Side-Effects of Drugs*, Karger, Basel, 1980.
47. **Schulz, K. H.,** Stellenwert und Aussagekraft von Testmethoden bei allergischen Arznei-Exanthemen, in *Fortschritte der praktischen Dermatologie und Venerologie. Bd. 9*, Braun-Falco, O. and Wolff, H. H., Eds., Springer, Berlin, 1979, 71.
48. **Kauppinen, I. and Stubb, S.,** Drug eruptions: causative agents and clinical types. A series of in-patients during a 10-year period, *Acta Derm. Venereol.*, 64, 320, 1984.
49. **Dukes, M. N. G., Ed.,** *Meyler's Side Effects of Drugs*, Elsevier, Amsterdam, 1988.
50. **Schöpf, E., Schulz, K. H., Kessler, R., Taugner, M., and Braun, W.,** Allergologische Untersuchungen beim Lyell-Syndrom, *Z. Hautkr.*, 50, 865, 1975.
51. **Goerz, G. and Ruzicka, T.,** *Lyell-Syndrom*, Grosse, Berlin, 1978.
52. **Roujeau, J. C., Dubertret, L., Moritz, S., Joualt, H., Heslan, M., Revuz, J., and Touraine, R.,** Involvement of macrophages in the pathology of toxic epidermal necrolysis, *Brit. J. Derm.*, 113, 425, 1985.
53. **Ring, J.,** Drug-induced Lyell's syndrome (toxic epidermal necrolysis), in *Prog. Allergy Clin. Immunol.*, Pichler, W. J., Stadler, B. M., Dahinden, C. A., Pécoud, A. R., Frei, P., Schneider, I. C. H., and de Weck, A., Eds., Huber, Bern, 1989, 455.
54. **Ring, J., Przybilla, B., and Gollhausen, R.,** Progressive pigmentary purpura provoked by a phytotherapeutic drug containing Echinacea extract, *Allergy Clin. Immunol. News*, 114, 108, 1989.
55. **Soter, N. A., Austen, K. F., and Gigli, I.,** Urticaria and arthralgias as manifestations of necrotizing angiitis (vasculitis), *J. Invest. Derm.*, 63, 485, 1974.
56. **Wolff, H. H. and Scherer, R.,** Allergic vasculitis, in *New Trends in Allergy*, Ring, J. and Burg, G., Eds., Springer, Berlin, 1981, 140.
57. **Burr, M. L., Ed.,** Epidemiology of clinical allergy, in *Monographs in Allergy*, 31, Karger, Basel, 1993.
58. **Ring, J., Ed.,** *Epidemiologie allergischer Erkrankungen*, MMV Medizin, München, 1991.
59. **Wüthrich, B.,** Epidemiology of allergic diseases: are they really on the increase?, *Int. Arch. Allergy Appl. Immunol.*, 90, 3, 1987.
60. **Kjellman, N.-I. M.,** Atopic disease in seven-year old children. Incidence in relation to family history, *Acta Pädiatr. Scand.*, 66, 465, 1977.

61. **Kunz, B. and Ring, J.,** Are allergies increasing?, in *New Trends in Allergy,* Ring, J., and Przybilla, B., Eds., Springer, Berlin, 1991, 3.

62. **Schlipköter, H. W., Krämer, U., Behrendt, H., Dolgner, R., Stiller-Winkler, R., Ring, J., and Willer, H. J.,** Impact of air pollution on children's health. Results from Saxony-Anhalt and Saxony as compared to Northrhine-Westphalia, in *Health and Ecological Effects, Critical Issues in the Global Environment,* Vol. 5, IU-A 2103, Air & Waste Management Association, Pittsburgh, 1992.

63. **Vieluf, D., Behrendt, H., Ring, J., Krämer, U., Kahle, S., Dolgner, R., Hinrichs, L., and Schlipköter, W.,** Epidemiologische Untersuchung zur Prävalenz von atopischem Ekzem bei 6jährigen Kindern in verschiedenen Städten Deutschlands, *Allergo J.,* 2 (Suppl. 1), (Abstr.), 3, 1993.

64. **Schäfer, T. and Ring, J.,** Epidemiology of urticaria, in *Epidemiology of Clinical Allergy,* Burr, M. L., Ed., Karger, Basel, 1993, 49.

65. **Frosch, P. J.,** Aktuelle Kontaktallergene., *Hautarzt,* 41 (Suppl. 10), 129, 1990.

66. **Agrup, G.,** Hand eczema and other hand dermatoses in South Sweden, *Acta Derm. Venerol.,* 49 (Suppl. 61), 1, 1969.

67. **Rea, J. N., Newhouse, M. L., and Halil, T.,** Skin disease in Lambeth, *Brit. J. Prev. Soc. Med.,* 30, 107, 1976.

68. **Johnson, M. L. T. and Roberts, J.,** Skin conditions and related need for medical care among persons 1–74 years. United States, 1971–1974, *USA: Vital Health Statistics,* 11, 1, 1978.

69. **Coenraads, P. J., Nater, J. P., and van der Lende, R.,** Prevalence of eczema and other dermatoses of the hands and arms in the Netherlands. Association with age and occupation, *Clin. Exp. Dermatol.,* 8, 495, 1983.

70. **Kavli, G. and Förde, O. H.,** Hand dermatoses in Tromsö, *Contact Dermatitis,* 10, 174, 1984.

71. **Meding, B. and Swanbeck, G.,** Prevalence of hand eczema in an industrial city, *Br. J. Dermatol.,* 116, 627, 1987.

72. **Miyamoto, V. S. and Takafuji, S.,** Environmental pollution and allergy, in *New Trends in Allery III,* Ring, J. and Przybilla, B., Eds., Springer, Berlin, 1991, 165.

73. **Ishizaki, T., Koizumi, K., Ikemori, R., Ishiyama, Y., and Kushibiki, E.,** Studies of prevalence of Japanese cedar pollinosis among the residents in a densely cultivated area, *Ann. Allergy,* 58, 265, 1987.

74. **Krämer, U., Behrendt, H., Dolgner, R., Kainka-Stänicke, E., Oberbarnscheidt, J., Sidaoui, H., and Schlipköter, H. W.,** Auswirkung der Umweltbelastung auf allergologische Parameter bei 6jährigen Kindern, in *Epidemiologie allergischer Erkrankungen,* Ring, J., Ed., MMV, München, 1991, 165.

75. **Behrendt, H., Krämer, U., Dolgner, R., Hinrichs, J., Willer, H., Hagenbeck, H., and Schlipköter, H. W.,** Elevated levels of total serum IgE in East German children: atopy, parasites, or pollutants?, *Allergo J.,* 2, 31, 1993.

76. **von Mutius, E., Fritzsch, C., Weiland, S. K., Röll, G., and Magnussen, H.,** Prevalence of asthma and allergic disorders among children in united Germany: a descriptive comparison, *Br. Med. J.,* 305, 1395, 1992.

77. **Takafuji, S., Suzuki, S., Koizumi, K., Tadokoro, K., Miyamoto, T., Ikemori, R., and Muranaka, M.,** Diesel-exhaust particulates inoculated by the intranasal route have an adjuvant activity for IgE production in mice, *J. Allergy Clin. Immunol.,* 79, 639, 1987.

78. **Riedel, F., Krämer, M., Scheibenbogen, C., and Rieger, C. H. L.,** Effects of SO_2 exposure on allergic sensitization in the guinea pig, *J. Allergy Clin. Immunol.,* 82, 527, 1988.

79. **Behrendt, H., Wieczorek, M., Wellner, S., and Winzer, A.,** Effect of some metal ions (Cd++, Pb++, Mn++) on mediator release from mast cells in vivo and in vitro, in *Environmental Hygiene,* Seemayer, N. and Hadnagy, W., Eds., Springer, Berlin, 1988, 105.

80. **Raulf, M. and König, W.,** In vitro effects of polychlorinated biphenyls on human platelets, *Immunology*, 72, 287, 1991.
81. **Behrendt, H. J., Friedrich, K. H., Kainka-Stänicke, E., Darsow, U., Becker, W. M., and Tomingas, R.,** Pollens and pollutants in the air. A complex interaction, in *New Trends in Allergy III*, Ring, J. and Przybilla, B., Eds., Springer, Berlin, 1991, 465.

Chapter 12

DIAGNOSIS AND TREATMENT OF HYPERSENSITIVITY REACTIONS OF THE SKIN

Johannes Ring

CONTENTS

I. INTRODUCTION

The final aim of allergy diagnosis comprises

1. Detection of the eliciting agent
2. Elucidation of the pathomechanism involved
3. Recommendations for prophylaxis and treatment
4. Information about safe alternatives

The classical sequence of different steps in allergy diagnosis[1,2] is

1. History
2. Skin test
3. *In vitro* allergy diagnosis
4. Provocation test

II. HISTORY

The careful history is the main part of allergy diagnosis; it is the domain of the experienced specialist, who will, in more than 50% of the cases, be able to establish a diagnosis by history. Specific allergy tests *in vitro* or *in vivo* are only indicated on the basis of a careful history. There is never an indication for a "blind screening." The allergological history can be supplemented by questionnaires, which are valuable together with the personal history.

Questions important in an allergy history are those that consider the time and space relations of complaints (in the house, out of the house), the seasonal periodicity (e.g., hayfever), and work related symptoms. Questions should always also cover possible indoor allergen sources (animals, molds, house dust mites), family history, drug treatment, and psychosocial background.

III. SKIN TEST PROCEDURES

There are a variety of skin test procedures for the diagnosis of several types of allergic reactions (Table 12-1).

A. EPICUTANEOUS TEST (PATCH TEST)

Since the introduction of the classical patch test at the end of the last century by Jadassohn,[3] this procedure has been the most practical and widely used test in the diagnosis of allergic contact dermatitis.[4-6] The suspected allergen usually is delivered in an appropriate dilution (in order to avoid toxic reactions, the proper concentration has to be determined for each allergen in separate studies in human volunteers) in petrolatum or some other indifferent vehicle. After 2 days the patch is removed and the test reaction is read after 48 and after 72 hr. The test result is graded according to the intensity of the dermatitis:

0 = negative (no reaction)
1+ = weakly positive (marked erythema)
2+ = positive (erythema and papules)
3+ = strongly positive (erythema, infiltration, vesicles, blisters, and erosions)

Patch test results should be read 20 to 60 min at the earliest after removal of the patch. In special cases, a third reading after 1 week or later may be indicated. In distinguishing toxic from allergic patch test reactions, the following criteria may be helpful:

TABLE 12-1
Diagnostic Value of Skin Test Procedure

Test		Reaction Type	Unspecific
Epicutaneous	Classic patch	IV	Tape sensitivity
	Open patch	I	"Angry back"
	Stripped patch	IV	Alkali neutralization
	Diseased skin	IV	Nitrazin yellow
	"Friction"	I	Dermographism
Cutaneous	Scratch	I	Histamine reaction
			Codeine reaction
	Prick	I	Saline reaction
	Intradermal	I, II, IV	Pharmacological response
			(e.g., acetylcholine)
Physical stimuli	Cold	(I)	Dermographism
	Heat		
	Pressure		
	Light:		Minimal erythema dose (MED)
	Monochromatic UV	(I)	
	Photopatch	IV	

Toxic reactions are usually clearly marked, reach their maximum at an earlier time (24 to 48 hr), and are positive in normal individuals (in a percentage of more than 20%).

False negative test results may be observed under systemic corticosteroid treatment of heavy UV-light exposure. Topical steroids should be avoided at least 5 days prior to patch tests. Antihistamines, ß-adrenergics, or theophylline do not influence classical patch test reactions.

False positive patch test reactions may occur due to irritancy (too high concentration of test substance), adhesive tape hypersensitivity, or concomitant to other very strong positive reactions. This phenomenon is called "excited skin syndrome" or "angry back."[7,8] The exact mechanism of this phenomenon is not yet clear; however, careful retesting is recommended when there are more than five positive patch test reactions (unrelated chemicals).[9]

When combined with UV-light, the patch test is useful in detecting photo-allergy; this procedure is called the photopatch test. Two samples of each individual allergen are placed upon the skin of the patient, and the tape is removed 24 hr. later. Then one series of allergens is exposed to UV-light (usually UV-A), and patch test reactions are read as usual after 48 and 72 hr.[10]

Patch test reactions are commonly performed in order to detect a type IV allergy according to the classification of Coombs and Gell.[10a] There are, however, modifications of epicutaneous testing which are suitable for demonstrating immediate type hypersensitivity, e.g., the open patch test, where the reaction is read after 20 min. This procedure is applied in the diagnosis of contact urticaria. Another modification is the "friction test"[11] where the allergen is rubbed gently upon the skin. This friction test correlates to a high degree with *in vitro* IgE-antibody measurement (RAST) or standard prick testing. It can be recommended in highly sensitized individuals. A new variant has been introduced as the "Atopy Patch Test," with the epicutaneous application of allergens known to induce IgE antibodies in patients with atopic eczema.[12]

B. CUTANEOUS TESTS

1. Prick Test

A drop of a solution of allergen extract is applied to the skin and is pricked with a needle or special prick instrument superficially (without bleeding). After 15 to 20 min, the test reaction is read.[13]

2. Scratch Test

The skin under the test substance (e.g., powder or suspension) is scratched superficially. This test is indicated when no proper allergen extracts or solutions are available.

3. Intradermal Test

In the intradermal test, 0.02 to 0.05 ml of an allergen solution (on the average, $^1/_{100}$ of the prick test concentration) are injected strictly intradermally with a fine needle. A small wheal will develop (up to 3 mm in diameter).

The test reaction of all cutaneous tests (prick, scratch, and intradermal) consists of a typical wheal and flare reaction and is read after 15 to 20 min. It is graded according to the diameter of the erythema and wheal, which may be recorded in mm or graded arbitrarily from 0 to 4+ reactions. In order to register late phase reactions,[14] the test sites should be re-read 4 to 8 hr after testing.

The diagnostic value of skin test procedures regarding both specific and unspecific reactivity is summarized in Table 12-1.

False positive reactions in cutaneous testing may occur with irritant solutions as well as in patients with factitious urticaria. False negative reactions can be observed, if the time interval after the allergic event is too short (in drug allergy testing), in certain skin or nervous diseases, if the antigen is too diluted, or under the influence of certain drugs, the most important of which are antihistamines, psychopharmaca, corticosteroids, or ß-adrenergics.[1,2]

At the time of an *in vivo* allergy test, the principles of allergen avoidance should be considered. During high pollen season or during intimate contact with animals at home, the allergen load of a skin test may lead to an exacerbation of symptoms (e.g., asthma, atopic eczema, or anaphylaxis).

C. SPECIAL SKIN TEST PROCEDURES

On the skin several physical tests are performed, especially in the diagnosis of physical urticarias with cold, heat, pressure, or UV-light (solar urticaria). Skin tests with UV-irradiation are indicated in patients with photosensitivity disease (photoallergic or phototoxic) (see photopatch test) and, according to the possible mechanism, also applied as a photo-prick- or photo-scratch-test. Skin tests not only give information about specific sensitizations, but also about unspecific parameters of the skin function, permeability, or irritability.

IV. PROVOCATION TEST

Provocation tests consist of the exposure of the relevant tissue to the allergen under controlled laboratory conditions.[1] Target organs used in the various modifications of provocation tests are the eye, the nasal mucosa, the bronchi, the gastrointestinal tract, or the subcutaneous tissue.

In the conjunctival provocation test, the erythema and inflammation of the eye after application of 1:100 or 1:10 dilutions of the allergen extract are evaluated. Similarly, nasal provocation can be done registering the subjective symptoms (rhinorrhea, sneezing, etc.). A more objective technique in nasal provocation is the use of a rhinomanometer, through which the reduction in air flow velocity can be demonstrated on a pressure-volume-diagram.

For bronchial provocation, the allergen solution is delivered as an aerosol to the asymptomatic patient. The increase in airway resistance or decrease in forced expiratory volume in one minute (FEV_1) is measured as a parameter of bronchial constriction. Apart from specific allergen, the unspecific irritability can be investigated with bronchial inhalation using methacholine in varying dilutions.

Provocation tests involving the gastrointestinal tract include sublingual provocation and oral provocation, as well as intragastral provocation under endoscopic control (IPEC).[15] Subcutaneous provocation tests are performed in some cases of drug hypersensitivity, e.g., local anesthetics.

All provocation tests are unpleasant for the patient and bear a considerable risk of serious, sometimes life-threatening reactions. In rare cases, fatalities have been reported even after skin testing.[16]

V. PASSIVE TRANSFER TEST

The passive transfer of reaginic antibodies, as originally demonstrated by Prausnitz and Küstner,[17] should not be performed as a routine diagnostic tool because of the potential risk of transferring infectious diseases (hepatitis, syphilis, HIV, etc.). Based on the same principle, the passive transfer of human serum to primate skin can be used for detecting IgE-dependent reactions in the so-called passive cutaneous anaphylaxis (PCA).[18]

VI. *IN VITRO* ALLERGY DIAGNOSIS

In vitro methods in the diagnosis of allergic diseases comprise both serological and cellular assays measuring either specific antibodies or sensitized cells or, unspecifically, concentrations of immunologically relevant factors.

TABLE 12-2
In Vitro Methods of Allergy Diagnosis

Antigen Specific	Nonspecific
Serological (quantitation of antibodies)	Quantitation of
Immunoprecipitation	Immunoglobulins (esp. IgE)
Immunodiffusion	complement factors
Hemagglutination	–chemical
Complement fixation	–functional
Immunofluorescence	Immune complexes
Radioimmunoassay (RIA)	Mediators in plasma
Enzyme-immunoassay (EIA)	
Cellular	Differential blood count
Release of mediators (e.g., histamine)	(Eosinophil, Basophil)
Basophil degranulation	Lymphocyte subpopulations
Chemiluminescence	Lymphocyte transformation (mitogens)
Lymphocyte transformation (antigen)	

A. SEROLOGICAL METHODS

In the diagnosis of immediate type allergic reactions, of course, the detection of IgE antibodies is of special importance. However, antibodies of other immunoglobulin classes also may be able to elicit allergic reactions (e.g., type III) or interfere as "blocking antibodies" with allergic phenomena, and are, therefore, of interest. For the detection of antibodies a variety of immunological techniques are useful, including immunodiffusion, hemagglutination, and complement fixation, as well as the more sensitive radioimmunoassays (RIA) or enzyme-immuno-assays (EIA) (Table 12-2).

Total serum IgE can be measured both by RIA and EIA. The theoretical principle of these test systems is very simple.[19,20] The marker of the anti-IgE may either be a radioisotope or an enzyme, or a chemiluminescence or fluorescence signal.

Figure 12-1 shows the principle of a paper-radio-immuno-sorbent test (PRIST) for the detection of total serum IgE and of the radio-allergo-sorbent test (RAST) for the detection of specific IgE antibodies. With the RAST, specific antibodies of the IgE class are detected.[20] The detection of specific IgG antibodies against allergens may be of interest in special cases of immune-complex allergic reactions as well as in the control of the efficacy of hyposensitization therapy. As a diagnostic parameter for food allergies, specific IgG antibodies have not proven to be useful.[21] In immunoblot techniques, the specific antibody reaction pattern against electrophoretically separated allergen extracts is detected by labeled anti-IgE or anti-IgG.

In the interpretation of RAST results, one has to carefully keep in mind that negative RASTs do not exclude allergic reactions since the half-life of circulating IgE antibodies in the plasma is rather short. With increasing time after

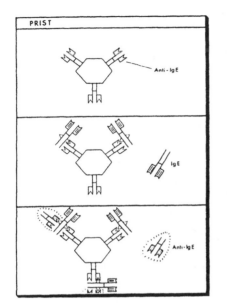

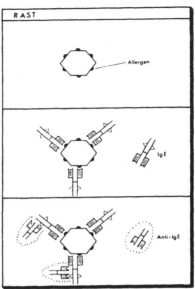

FIGURE 12-1. Schematic diagram of the principle of paper-radio-immuno-sorbent test (PRIST) and radio-allergo-sorbent test (RAST).

the allergen contact, specific IgE antibodies will decrease in concentration in the serum.

B. CELLULAR ASSAYS

Apart from serological methods, there are several cellular assays useful in the diagnosis of allergic diseases, namely the *in vitro* histamine release or basophil degranulation for IgE-mediated reactions and the lymphocyte transformation test for cellular hypersensitivity.

In vitro histamine release[22,23] is a suitable method and correlates well with RAST and skin test results. The assay, however, is time-consuming and not yet widely used in large clinical routine. Briefly, peripheral leukocytes of the patients are washed and incubated *in vitro* with varying dilutions of the allergen. The histamine content of the supernatant is measured and compared with the total histamine content of the cell suspension expressed as 100 percent. This test is particularly useful in situations where there is no RAST available or skin tests are too dangerous or not possible. Apart from the IgE antibodies on the basophil surface, it reflects the nonspecific parameter of "releasability," which might be of interest in certain atopic diseases.[24]

Recently an *in vitro* test for detection of sulfidopeptide leukotriene secretion from primed basophils (e.g., by interleukin 3) after allergen stimulation has been described.[25]

The direct demonstration of mediator substances like histamine during allergic reactions is not only of scientific interest, but also of possible practical

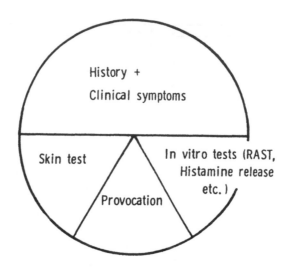

FIGURE 12-2. Synopsis of different diagnostic procedures in allergy. In doubtful cases the history is the decisive factor.

importance regarding antagonistic therapy. Studies regarding other mediators of allergic reactions would be of great interest, especially with regard to the measurement of eosinophil cationic protein (ECP), mast cell tryptase, or neutrophil myeloperoxidase.[26,27]

VII. COMPARISON OF DIFFERENT TECHNIQUES IN ALLERGY DIAGNOSIS

In the comparison of the different procedures (skin test, *in vitro* diagnosis, provocation test), one has to keep in mind that the different tests detect different factors of the allergic response. In the skin test, sensitized mast cells in the skin will be detected; in the RAST, circulating specific IgE antibodies are demonstrated; and in the *in vitro* histamine release, IgE antibodies on the surface of basophil leukocytes. Figure 12-2 shows a synopsis of the different diagnostic procedures in allergy diagnosis.

A positive allergy test is no proof of allergy but only of sensitization! The most difficult problem is the interpretation of the test results with regard to the clinical relevance. Only if this is proven (either by a clear-cut history or by provocation test) can the diagnosis of actual allergy be made. Therefore, allergy diagnosis requires special knowledge and experience and should only be performed by specially trained physicians (preferably allergists).

VIII. THERAPY OF ALLERGIC SKIN DISEASES

A. ALLERGEN AVOIDANCE

The most effective and simple treatment of allergic diseases consists in elimination of the eliciting allergen. This shows the primary importance of

TABLE 12-3
Pharmacotherapeutic Approaches in Allergic Diseases

Inhibition of mediator synthesis
Flavonoids, Tritoqualin
Mast-cell-blockers
Dinatriumcromoglicicum, Ketotifen, Oxatomide, ß-Agonists
Receptor blockade
Histamine antagonists (H$_1$, H$_2$, or combined)
  classical and non-sedating
Anti-inflammatory treatment
Glucocorticosteroids (topical, systemic)
Therapeutics with end organ specificity
ß-Agonists
Xanthine-derivatives (bronchi)
Anticholinergics
Secretolytics
Physical therapy
Substance under study or experimental protocols
Cytokines (Interferon)
Phytotherapeutics (Ginkgo biloba-extracts, Flavonoides)
Unsaturated fatty acids
Substances with doubtful efficacy
(Calcium, Homöopathy, bacterial vaccines)
Immunosuppressives

correct allergy diagnosis in these conditions. There is almost no other field in medicine where diagnosis and therapy are so closely connected.

In many, though not all, cases, allergen avoidance is possible. There are ubiquitous allergens (like pollen in the air) that cannot be avoided under normal conditions. Climatological changes may be helpful (high altitude, sea, desert). For many patients with occupational allergies, allergen avoidance implies a change of profession or at least a change of daily activity in their professional lives.

B. HYPOSENSITIZATION

Specific hyposensitization is the repeated application of increasing doses of the relevant allergen up to the tolerance level (i.e., disappearance of symptoms or maintenance dose). Hyposensitization is widely and successfully used in the treatment of the IgE-mediated allergic respiratory diseases.[28] The use of hyposensitization in allergic skin diseases is controversial and the subject of scientific investigation but not clinical routine.[29-31] There have been reports of successful hyposensitization in allergic contact dermatitis after oral application of the relevant antigens such as urushiol or nickel.[30,31]

C. PHARMACOTHERAPY OF ALLERGIC SKIN DISEASES

There are several steps of pharmacological intervention within the cascade of events of an allergic effector phase reaction (Table 12-3).

Apart from inhibitors of mediator synthesis, mast cell blockers, and mediator antagonists (e.g., antihistamines), anti-inflammatory agents such as glucocorticosteroids are often unavoidable in treating allergic skin diseases. For the majority of cases of atopic eczema or allergic contact dermatitis, topical glucocorticosteroids in the acute phase are mandatory. Rarely, systemic glucocorticosteroids will have to be given.

Antihistamines are of limited value in the treatment of eczematous dermatitis. However, they may act as antipruritics. In the treatment of urticaria and angioedema, however, antihistamines (H_1-antagonists) are effective in many cases.

In some situations the combined use of H_1- and H_2-antagonists has been shown to be more effective than H_1-antagonists alone. Major progress has been made in the development of nonsedating H_1-antagonists like terfenadine, astemizole, loratadine, and cetirizine.[32,33]

There are many approaches with new therapeutic agents which are either currently under scientific investigation (e.g., interferon in the treatment of atopic eczema) or of doubtful efficacy (so-called alternative medicine). Generally, the therapeutic arsenal in allergic diseases has been enriched due to the progress in experimental immunology and allergology during the last decades. There are, however, still many patients with severe allergic diseases and poor response to the above-mentioned modalities except for rather high doses of systemic glucocorticosteroids or immunosuppressives (e.g., cyclosporin A). Therefore, new approaches of antiallergic treatment are desirable; there is some hope for the development of new strategies by continued experimental and clinical research.

REFERENCES

1. **Middleton, E., Read, C. E., and Ellis, E. F., Eds.,** *Allergy: Principles and Practice*, 3rd Ed., Mosby, St. Louis 1988.
2. **Ring, J.,** *Angewandte Allergologie*, 2. Aufl., MMW Medizin, Munich, 1988.
3. **Jadassohn, J.,** *Verhandl. Dtsch. Derm. Gesellschaft*, 5. Kongreß, 1895, 103, 1896.
4. **Bandmann, H. J. and Fregert, S.,** Epicutantestung: Einführung in die Praxis, 2nd Ed., Springer, Berlin, 1982.
5. **Cronin, E.,** *Contact Dermatitis*, Churchill, Livingstone, London, 1980.
6. **Fisher, A. A.,** *Contact Dermatitis*, 3rd Ed., Lea and Febiger, Philadelphia, 1986.
7. **Maibach, H. I.,** The ESS, excited-skin syndrome (alias the "angry-back"), in *New Trends in Allergy*, Ring, J. and Burg, G., Eds., Springer, Berlin, 1981, 208.
8. **Mitchell, J. C.,** The angry back syndrome: eczema creates eczema, *Contact Dermatitis*, 1, 193, 1975.
9. **Luderschmidt, C., Heilgemair, G., Ring, J., and Burg, G.,** Polyvalente Kontaktallergie vs. "Angry-back-Syndrom," *Allergologie*, 5, 262, 1982.
10. **Hölzle, E., Plewig, G., Hofman, L., and Braun-Falco, O.,** Photopatchtestung, *Ztrlbl. Haut Geschlkr.*, 151, 361, 1985.

10a. **Coombs, R. R. A. and Gell, P. G. H.,** The classification of allergic reactions underlying disease, in *Clinical Aspects of Immunology,* Gell, P. G. H. and Coombs, R. R. A., Eds., Davis, Philadelphia, 1963, 317.

11. **Gronemeyer, W. and Debelic, M.,** Der sog. Reibtest, seine Anwendung und klinische Bedeutung, *Dermatologica,* 134, 208, 1967.

12. **Vieluf, D., Kunz, B., Bieber, T., Przybilla, B., and Ring, J.,** "Atopy Patch Test" with aeroallergens in patients with atopic eczema, *Allergo J.,* 2, 9, 1993.

13. **Dreborg, S. and Trew, A.,** Position paper. Allergen standardization and skin test, *Allergy,* 14 (Suppl. 1), 49, 1993.

14. **Dorsch, W. and Ring, J.,** Induction of late cutaneous reaction (LCR) by skin blister fluid (SBF) from allergen-tested and normal skin, *J. Allergy Clin. Immunol.,* 67, 117, 1981.

15. **Reimann, H. J., Ring, J., Ultsch, B., and Wendt, P.,** Intragastral provocation under endoscopic control (IPEC) in food allergy: mast cell and histamine changes in gastric mucosa, *Clin. Allergy,* 15, 195, 1985.

16. **Lockey, R. F., Benedict, L. M., Turkeltaub, P. C., and Bukantz, S. C.,** Fatalities from immunotherapy (IT) and skin testing (ST), *J. Allergy Clin. Immunol.,* 79, 660, 1987.

17. **Prausnitz, C. and Küstner, H.,** Studien über die Empfindlichkeit, *Zbl. Bakt. I Orig.,* 86, 160, 1921.

18. **Ovary, Z.,** Cutaneous anaphylaxis in the albino rat, *Int. Arch. Allergy,* 3, 293, 1952.

19. **Ceska, M., Eriksson, R., and Varga, J. M.,** Radioimmunosorbent assay of allergens, *J. Allergy Clin. Immunol.,* 49, 1, 1972.

20. **Wide, L., Bennich, H., and Johansson, S. G. O.,** Diagnosis of allergy by an in vitro test for allergen antibodies, *Lancet II,* 1105, 1967.

21. **Aalberse, R. C., Gaag, van der R., and Leeuwen, J.,** Serologic aspects of IgG4 antibodies. I. Prolonged immunization results in an IgG4-restricted response, *J. Immunol.,* 130, 722, 1983.

22. **Lichtenstein, L. M. and Osler, A. G.,** Studies on the mechanism of hypersensitivity phenomena. IX. Histamine release from human leukocytes by ragweed pollen allergen, *J. Exp. Med.,* 120, 507, 1967.

23. **Siraganiarin, R. P. and Hook, W. A.,** Complement-induced histamine release from human basophils. II. Mechanism of the histamine release action, *J. Immunol.,* 116, 639, 1976.

24. **Conroy, M. C.,** Releasability — a new dimension in basophil and mast cell reactivity, in *New Trends in Allergy,* Ring, J. and Burg, G., Eds., Springer, Berlin, 1981, 40.

25. **de Weck, A. L., Dahinden, C. A., Furukawa, K., and Maly, F. E.,** A new cellular assay for the diagnosis of allergy, in *Progr. Clin. Immunol. Allergy,* Vol. 2, Miyamoto, T. and Okuda, M., Eds., Huber, Bern, 1992, 197.

26. **Juhlin, L. and Venge, P.,** Eosinophilic cationic protein (ECP) in skin disorders, *Acta Derm. Venerol.,* 71, 491, 1991.

27. **Jakob, T., Hermann, K., and Ring, J.,** Eosinophilic cationic protein in atopic eczema, *Arch. Derm. Res.,* 283, 5, 1991.

28. **Norman, P. S.,** An overview of immunotherapy, *J. Allergy Clin. Immunol.,* 65, 87, 1980.

29. **Ring, J.,** Hyposensitization and atopic eczema, *Allergo J.,* 2, 1, 1993.

30. **Epstein, W. L., Byers, V. S., and Frankart, W.,** Induction of antigen specific hyposensitization to poison oak in sensitized adults, *Arch. Dermatol.,* 118, 630, 1982.

31. **Sjövall, P. and Christensen, O. B.,** Oral hyposensitization in allergic contact dermatitis, *Seminars Dermatol.,* 9, 206, 1990.

32. **Rimmer, S. J. and Church, M. K.,** The pharmacology and mechanisms of action of histamine H1-antagonists, *Clin. Exp. Allergy,* 20 (Suppl. 2), 3, 1990.

33. **Schmutzler, W.,** Antihistamines, in *Handbook of Atopic Eczema,* Ruzicka, Th., Ring, J., and Przybilla, B., Eds., Springer, Berlin, 1991, 396.

Chapter 13

PREDICTIVE TESTING FOR SKIN ALLERGY

Thomas Maurer

CONTENTS

I. INTRODUCTION

A. CONTACT ALLERGY

Predictive tests for the evaluation of skin sensitizing effects of chemicals have existed for nearly 50 years. The Draize test, published in 1944,[1] was included in the first guideline in 1959.[2] The test was designed to screen out potent sensitizers. Negative results with known human sensitizers, such as benzocaine, neomycin, or nickel sulfate,[3] and the simulation of the regular routes of exposure in man were the main reasons to develop other protocols.

A considerable change in the development of tests was induced by the introduction of Freund's complete adjuvant (FCA) injections during the induction phase. FCA enhances the immunological responses of the animals and it functions as a depot for the distribution of the allergen.[4] With the introduction of FCA, it was possible to detect not only potent, but also moderate and weak, contact allergens (Section III.B).

Guidelines for the testing of sensitizing effects were published in 1981 by the OECD[5] and in 1984 by the EEC.[6] Until recently, both guidelines included seven methods, four protocols using adjuvant in the induction phase, and three tests which do not use FCA. A new OECD[7] guideline with a reduced number of methods was accepted by the council in July 1992, and a new EEC[92] guideline in December 1992. They will be discussed later. Independent of the mentioned guidelines, additional protocols have been published for special purposes which are reviewed by Maurer[8] and Andersen and Maibach.[9]

B. PHOTOALLERGY

Predictive methods to evaluate the photoallergenic potential of a chemical or formulation have also existed for many years. But the importance attributed to photoallergic side effects of chemicals was never as high as that of contact allergy. Environmental factors may change the importance of photoallergic side effects in the future. As in contact allergy testing, adjuvant injections during the induction were introduced to increase the sensitivity of the methods. Some years ago the initiative was taken to propose an OECD guideline for predictive photoallergenicity testing. The proposed method is in discussion.

II. CONTACT ALLERGENICITY TESTS INCLUDED IN OECD AND EEC GUIDELINES

A. NON-ADJUVANT TESTS: DRAIZE TEST, BÜHLER TEST, OPEN EPIDERMAL TEST

As already mentioned, the main advantage of the Draize test[1,2] is the great amount of experience in use as a result of its early development. The disadvantages are the lack of full standardization and its insensitivity to moderate or weak sensitizers (Table 13-1). Modifications of the Draize test, such as an increase of test concentrations[10] or repetition of the induction procedure[11] have been made, but without a substantial increase in the sensitivity of the method.

The Bühler test[12-14] was developed to screen strong and moderate sensitizers prior to testing in man. According to Bühler,[12] the test is more sensitive than the intradermal method developed by Draize and, when properly performed, comparably sensitive to that of an adjuvant-type test.[14] One of the advantages is that the route of application is similar to that encountered in human use. The major disadvantages of the test were the use of different protocols and the very low incidences of positive animals obtained with known sensitizers (Table 13-2).

TABLE 13-1
Results Obtained According to Draize Protocol and Modifications

Compound	Bühler[12]	Magnusson[3]	Klecak[17]	Sharp[11]
TCSA	0/10	2/25		
PPD	0/10			
Benzocaine	0/10	0/25		
Pot. dichr.	1/10	3/20		
Formaldehyde	1/10	1/20		
Cinn. ald.			+	+
Citral			+	−
Citronellal			−	−
Salicylates			−	−

Note: TCSA = tetrachlorosalicylanilide, PPD = p-phenylene diamine, Pot. dichr. = potassium dichromate, Cinn. ald. = cinnamic aldehyde.

The latter could be a result of not performing the test properly. However, a false negative result has been published by Bühler with CPY,[13] a pyrazoline derivate, which induced clinical cases in a short period after its introduction in washing powders.[15] The ability of the test to detect known allergens has been improved using an occlusive application device, the Hill Top Chamber.®[16]

The Open Epidermal Test (OET) uses epidermal open applications for induction and challenge.[17] This is the only test using different concentrations of the test compound for induction and challenge. According to Klecak et al.[17] it is possible to define the minimal sensitizing and elicitation concentration for each compound. Most experience with the test has been gained testing fragrances. Consequently, its sensitivity for predictions of the sensitizing potential of industrial chemicals is not known.

TABLE 13-2
Results Obtained According to Bühler Protocol

		Polikandritou[16]	
Compound	Bühler[12,13]	HTC	WP
TCSA	8/10		
PPD	10/10	9/10	5/10
Benzocaine	2/10		
Pot. dichr.	1/10	6/10	3/10
Formaldehyde	3/10		
Neomycin		7/10	2/10
CPY	0/20		

Note: TCSA = tetrachlorosalicylanilide, PPD = p-phenylene diamine, Pot. dichr. = potassium dichromate, CPY = 1-(3-chlorophenyl)-3-(4-chlorophenyl)-2-pyrazoline, HTC = Hill Top Chamber®, WP = webril patch.

TABLE 13-3
Results Obtained According to Maximization Protocol

Compound	Magnusson/Kligman[3]	Maurer/Hess Standard Protocol[18]	Modified Protocol
TCSA	18/25		
Benzocaine	7/25	6/10	5/10
Pot. dichr.	18/24		8/10
Formaldehyde	10/20	9/10	7/10
Neomycin	18/25		
DNCB		20/20	20/20
PPD		20/20	10/10
CPY		20/20	10/10

Note: TCSA = tetrachlorosalicylanilide, Pot. dichr. = potassium dichromate, DNCB = dinitrochlorobenzene, PPD = p-phenylene diamine, CPY = pyrazoline derivate.

B. ADJUVANT TESTS: MAXIMIZATION TEST, SPLIT ADJUVANT TEST, FCA TEST, OPTIMIZATION TEST

The major steps introduced to enhance the sensitivity of the guinea pig test in the maximization test of Magnusson and Kligman[3] were (1) injections of FCA during the induction phase, (2) the use of maximal tolerated concentrations, and (3) irritation of the application site before the epidermal induction treatment, for those for which it is not itself the irritant.

The sensitivity of the test is high and all known sensitizers which are negative in the Draize test are positive in the maximization test (Table 13-3). A modification of the protocol for the use of noninjectable compounds was shown to be as sensitive as the original method.[18]

Maguire and Chase[19-21] also used the combination of adjuvant injections and epidermal application of the test compound in the *split adjuvant test.* Positive results with benzocaine indicate that the method is sensitive. It is claimed that the method of exposure to the test compound is more "natural" than that used in the maximization test. However, it is questionable because the application site has to be shaved before application "till glistening" or pretreated with dry ice.

Similar to the OET, not much experience with compounds other than fragrances tested in the *FCA test* is published.[9,17] The induction consists of five injections of the compound in adjuvant and epidermal open challenge application. The aim when developing the *optimization test*[22,23] was to provide a test which was as sensitive as the maximization test, which used standard concentrations for the intradermal injections to permit easy comparison of the sensitizing potential of various compounds, and in which skin reactions could be evaluated objectively. The experience with this method is based on many different chemical classes (Table 13-4).[9] The limited use in other laboratories

TABLE 13-4
Results Obtained According to Optimization Protocol

Compound	Maurer[9] Incidence of Positive Animals i.d. chall.	epid. chall.
Dinitrochlorobenzene	20/20	20/20
Formaldehyde	20/20	10/20
Tetrachlorosalicylanilide	11/20	19/19
p-Phenylene diamine	20/20	13/20
Potassium dichromate	19/20	13/20
Pyrazoline derivate (CPY)	19/20	17/20
Neomycin	11/20	2/20
Methyl paraben	3/20	4/20
Propyl paraben	10/20	0/20
Phenoxy ethanol	1/20	0/20
	13/20[a]	0/20
Phenyl propanol	2/20	0/20
	14/20[a]	3/20

Note: i.d. chall. = intradermal challenge, epid. chall. = epidermal occlusive challenge.

[a]   2% induction concentration.

is probably a consequence of the longer duration of the test, because separate intradermal and epidermal challenge applications are regularly performed. The two routes of challenge applications makes it possible to get information on the effects of a substance on intact and diseased skin. This may be especially important when testing the active ingredients of dermatics or transdermal therapeutic systems.

C. CHANGES IN GUIDELINES

The OECD guideline of 1981 listed all seven methods without giving preference to a certain method. The EEC favored the use of the maximization test. Both guidelines described the test in tabular form and many modifications of the original protocol were possible. Approximately 5 years ago, a working group of the Federal Health Office in Germany started to discuss reasonable changes to the existing guidelines. A limited number of tests were proposed for a new guideline (two adjuvant tests, two non-adjuvant tests), and special preference was given for the maximization protocol.[24] The new EEC guideline[92] includes only the maximization and the Bühler test.

A review of predictive animal and human tests was published in 1990 by the European Industrial Ecology Toxicology Center.[25] From a survey in Europe and the U.S. it was apparent that the maximization test is the most frequently used test in Europe, but that the Bühler test was most common in the U.S. Therefore, ECETOC proposed that preference be given to these two tests. They recommended that the tests be performed according to the original literature and that moderate and/or weak allergens be used for validation in each laboratory.

In May 1991 an OECD expert meeting was held in Paris to discuss skin sensitization testing. The major changes suggested for a new guideline, which was accepted by the council on July 17, 1992 are[7]:

1. The protocol of Magnusson and Kligman and that of Bühler are primarily recommended. Both protocols are described in detail.
2. If there are special reasons, other protocols can be used.
3. New reference compounds for internal validation are recommended (hexyl cinnamic aldehyde, 2-mercaptobenzothiazole, bezocaine) which have been proven to be sensitizers in two laboratories recently.[26]
4. If a positive result in an early screening of compounds has been obtained in a mouse test (MEST or LLNA), no additional guinea pig test is necessary for registration of a new chemical. In contrast, a negative result in a mouse test must be confirmed in a guinea pig test, because the threshold of sensitivity of the mouse tests is not yet fully known.

The new EEC guideline,[92] in contrast, does not include mouse tests.

III. FACTORS INFLUENCING PREDICTIVE GUINEA PIG TEST RESULTS

Besides the choice of protocol, there are several factors which influence the outcome of a sensitization study. They have been discussed in detail by Magnusson and Kligman[27,28] and Maurer.[8] Only two important factors are discussed here.

A. SELECTION OF INDUCTION CONCENTRATIONS

Magnusson and Kligman[27] found that the number of positive animals is dependent on the induction concentrations used. They recommended concentrations up to 5% for the intradermal injections and 25% for epidermal treatment if the compound was not a local irritant and was systemically well tolerated. With strong allergens like dinitrochlorobenzene (DNCB) much lower concentrations are already sufficient to induce sensitization. Is it reasonable to use even higher concentrations than Magnusson recommended to avoid any false negative response? To use one concentration in a larger group of animals has generally given priority over the use of several doses in smaller groups of animals. If maximal concentrations are used in every step and strong local effects are induced, a diminished response rate may be possible. Schäfer et al.[29] studied the induction of sensitization to para-substituted benzenes using various concentrations. They concluded from the studies with paraphenylene diamine that a concentration of 1 to 2% is optimal for sensitization and that higher concentrations have an inhibitory effect.

Roberts and Williams[30] composed a parameter called relative alkylation index (RAI) based on *in vitro* chemical reactivity, lipophilicity (partition coefficient), and induction dose. They could show with various sultones that with increasing induction doses, the frequency of reactions can only be increased to a certain level, after which the frequency decreases (overload region). Similar results were obtained with alkyl transfer agents[31] tested in the single adjuvant injection test.

A nonlinear relationship between concentration and response was also found by Andersen[32] for formaldehyde in the maximization test. Linear and nonlinear relations were found by Rohold et al.[33] for nickel sulfate in the maximization test. The responses at the 48-hr evaluation were linearly related to the intradermal induction dose. The readings at 72-hr, in contrast, showed a nonlinear dose relationship.

B. REACTION ASSESSMENT

A repeated point of discussion is the type of reaction evaluation. The visual assessment of erythema reactions and the assessment of the skinfold thickness increase by palpation according to grading scales is subjective and dependent on the person performing the assessment. But with a careful selection of the challenge concentration, avoiding any irritation in all controls, and good training of the technicians, reproducible results are obtained from repeated studies with reference compounds.

Objective assessments are possible, but they are often time-consuming and do not always clearly distinguish between allergic and irritant reactions. An easy, objective way to measure reactions after intradermal injections is the measurement of the reaction diameter and the skinfold thickness with a caliper as it is performed in the optimization test for the induction and challenge reactions.[9,23] To use several parameters instead of skinfold thickness alone has been proposed by Scheper et al.,[34] because the influence on the parameters evaluated is compound related. Skinfold thickness measurements after occlusive application of the test compound may already give difficulties, because occlusion itself has an acanthotic effect.

Another alternative is histological assessment. This is time-consuming, and conflicting results have been published on the possibility of differentiating between allergic and irritant reactions. Reitamo et al.[35] used α-naphthyl acetate esterase and endogenous peroxidase as markers for inflammatory cells and could not find a statistical difference between irritant and allergic reactions in patients. Kanerva et al.[36] examined the distribution pattern of the Langerhans cell (LC) and their contact with mononuclear cells in sequential biopsies from allergic and irritant reactions. They found no difference in 28 patients tested. Scheynius and Fischer[37] examined allergic reactions to nickel and cobalt, and irritant effects to sodium lauryl sulfate in nine patients. The expression of HLA-DR on keratinocytes was only found in 9 of 14 allergic reactions, but not in irritant ones. The evaluation period was 4 to 20 days after treatment. A difference in the distribution of LC in the epidermis and dermis in irritant and allergic reactions was found by Marks et al.[38] In allergic reactions, LC in the dermis were predominately perivasculary 2 to 14 days after treatment, in irritant reactions the LC were widely distributed in the dermis 2 days after treatment, but the number of LC was markedly reduced 4 to 7 days afterwards.

Much less work has been done in experimental animals. A useful tool for the differentiation of weak to moderate allergic reactions from irritant reactions in guinea pigs was found by Robinson et al.[39] to be the cutaneous basophil hypersensitivity response. Guinea pigs sensitized with oxazolone, citronellal, or cinnamic aldehyde showed a clear quantitative difference in basophil counts to animals treated the first time with the same compounds or with sodium lauryl sulfate. In most cases the values obtained 24 hr after challenge were higher than those obtained after 72 hr.

Maurer et al.[40] compared standard histology and immunohistological evaluation of LC in irritant and allergic reactions to DNCB in guinea pigs. With standard histology, it was impossible to differentiate between irritant and allergic response. The immunohistochemical quantification of LC in epidermal sheets or in tissue sections revealed contradictory results and it was concluded that no clear differentiation between an irritant and allergic reaction can be made.

Various publications have demonstrated difficulties in differentiating irritant and allergic reactions by histological methods. There are differences in the process of the two modes of reactions. But in predictive testing, such time-dependent evaluation is not possible and, therefore, histology is not an ideal tool in routine sensitization testing.

Noninvasive techniques have been used quite frequently in man in recent years to evaluate objectively skin responses and to differentiate between weak and moderate, or doubtful and weak reactions. As mentioned by Li et al.,[41] much less experience is available in animals. They used laser doppler flowmetry in guinea pigs sensitized with DNCB. Based on the results presented, it was possible to distinguish between negative and positive response, but not between different positive intensities.

Evaluation of reactions by a colorimetric method was used by Rohold et al.[42] in guinea pigs sensitized with nickel sulfate. The variation in responses in negative and weakly positive skin sites made it impossible to differentiate between negative and positive guinea pigs.

Objective skin reaction assessment is still difficult or time-consuming. Careful selection of challenge concentrations is still an important factor. If doubtful reactions are seen in the majority of reacting animals, repetition of the challenge procedure is recommended after an additional rest period. During a 1-year period, the erythema reactions of 17 out 100 tested compounds were very weak and the incidence lower than 50%. After a second challenge, the results were much clearer; in 8 out of 17 compounds the incidence of positive animals significantly increased and in 9 of 17 cases the reactivity was clearly diminished.[18]

IV. RELEVANCE OF GUINEA PIG DATA FOR MAN

A positive result in a guinea pig test shows the skin sensitizing potential of the test compound. Magnusson and Kligman[3] classified the compounds in five classes of potency. Even in the case of a negative result in a guinea pig maximization test, the compound is classified as a weak sensitizer. The background for this classification is the long-lasting experience in occupational dermatology of the two clinicians and the experience that every compound may at some time induce an allergic reaction in man.

In the early phase after introduction of adjuvant tests for predictive screening in my laboratory, some of the compounds were additionally tested in repeated insult patch tests (RIPT) in man. The results obtained showed a good correlation between positive effects after epidermal application in guinea pigs and positive effects in man.[9] It was also apparent that the number of volunteers is a crucial factor for the relevance of a RIPT and that more than 50 volunteers per group should be tested to get an equally good prediction, as in guinea pigs.

The correlation between experimental animal and human studies is also dependent on the protocol used. The difference in sensitivity of adjuvant and non-adjuvant tests was clearly demonstrated by Marzulli and Maguire (Table 13-5).[64] This difference has been taken into account for the labeling of compounds according to Commission Directive.[43] Any compound inducing at least 30% positive animals in an adjuvant test must be labeled with the risk phrase R43. In the case of a non-adjuvant tests 15% are sufficient.

As mentioned before, not all compounds positive in animals are sensitizers in man. The risk of sensitization is not only dependent on the sensitizing potential of the compound, but also on the frequency of contact, the intensity of contact, the type of skin condition, the genetic background, and environmental factors. A few examples are discussed below.

Nickel sulfate is generally a weak sensitizer in guinea pigs and one of the most frequent sensitizers in man (Section III.B). The reason for the high number of sensitized people is the wide distribution. Similar experience is known from the parabens used as preservatives. They have a low sensitizing potential in the optimization test (Table 13-4), but due to their wide use in cosmetics and dermatics, the parabens are in the list of standard allergens which are used for diagnostic patch tests. Substitution of parabens as preservatives must be made very carefully. Alcohols and glycols showed no sensitizing potential when tested with the standard concentration of 0.1%, but showed a higher sensitizing potential than the parabens when they were tested at their approximate use concentrations (Table 13-4). Replacing the parabens by these compounds would only reduce the number of sensitized people for a short time, and sensitization to the substitutes would increase with time and frequency of use. A good example is Kathon CG which is a very strong sensitizer in guinea pigs and which induced sensitization in man shortly after introduction of the compound, even if the compound is used in ppm concentrations.[44,45]

TABLE 13-5
Correlation Between Predictive Animal and Human Tests

Guinea Pig Protocol	Assessment with Predictive Human Tests		
	Agreement	False Pos.	False Neg.
Draize	14	1	15
Bühler	10	0	20
Maximization	29	0	1
Split adjuvant T.	22	2	6

Adapted from Marzulli, F. and Maguire, H. C., in *Dermatoxicology*, Hemisphere, Washington, 1987. With permission.

An example that a negative result in guinea pigs does not mean a negative result in humans is the drug clonidine, which only induces contact sensitization in patients when used as TTS devices. The frequency of allergic reactions increases with the duration of therapy.[46] Scheper et al.[47] used the cumulative contact enhancement test in guinea pigs and could not induce allergic responses. Even after cyclophosphamide and/or croton oil pretreatment, only a few of the treated animals showed skin reactions.

The influence of the type of skin is well demonstrated in the frequency of responses in the general population and in the high risk group of leg ulcer patients (Table 13-6). The frequency of positive reactions to nickel and chromate demonstrates that the response rate is not generally elevated in leg ulcer patients. The frequency to parabens and balsam of Peru, in contrast, shows how important generally weak sensitizers may be in special risk groups with long lasting treatment under occlusion.

It can be concluded that not all positive compounds in animals have the same relevance for man. Knudsen et al.[93] proposed criteria for the classification of allergens based on their experimental potential in animals and clinical experience.

V. ALTERNATIVE TESTING

A. MOUSE TESTS
The guinea pig replaced the mouse as the animal of choice many years ago and was the only accepted species for a long time. The latest predictive *in vivo* methods established use the mouse again.

Maisey and Miller[48] used a higher vitamin A content in the pellets to enhance the response of mice. Positive results were obtained with benzocaine, formaldehyde, or citral, all of which had not previously been positive in mice.

Other methods used the combination of adjuvant injections and epidermal compound application during induction and epidermal treatment of the ear for challenge, such as in the mouse ear swelling test (MEST) by Gad et al.[49] or mouse ear sensitization assay.[50] The ear thickness increase was used as an

TABLE 13-6
Epidemiology in Risk Groups

Compound	General Population	Patients with Leg Ulcers
Nickel sulfate	10.0%	5.5%
Potassium chromate	11.0%	6.1%
Neomycin sulfate	4.1%	15.3%
Parabens	2.1%	11.5%
Balsam of Peru	5.3%	25.4%

Adapted from Maurer, Th., *A Manual of Predictive Test Methods*, Marcel Dekker, New York, 1983. With permission.

objective parameter for the reaction assessment. The measurements with a caliper are objective, but require the extensive experience of the technicians to obtain reproducible results. The reproducibility of measurements in nonanesthetized animals is questioned. Dunn et al.[51] tried to reproduce some of the results obtained by Gad et al.[49] Based on negative results with moderate sensitizers, such as nickel sulfate and methyl methacrylate, they concluded that the mouse ear swelling test is not sensitive enough to detect moderate and weak sensitizers.

The mouse test developed by Kimber et al.,[52,53] the local lymph node assay (LLNA), is based on a different approach. Mice are treated on the ear daily for 3 days and the proliferation of the local lymph node is measured by assessing the incorporation of radioactive thymidine. In the first protocol, the thymidine incorporation was measured *ex vivo* in lymphocyte cultures; in a second protocol the thymidine was given intravenously to the animals and the lymph node was subsequently excised. The advantages of this method are the short time needed and the objective parameter measured. The disadvantage is the necessity of using radioactive thymidine.

Much work was put into early international validation of the LLNA. The results showed that the LLNA is reproducible in other laboratories,[54-56] and the method can be classified as validated. Quite close concordance was also found between the *in vivo* Bühler test and the LLNA.[57] Good correlation in assessment of various sensitizing potencies was also found between the guinea pig maximization test and the LLNA in guinea pigs.[58] The results with the LLNA in guinea pigs confirm the general mechanism of lymph node proliferation after allergen exposure. However, differences in the response rates to the various chemicals have been found in the two species compared (Table 13-7). In a comparison between guinea pig maximization test and mouse LLNA with 40 chemicals, Basketter and Scholes[59] found a good correlation with most chemicals showing a moderate to strong potential in the maximization test. Some chemicals which are mild sensitizers in the maximization test, such as benzocaine, were negative in the LLNA.

TABLE 13-7
Results Obtained with LLNA

Compound	LLNA Mouse PI Range (Basketter[59])	Max. Guinea Pig Incidence (Basketter[59])	LLNA Guinea Pig Max. PI (Maurer[58])
DNCB	6.2–24.0	100%	6.4
Formaldehyde	3.7–5.8	90%	2.02
PPD	12.8–23.3	100%	7.06
Pot. dichr.	3.5–10.4	90%	N.D.
Benzocaine	0.9–2.0	50%	2.41
Cinnamic ald.	12.5–18.4	100%	11.5
Citral	2.1–9.3	50%	N.D.
Nickel sulfate	1.1–1.5	30%	N.D.

Note: PI range = proliferation index range in 3 experiments, max. PI = maximal value obtained with induction concentrations up to 5%, DNCB = dinitrochlorobenzene, PPD = p-phenylene diamine, N.D. = not done

B. *IN VITRO* TESTS

The complex interactions between different cell types for induction and challenge and the difficulties of culturing LC in an active form are the main reasons why a predictive *in vitro* method is not yet available.

Parish[60] reviewed various possible approaches for an *in vitro* method based on cellular interactions during induction and challenge. Von Blomberg and Scheper[61] reviewed the *in vivo* mechanisms of induction and challenge and looked for *in vitro* correlates. Special emphasis was noted on the LLNA and its *in vitro* lymphocyte proliferation assessment. No real *in vitro* system was described.

Becker et al.[62] studied the *in vitro* endocytotic activation of murine LC after exposure to solvents, irritants, or contact allergens. The results of this research project are very interesting, but as a predictive method, more compounds must be tested and validation in other laboratories is needed. Other projects are in progress, e.g., supported by the Center for Alternatives to Animal Testing (CAAT) in Baltimore or by the EEC.[63] Results are not yet available.

VI. PHOTOALLERGENICITY TESTS

A. GENERAL COMMENTS

The clinical and epidemiological aspects of phototoxic and photoallergic reactions are reviewed in Chapter 11. Photosensitivity reactions due to drugs have been summarized, for example, by Epstein and Wintroub.[65] Clinical reactions may be observed after local or systemic exposure of humans to chemicals or drugs. Nearly all predictive methods, however, are based on local exposure of test animals, even for compounds which are used systemically in humans. Epstein stated in his publication: "however, to date (1985), no animal models have proved to be predictive for screening the potential for chemicals which had not previously been identified to produce photoallergic contact reactions."

Before and since 1985, many chemicals have been tested for their photoallergenic potential and the question should be: have compounds not been detected as photoallergens in screening tests which have induced problems in humans later?

The first predictive method was published by Vinson and Borselli[66] and halogenated salicylanilides were the main reference compounds used. Similar to the development of methods for the evaluation of contact allergenic potentials, attempts were made to increase the sensitivity by stimulation of the immune system with intradermal adjuvant injections or by enhancing the penetration with chemical or mechanical irritation of the application site. The guinea pig was the animal of choice for many years, but the mouse has gained importance. No international guideline exists yet for industrial chemicals.

Factors which play no role in predictive contact allergy testing are important in predictive photoallergenicity testing, such as light source and emission spectrum, UV-dose and measurement of UV-intensities. These factors are reviewed in detail by Harber et al.[67,68] and Maurer.[8]

At the beginning of photosensitivity testing, carbon arc lamps were used for the simulation of sunlight. Their instability in light intensity was one of the biggest disadvantages of this type of burner. Heat was the major discomfort of the xenon arc light sources used later, even when the light source was cooled by water. New metal halonid burners are now the best choice for solar light simulation. Fluorescent tubes with defined UV-B or UV-A emissions have also been used. For many years they were the only light sources suitable for *in vitro* testing. The main advantages are the low costs of tubes and equipment. The disadvantage is the limited range of spectrum compared to sunlight.

As important as the type of light source is the control of the emission of the light source, and the control of the actual intensity for each test. Examples of types of instruments are given in reviews by Forbes et al.[69] or Parrish et al.[70] The range of sensitivity of the detectors and the main emission range of the light source must correlate, and regular calibration of the instrument should be guaranteed. The light intensity must be controlled for every test, because the intensity of the light source changes with age and frequency of ignitions. A change in the output in the UV-B range may also result from a change in the transmission of the filters used.

To classify reactions as photoallergic at the end of a test is only possible when various controls have been included in the testing, because chemicals may have a phototoxic and photoallergenic or a contact and photocontact allergenic potential simultaneously. A schema of controls is given by Parker et al.[71]

B. GUINEA PIG TESTS

A selection of different photoallergenicity protocols is summarized in Table 13-8. It shows that a wide range of induction and challenge concentrations and a number of treatments have been used in the different methods. Open epidermal application of the test compound was used in many cases. A combination of occlusive application and UV irradiations was also used.[72] The advantage of adjuvant injections during the induction phase, as in contact sensitization testing, was first published by Maurer et al.[73] and was subsequently used in many other methods.[72,74-76]

During the development of the method of Maurer et al.,[73] it was observed that the induction concentration is an important factor and that the highest concentration may not always be optimal. For example, 1% solutions of halogenated salicylanilides induced fewer positive responses than 0.1% solutions of the same compounds. Photoallergenic compounds must absorb light to induce photoreactions. The combined effect of chemical and light can only take place when sufficient chemical and light are reaching the site of action in the skin. High concentrations may induce higher penetration rates into the skin, but also lower penetration of light due to stronger absorption on the surface (filter effect).

Similar observations were made by Ramsay and Pellett[77] in studies with TCSA. They observed not only a concentration dependency but also a UV-dose dependency of induction and challenge. Lovell and Sanders[76] confirmed the concentration dependency with different musks and the UV dose dependency in studies with TCSA.[78] They even showed that the number of positive animals decreased when high UV doses were used. It is known that parts of the UV range have immunosuppressive effects; Miyachi and Takigawa[79] demonstrated it, for example, in a mouse model.

C. ALTERNATIVE TESTS

One of the first photoallergenicity tests in the mouse was published in 1982 by Maguire and Kaidbey.[80] In many tests the induction was performed on the flank or the abdomen of the animals and the challenge was conducted with a single application on the ear. The ear thickness increase after challenge was proposed as an objective measure of the edema formation, in contrast to subjective graduation of erythema intensities. A selection of protocols is summarized in Table 13-9.

Many studies have been performed with TCSA which induced reactions in the mouse, but not as easily as in guinea pigs. Various enhancement treatments have been used; cyclophosphamide was the most effective one in the mouse. Brown et al.[81] showed that the reactivity can be optimized by a careful selection of chemical concentration and UV dose. With TCSA, concentration and UV dependency of induction was found in the mouse, too.

Whether UV-B irradiation is necessary or beneficial during induction in guinea pigs has been discussed several times. Gerberick and Ryan[82] showed that in the mouse, the combination of UV-A and UV-B is beneficial for the induction in the case of 6-MC, but not for TCSA or musk ambrette.

Miyachi and Takigawa[79] showed that LC are necessary for the induction of photoallergic reactions as in contact sensitization, and that UV-B has an

TABLE 13-8
Guinea Pig Photoallergenicity Protocols

Authors	Test Concentration	No. of Treatments	Results
Vinson[66]	Induction 2%	5	TCSA+, TriCSA+,
	Challenge 0.1–0.2%	2–3	TBS–, Bith.+,
			TCC–
Morikawa[90]	Induction 2%	10	TBS+, DBS+,
	Challenge 0.1–1%	1	MBS+, TCSA+
			Bith.(+), Hex.–
Kochevar[91]	Induction 5–10%	4	TCSA+, Musk
	Challenge 2.5–7%	1	ambrette+
Maurer[73]	Induction 0.1%	12	TCSA+, TBS+,
	Challenge 0.1%	3	Bith.+, TCC–,
			CPZ+, Musk
			ambrette+,
			6-MC–, etc.
Gerberick[72]	Induction 0.1–10%	6	Musk ambrette+,
	Challenge 0.01–0.1%	1	TCSA+,
			PABA+,
			Eusolex 8020–
Lovell[76]	Induction 0.1–10%	2	Musk ambrette+,
	Challenge 0.1–1.5%	1	musk moskene+,
			musk xylene–
Parker[71]	Induction 10%	9	Musk ambrette+,
			musk ketone–,
			musk xylene–,
			musk moskene–

immunosuppressive effect. Gerberick et al.[83] demonstrated that dendritic cells move to the regional lymph node and that TCSA induces lymph node proliferation in a manner similar to contact allergens in the local lymph node assay. In a larger experiment with different chemicals, Scholes et al.[84] confirmed the results of Gerberick with TCSA and also got positive results with fentichlor; on the other hand negative results were obtained with 6-MC and nonreproducible results with musk ambrette. Based on reactions obtained with photoirritants such as acridine and anthracene, Scholes et al.[84] concluded that the photoLLNA may only screen out strong or moderate photoallergens and that the specificity of the reactions may be questionable.

In the case of pharmaceuticals, induction after systemic exposure is clinically relevant. No animal model demonstrating successful induction of photoallergy after peroral exposure is available, although this is the most relevant exposure in man. A small amount of experience has been obtained from animal studies after i.p. injections. Wirestrand and Ljunggren[85] could demonstrate photoallergic reactions to quinidine after i.p. induction and challenge with cyclophosphamide pretreatment. With a similar method, Giudici et al.[86] induced positive reactions to sulfanilamide and chlorpromazine.

TABLE 13-9
Mouse Photoallergenicity Protocols

Authors	Test Concentration	No. of Treatments	Results
Maguire[80]	Induction 1%	2	TCSA+, 6-MC?,
	Challenge 1%	1	CPZ?
Brown[81]	Induction 0.5–6%	2	TCSA+
	Challenge 0.5%	1	
Gerberick[82]	Induction 1–20%	3	TCSA+,
	Challenge 1–10%	1	Musk ambrette+
Wirestrand[85]	Induction 100 mg/kg	2	Quinidine
	Challenge 100 mg/kg	1	

Little information is available on *in vitro* photoallergenicity tests. Maguire[87] could show that photoallergens can be coupled *in vitro* with lymphoid cells and that photoallergic reactions can be elicited when such cells are transferred to recipient mice. A lymphocyte proliferation response was induced *in vitro* by Gerberick et al.,[88] but the cells were obtained from *in vivo* treated animals.

The only real *in vitro* study was performed by Pendlington and Barrat.[89] They found a good correlation between the photochemical binding of photoallergens with human serum albumin and the known photoallergenic potential of the compounds in man. A sunscreen was included as a negative reference compound in the test series and showed no photobinding with serum albumin.

VII. DISCUSSION AND RECOMMENDATIONS

A. CONTACT ALLERGY

Since the experiments of Draize, predictive contact allergenicity testing in guinea pigs has much improved and the knowledge on the reaction mechanism has increased. For screening purposes, control of the guinea pig strain and the performance of the test system (validation with reference compounds) is probably more important than whether an adjuvant protocol or the Bühler test is used. The sensitivity and specificity of the tests is generally high. How many false positive reactions are produced is not known, because many sensitizing chemicals in guinea pigs do not reach the market.

Even the number of major methods listed in OECD and EEC guidelines has been reduced, and there are still several ways to evaluate the contact allergenicity potential of a chemical and to assess the sensitizing risk for man.

A single adjuvant test may be enough to evaluate the sensitizing potential for labeling purposes. However, for formulations it may be advantageous to test the active chemical and the formulation. The dilution of a compound in a formulation and the interaction with other ingredients may alter the sensitizing potential of the compound.

Adjuvant methods are sensitive and should yield reproducible results with known sensitizers. The Bühler test, however, is more dependent on the quality of performance and, therefore, the internal validation with reference compounds is very important.

The importance of internal validation is underlined in a recent publication by Basketter et al.[26] on the newly recommended reference compounds.

New mouse tests are now accepted in the OECD guideline, and subsequently, a new testing strategy is possible. Information on the sensitizing potential can be obtained in an early stage of development with a small amount of compound and in a short time. A reduced number of guinea pig tests may be needed and the protocols can be better designed in a later stage of development and related to the expected future use.

As guidance for further work, the following recommendations can be made:

- Further validation of the mouse test is needed to clarify the threshold of sensitivity of each method compared to guinea pig tests.
- Because comparison of our own data and literature data is difficult, in-house validation with allergens of different potentials is highly recommended.
- Research work should be supported to evaluate possibilities for an *in vitro* predictive method of contact sensitization.
- Comparisons of classification of allergens from guinea pig tests and from clinical studies should be encouraged to get a better estimation of the prediction of contact allergenic potential from animal experiments.

B. PHOTOALLERGY

The number of photoallergy experiments is much lower than in the case of contact allergy. The number of tested chemicals is much lower and they are from a limited number of chemical classes. No guideline is available and comparable results with a certain method are not available from various laboratories. Standardization is necessary and interlaboratory validation is needed. Such a validation could be done with a proposed protocol for the OECD by a working group initiated by R. Nielsson (Sweden).

International validation would be of great importance because the performance of a photoallergenicity test is much more complex than a regular sensitization test. The selection of chemical concentration and UV dose is important, as false negative results are possible. Additional difficulties result from the fact that photoallergens may often act also as phototoxins or contact allergens. Controls for a good distinction between the three modes of action are important.

A combination of the evaluation of phototoxins and photoallergens in the same test is, however, not recommended, because generally only one concentration is used for the induction of photoallergy and because phototoxicity is much more dose-dependent and several concentrations should therefore be tested.

After activation of a chemical by light, the mechanism of sensitization in photoallergy and contact allergy are comparable. However, the photoLLNA was neither as sensitive nor as specific as the LLNA for contact allergens. Consequently, further work is required to develop more refined methods in this area.

The recommendations for photoallergy are, therefore, as follows.

- Standardization and international validation of an animal method is needed.
- The guinea pig methods available today using topical application of the test compounds and predictions are made for compounds active in man after topical or systemic administration. The relevance of results from topical treatment in animals for systemic exposure in man must be further evaluated.

REFERENCES

1. **Draize, J. H., Woodward, G., and Calvery, H. O.,** Methods for the study of irritation and toxicity of substances applied topically to the skin and mucous membranes, *J. Pharm. Exp. Therap.*, 82, 377, 1944.
2. **Draize, J. H.,** Intracutaneous sensitisation test on guinea pigs, in *Appraisal of the Safety of Chemicals in Food and Cosmetics,* Association of Food and Drug Officials of the United States, Austin, Texas, 1959.
3. **Magnusson, B. and Kligman, A. M.,** The identification of contact allergens by animal assay. The guinea pig maximization test, *J. Invest. Dermatol.*, 52, 268, 1969.
4. **Waksman, B. H.,** Adjuvants and immune regulation by lymphoid cells, *Springer Semin. Immunopath.*, 2, 5, 1979.
5. **OECD,** Organisation for Economic Co-operation and Development — OECD Guideline for Testing of Chemicals, No. 406, Skin Sensitization, 1981.
6. **EEC,** Commission Directive of 25 April 1984 adapting to technical progress for the sixth time Council Directive 67/548/EEC on the laws, regulations and administrative provisions relating to the classification, packaging and labelling of dangerous substances, *Off. J. Eur. Commun.*, L 251, 27, 113, 1984.
7. **OECD,** Guideline for Testing of Chemicals, Skin Sensitization, Adopted by the Council on July 17, 1992.
8. **Maurer, Th.,** Contact and Photocontact Allergens, in *A Manual of Predictive Test Methods,* Marcel Dekker, New York, 1983.
9. **Andersen, K. E. and Maibach, H. I., Eds.,** *Contact Allergy, Predictive Tests in Guinea Pigs,* Karger, Basel, 1985.
10. **Vos, J. G.,** Skin sensitization by mercaptans of low molecular weight, *J. Invest. Dermatol.*, 31, 273, 1958.

11. **Sharp, D. W.,** The sensitization potential of some perfume ingredients tested using a modified Draize procedure, *Toxicology*, 9, 261, 1978.

12. **Bühler, E. V.,** Delayed contact hypersensitivity in the guinea pig, *Arch. Dermatol.*, 91, 171, 1965.

13. **Bühler, E. V. and Griffith, J. F.,** Experimental skin sensitization in the guinea, pig and man, in *Animal Models in Dermatology*, Maibach, H. I., Ed., Churchill Livingstone, Edinburgh, 1975, 56.

14. **Robinson, M. K., Nusair, T. L., Fletcher, E. R., and Ritz, H. L.,** A review of the Bühler guinea pig skin sensitization test and its use in a risk assessment process for human skin sensitization, *Toxicology*, 61, 91, 1990.

15. **Osmundsen, P. E. and Alani, M. D.,** Contact allergy to an optical whitener "CPY" in washing powders, *Brit. J. Dermatol.*, 85, 61, 1971.

16. **Polikandritou, M. and Conine, D.,** Enhancement of the sensitivity of the Bühler method by use of the Hill Top Chamber®, *J. Soc. Cosmet. Chem.*, 36, 159, 1985.

17. **Klecak, G., Geleick, H., and Frey, J. R.,** Screening of fragrance materials for allergenicity in the guinea pig, comparison of four testing methods, *J. Soc. Cosmet. Chem.*, 28, 53, 1977.

18. **Maurer, T. and Hess, R.,** The maximization test for skin sensitization potential — updating the standard protocol and validation of a modified protocol, *Fd. Chem. Toxicol.*, 27, 807, 1989.

19. **Maguire, H. C.,** The bioassay of contact allergens in the guinea pig, *J. Soc. Cosmet. Chem.*, 24, 151, 1973.

20. **Maguire, H. C.,** Estimation of the allergenicity of prospective human contact sensitizers in the guinea pig, in *Animal Models in Dermatology*, Maibach, H. I., Ed., Churchill Livingstone, Edinburgh, 1975, 67.

21. **Maguire, H. C. and Chase, M. W.,** Studies on the sensitization of animals with simple chemical compounds, *J. Exp. Med.*, 135, 357, 1972.

22. **Maurer, T., Thomann, P., Weirich, E. G., and Hess, R.,** The optimization test in the guinea pig, *Agents Actions*, 5, 174, 1975.

23. **Maurer, T., Weirich, E. G., and Hess, R.,** The optimization test in the guinea pig in relation to other predictive sensitization methods, *Toxicology*, 15, 163, 1980.

24. **Schlede, E., Maurer, T., Potokar, M., Schmidt, W. M., Schulz, K. H., Roll, R., and Kayser, D.,** A differentiated approach to testing skin sensitization, *Arch. Toxicol.*, 63, 81, 1989.

25. **ECETOC,** European Chemical Industry Ecology and Toxicology Centre, Monograph No. 14, Skin Sensitization, March 9, 1990.

26. **Basketter, D. A., Selbie, E., Scholes, E. W., Lees, D., Kimber, I., and Botham, P. A.,** Results with OECD recommended positive control sensitizers in the maximization, Buehler and local lymph node assays, *Fd. Chem. Toxicol.*, 31, 63, 1993.

27. **Magnusson, B. and Kligman, A. M.,** *Allergic Contact Dermatitis in the Guinea Pig*, Charles C. Thomas, Springfield, 1970.

28. **Magnusson, B. and Kligman, A. M.,** Factors influencing allergic contact sensitization, in *Dermatotoxicology*, 3rd ed., Marzulli, F. N. and Maibach, H. I., Eds., Hemisphere, Cambridge, 1987.

29. **Schäfer, U., Metz, J., Pevny, I., and Röckl, H.,** Sensibilisierungsversuche an Meerschweinchen mit fünf parasubstituierten Benzolderivaten, *Arch. Derm. Res.*, 261, 153, 1978.

30. **Roberts, D. W. and Williams, D. L.,** The derivation of quantitative correlations between skin sensitisation and physico-chemical parameters for alkylating agents, and their application to experimental data for sultones, *J. Theor. Biol.*, 99, 807, 1982.

31. **Roberts, D. W. and Basketter, D. A.,** A quantitative structure activity/dose response relationship for contact allergic potential of alkyl group transfer agents, *Contact Dermatitis*, 23, 331, 1990.

32. **Andersen, K. E.,** Testing for contact allergy in experimental animals, *Pharmacol. Toxicol.,* 61, 1, 1987.
33. **Rohold, A. E., Nielsen, G. D., and Andersen, K. E.,** Nickel-sulfate-induced contact dermatitis in the guinea pig maximization test: a dose-response study, *Contact Dermatitis,* 24, 35, 1991.
34. **Scheper, R. J., Noble, B., Parker, D., and Turk, J. L.,** The value of an assessment of erythema and increase in thickness of the skin reaction for a full appreciation of the nature of delayed hypersensitivity in the guinea pig, *Int. Arch. Allergy Appl. Immunol.,* 54, 58, 1977.
35. **Reitamo, S., Tovanen, E., Konttinen, Y. T., Käyhkö, K., Förström, L., and Salo, O. P.,** Allergic and toxic contact dermatitis: inflammatory cell subtypes in epicutaneous test reactions, *Brit. J. Dermatol.,* 105, 521, 1981.
36. **Kanerva, L., Ranki, A., Mustakallio, K., and Lauharanta, J.,** Langerhans cell-mononuclear cell contacts are not specific for allergy in patch tests, *Brit. J. Dermatol.,* 109, (Suppl. 25), 64, 1983.
37. **Scheynius, A. and Fischer, T.,** Phenotypic difference between allergic and irritant patch test reactions in man, *Contact Dermatitis,* 14, 297, 1986.
38. **Marks, J. G., Zaino, R. J., Bressler, M. F., and Williams, J. V.,** Changes in lymphocyte and Langerhans cell populations in allergic and irritant contact dermatitis, *Int. J. Dermatol.,* 26, 354, 1987.
39. **Robinson, M. K., Fletcher, E. R., Johnson, G. R., Wyder, W. E., and Maurer, J. K.,** Value of the cutaneous basophil hypersensitivity (CBH) response for distinguishing weak contact sensitization from irritation reactions in the guinea pig, *J. Invest. Dermatol.,* 94, 636, 1990.
40. **Maurer, T., Germer, M., and Krinke, A.,** Does the immunohistochemical detection of Langerhans cell help in the differential diagnosis of irritative and allergic skin reactions?, *Prog. Histo. Cytochem.,* 23, 256, 1991.
41. **Li, Q., Aoyama, K., and Matsushita, T.,** Evaluation of contact allergy to chemicals using laser Doppler flowmetry (LDF) technique, *Contact Dermatitis,* 26, 27, 1992.
42. **Rohold, A. E., Nielsen, G. D., and Andersen, K. E.,** Colorimetric quantification of erythema in the guinea pig maximization test, *Contact Dermatitis,* 24, 373, 1991.
43. **EEC,** Commission Directive of 29 July 1983 adapting to technical progress for the fifth time Council Directive 67/648/EEC on the approximation of the laws, regulations and administrative provisions relating to the classification, packaging and labelling of dangerous, *Off. J. Eur. Comm.,* L 257, 1, 1983.
44. **Björkner, B., Bruze, M., Dahlquist, I., Fregert, S., Gruvberger, B., and Persson, K.,** Contact allergy to the preservative Kathon® CG, *Contact Dermatitis,* 14, 85, 1986.
45. **Menne, T., et al.,** Contact sensitization to 5-chloro-2-methyl-4-isothiazolin-3-one and 2-methyl-4-isothiazlin-3-one (MCI/MI), a European multicentre study, *Contact Dermatitis,* 24, 334, 1991.
46. **Maibach, H. I.,** Clonidine: irritant and allergic contact dermatitis assays, *Contact Dermatitis,* 12, 192, 1985.
47. **Scheper, R. J., von Blomberg, B. M. E., De Groot, J., Goeptar, A. R., Oostendorp, R. A. J., Bruynzeel, D. P., and van Tol, R. G. L.,** Low allergenicity of clonidine impedes studies of sensitization mechanisms in guinea pig models, *Contact Dermatitis,* 23, 81, 1990.
48. **Maisey, J. and Miller, K.,** Assessment of the ability of mice fed on vitamin A supplemented diet to respond to variety of potential contact sensitizers, *Contact Dermatitis,* 15, 17, 1986.
49. **Gad, S. C., Dunn, B. J., Dobbs, D. W., Reilly, C., and Walsh, R. D.,** Developement and validation of an alternative dermal sensitization test: the mouse ear swelling test (MEST), *Toxicol. Appl. Pharmacol.,* 84, 93, 1986.
50. **Descotes, J.,** Identification of contact allergens: the mouse ear sensitization assay, *J. Toxicol. Cut. Ocular Toxicol.,* 7, 263, 1988.

51. **Dunn, B. J., Rusch, G. M., Siglin, S. C., and Blaszcak, D. L.,** Variability of a mouse ear swelling test (MEST) in predicting weak and moderate contact sensitization, *Fund. Appl. Toxicol.,* 15, 242, 1990.
52. **Kimber, I. and Weisenberger, C.,** A murine local lymph node assay for the identification of contact allergens. Assay development and results of an initial validation study, *Arch. Toxicol.,* 63, 274, 1989.
53. **Kimber, I., Hilton, J., and Weisenberger, C.,** The murine local lymph node assay for identification of contact allergens: a preliminary evaluation of in situ measurement of lymphocyte proliferation, *Contact Dermatitis,* 21, 215, 1989.
54. **Kimber, I., Hilton, J., Botham, P. A., Basketter, D. A., Scholes, E. W., Miller, K., Robbins, M. C., Harrison, P. T., Gray, T. J. B., and Waite, S. J.,** The murine local lymph node assay: results of an inter-laboratory trial, *Toxicol. Lett.,* 55, 203, 1991.
55. **Basketter, D. A., Scholes, E. W., Kimber, I., Botham, P. A., Hilton, J., Miller, K., Robbins, M. C., Harrison, P. T. C., and Waite, S. J.,** Inter-laboratory evaluation of the local lymph node assay with 25 chemicals and comparison with guinea pig test data, *Toxicol. Methods,* 1, 30–43.
56. **Scholes, E. W., et al.,** The local lymph node assay: results of a final inter-laboratory validation under field conditions, *J. Appl. Toxicol.,* 12, 217, 1992.
57. **Kimber, I., Hilton, J., and Botham, P. A.,** Identification of contact allergens using the murine local lymph node assay: comparisons with the Buehler occluded patch test in guinea pigs, *J. Appl. Toxicol.,* 10, 173, 1990.
58. **Maurer, T. and Kimber, I.,** Draining lymph node cell activation in guinea pigs: comparisons with the murine local lymph node assay, *Toxicology,* 69, 209, 1991.
59. **Basketter, D. A. and Scholes, E. W.,** Comparison of the local lymph node assay with the guinea pig maximization test for the detection of a range of contact allergens, *Fd. Chem. Toxicol.,* 30, 65, 1992.
60. **Parish, W. E.,** Evaluation of in vitro predictive tests for irritaation and allergic sensitization, *Fd. Chem. Toxicol.,* 24, 481, 1986.
61. **Von Blomberg-van der Flier, B. M. E., and Scheper, R. J.,** In vitro tests with sensitized lymphocytes — relevance for predictive allergenicity testing, *Toxicol. In Vitro,* 4, 246, 1990.
62. **Becker, D., Kolde, G., Reske, K., and Knop, J.,** An in vitro test for endocytotic activation of murine epidermal Langerhans cells under the influence of contact allergens, *J. Immunol. Meth.,* 169, 195, 1994.
63. **EEC,** Commission of the European Communities Directorate-General XII, Biotechnology Research for Innovation, Development and Growth in Europe (1990–1993), Development of a predictive in vitro test for detection of sensitizing compounds, Contract No. BIOT CT-900186, 1992.
64. **Marzulli, F. and Maguire, H. C.,** Validation of guinea pig tests for skin hypersensitivity, in *Dermatotoxicology,* 3rd ed., Marzulli, F. N. and Maibach, H. I., Eds., Hemisphere, Washington, 1987.
65. **Epstein, J. H. and Wintroub, B. U.,** Photosensitivity due to drugs, *Drugs,* 30, 42, 1985.
66. **Vinson, L. J. and Borselli, V. F.,** A guinea pig assay of the photosensitizing potential of topical germicides, *J. Soc. Cosmet. Chemists,* 17, 123, 1966.
67. **Harber, L. C.,** Current status of mammalian models for predicting drug photosensitivity, *J. Invest. Dermatol.,* 77, 65, 1981.
68. **Harber, L. C., Armstrong, R. B., and Ichikawa, H.,** Current status of predictive animal models for drug photoallergy and their correlation with drug photoallergy in humans, *J. Natl. Cancer Inst.,* 69, 237, 1982.
69. **Forbes, P. D., Davies, R. E., D'Alosio, L. C., and Cole, C.,** Emission spectrum differences in fluorescent blacklight lamps, *Photochem. Photobiol.,* 24, 613, 1976.
70. **Parrish, J. A., Anderson, R. R., Urbach, F., and Pitts, D.,** *UV-A.* John Wiley & Sons, New York, 1978.

71. **Parker, R. D., Buehler, E. V., and Newmann, E. A.,** Phototoxicity, photoallergy, and contact sensitization of nitro musk perfume raw materials, *Contact Dermatitis*, 14, 103, 1986.
72. **Gerberick, G. F. and Ryan, C. A.,** Contact photoallergy testing of sunscreens in guinea pigs, *Contact Dermatitis*, 20, 251, 1989.
73. **Maurer, Th., Weirich, E. G., and Hess, R.,** Predictive animal testing for photocontact allergenicity, *Brit. J. Dermatol.*, 103, 593, 1980.
74. **Ichikawa, H., Armstrong, R. B., and Harber, L. C.,** Photoallergic contact dermatitis in guinea pigs: improved induction technique using Freund's complete adjuvant, *J. Invest. Dermatol.*, 76, 498, 1981.
75. **Guillot, J. P., Gonnet, J. F., Loquerie, J. F., Martini, M. C., Convert, P., and Cotte, J.,** A new method for the assessment of phototoxic and photoallergic potentials by topical applications in the albino guinea pig, *J. Toxicol. Cut. Ocular Toxicol.*, 4, 117, 1985.
76. **Lovell, W. W. and Sanders, D. J.,** Photoallergic potential in the guinea-pig of the nitromusk perfume ingredients musk ambrette, musk moskene, musk xylene, musk ketone, and musk tibetene, *Int. J. Cosm. Sci.*, 10, 271, 1988.
77. **Ramsay, C. A. and Pellett, F.,** Ultraviolet radiation and chemical requirements for experimental contact photosensitivity, *Photodermatology*, 3, 41, 1986.
78. **Lovell, W. W. and Sanders, D. J.,** Dose-response study of ultraviolet radiation for induction of photoallergy to tetrachlorosalicylanilide in guinea pigs, *Photoderm. Photoimmunol. Photomed.*, 7, 192, 1990.
79. **Miyachi, Y. and Takigawa, M.,** Mechanisms of contact photosensitivity in mice: II. Langerhans cells are required for successful induction of contact photosensitivity to TCSA, *J. Invest. Dermatol.*, 78, 363, 1982.
80. **Maguire, H. C. and Kaidbey, K.,** Experimental photoallergic contact dermatitis: a mouse model, *J. Invest. Dermatol.*, 79, 147, 1982.
81. **Brown, W. R., Furukawa, R. D., Shivji, G. M., and Ramsay, C. A.,** Optimization of tetrachlorosalicylanilide and ultraviolet A doses at sensitization and challenge for contact photosensitivity in the mouse, *Arch. Dermatol. Res.*, 281, 351, 1989.
82. **Gerberick, G. F. and Ryan, C. A.,** Use of UV-B and UV-A to induce and elicit contact photoallergy in the mouse, *Photodermatol. Photoimmunol. Photomed.*, 7, 13, 1990.
83. **Gerberick, G. F., Ryan, C. A., Fletcher, E. R., Howard, A. D., and Robinson, M. K.,** Increased number of dendritic cells in draining lymph nodes accompanies the generation of contact photosensitivity, *J. Invest. Dermatol.*, 96, 355, 1991.
84. **Scholes, E. W., Basketter, D. A., Lovell, W. W., Sarll, A. E., and Pendlington, R. U.,** The identification of photoallergic potential in the local lymph node assay, *Photodermatol. Photoimmunol. Photomed.*, 8, 249, 1991.
85. **Wirestrand, L. E. and Ljunggren, B.,** Photoallergy to systemic quinidine in the mouse: dose-response studies, *Photodermatology*, 5, 201, 1988.
86. **Giudici, P. A. and Maguire, H. C.,** Experimental photoallergy to systemic drugs, *J. Invest. Dermatol.*, 85, 207, 1985.
87. **Maguire, H. C.,** A general method for photohaptenization, *Contact Dermatitis*, 22, 57, 1990.
88. **Gerberick, G., Ryan, C. A., Von Bargen, E. C., Stuard, S. B., and Ridder, G. M.,** Examination of tetrachlorosalicylanilide (TCSA) photoallergy using in vitro photohapten-modified Langerhans cell-enriched epidermal cells, *J. Invest. Dermatol.*, 97, 210, 1991.
89. **Pendlington, R. U. and Barrat, M. D.,** Photochemical binding of photoallergens to human serum albumin: a simple *in vitro* method for screening potential photoallergens, *Toxic. In Vitro*, 4, 307, 1990.
90. **Morikawa, F., et al.,** Technique for evaluation of phototoxicity and photoallergy in laboratory animals and man, in *Sunlight and Man*, Fitzpatrick, T. B., et al., Eds., University of Tokyo Press, 1974.

91. **Kochevar, I. E., Zalar, G. L., Einbinder, J., and Harber, L. C.,** Assay of contact photosensitivity to musk ambrette in guinea pigs, *Contact Dermatitis*, 73, 144, 1979.

92. **EEC,** Commission Directive of 29 December 1992 adapting to technical progress for the 17th time council Directive on laws, regulations and administrative provisions relating to classification, packaging and labeling of dangerous substances, *Off. J. Eur. Commun.*, L 383 A, 131, 1994.

93. **Knudsen, B. B., Wahlberg, J. E., Andersen, I., and Menne, T.,** Classification of contact allergens, *Dermatosen*, 41, 5, 1993.

Chapter 14

SKIN ALLERGY: EXPOSURES AND DOSE–RESPONSE RELATIONSHIPS

M.-A. Flyvholm, T. Menné, and H. I. Maibach

CONTENTS

I. INTRODUCTION

There are millions of naturally occurring and man-made chemicals (CAS Registry Numbers, 1991). Many thousands of new chemicals are synthesized yearly. The skin may be exposed to chemicals either by direct contact or by airborne exposure. Exposure may be occupational or nonoccupational. This distinction has important legal implications, but often the same allergens appear both in the domestic and the occupational environment.

Chemicals dealt with here were restricted to chemicals known as contact allergens, which are substances capable of inducing allergic contact sensitization or type IV hypersensitivity. According to the textbooks, thousands of substances have been suggested to be contact sensitizers.[1] To provide hazard

TABLE 14-1
Sources of Information that May Be Used in the Evaluation of Chemicals as Skin Sensitizers

Human evidence
   1   Patch test data from dermatological clinics (supported with information on exposure).
   2   Epidemiological studies of contact dermatitis (supported with information on exposure).
   3   Data from human experiments and pre-market use tests, consumer complaints, etc.
Evidence from animal experiments
   4   Data from guinea pig experiments.
   *   Data from local lymph node assay (LLNA).
   *   Data from mouse ear swelling test (MEST).
Other supporting evidence
   5   Data from experiments performed in other animal species, i.e., mice.
   6   *In vitro* experiments.
   7   Quantitative structure-activity relationship (QSAR), etc.

*   Modified by the authors.

Adapted from Knudsen, B. B., Wahlberg, J. E., Andersen, I., and Menné, T., *Dermatosen*, 41, 5, 1993.

identification with respect to contact allergy, an operational classification of contact allergens is necessary. Knudsen et al. (1993)[2] recently suggested criteria for classification of contact allergens based on the principles used by IARC (International Agency for Research on Cancer) to evaluate data on the carcinogenicity of chemicals. The basic information needed to achieve such classification is summarized in Table 14-1. However, the majority of contact allergies are considered to be caused by about 100 allergens, i.e., chemicals or naturally occurring substances. The most common contact allergens belong to the groups of substances listed in Table 14-2.

The introduction of patch testing with Standard series by Bonnevie[3] and the ICDRG (International Contact Dermatitis Research Group)[4] is the basis for our extensive knowledge on contact sensitizing substances. Standard series are under continuous development[5,6] by the different international contact dermatitis research groups.[7] These series include substances yielding positive patch test reactions in more than 1% of the tested patients or substances difficult to suspect from the patients' history.

TABLE 14-2
Most Common Contact Allergens

Metals
Preservatives
Rubber additives
Perfumes
Medicaments
Miscellaneous

Among eczema patients tested in European dermatological clinics, 39 to 55% had positive patch test reactions to allergens included in the European Standard series or supplementary patch test series. About 77 to 95% of the patients with positive patch test reactions reacted to allergens from the European Standard series.[8]

A Danish population study, patch testing 567 persons between 15 and 69 years old, demonstrated that about 15% in a nonselected population has positive patch test reactions to allergens included in a modified version of the European Standard series (formaldehyde and primin excluded; thiomersal included).[9]

The frequency of positive patch test reactions in the general population[9] and in eczema patients tested at a dermatological clinic in the same area of greater Copenhagen, Denmark, is shown in Table 14-3. The allergens most often causing positive reactions in eczema patients were nickel, fragrance mix, cobalt chloride, colophony, and balsam of Peru. For the general population, nickel and thiomersal were the most common causes of positive patch test reactions. Generally, contact allergy is more frequent among patients investigated at dermatological centers as compared to the general population.

II. EXPOSURE

Assessment of exposure to contact allergens is of major importance for prevention and treatment of allergic contact dermatitis (primary and secondary prevention), as information on occurrence of the offending agent is critical for the prognosis of eczema patients with contact allergy.[10-12]

A. EXPOSURE TO CONTACT ALLERGENS

Important contact allergens are the following compounds and chemicals: the metals, nickel, cobalt, and chromate; preservatives, particularly formaldehyde and formaldehyde-releasing substances; and isothiazolinones. Further, rubber additives, perfumes, medicaments, plant allergens, and plastic products are contact allergens.

Sources of exposure to contact allergens can be divided into groups of substances, products, or use categories. Exposure to allergens can be divided into five different situations: (a) occupational, (b) domestic work, (c) hobby and leisure time activities, (d) topical medicaments, and (e) cosmetics, personal care products, clothing, shoes, etc.

Based on textbooks of dermatology and contact dermatitis[13-23] examples of allergens for each of these exposures situations are described below and listed in Tables 14-4 to 14-8.

A variety of product categories can cause occupational exposure to allergens (see Table 14-4). Among the most common mentioned are metals, resins, cosmetics/toiletries, metal working fluids, and medicaments. Allergens such as plastics and preservatives may occur in many different product categories.

TABLE 14-3
Comparison of Frequencies of Positive Patch Test Reactions in the General Population,* and in Eczema Patients at a Dermatological Clinic in the Same Area of Greater Copenhagen in 1990[a]

Test Substances	General Population[b] % Positive of Tested			Dermatological Clinic[c] % Positive of Tested		
	Men n = 279	Women n = 288	Total n = 567	Men n = 262	Women n = 410	Total n = 672
Potassium dichromate	0.7	0.3	0.5	1.9	2.7	2.4
Neomycin sulfate	0.0	0.0	0.0	3.4	3.7	3.6
Thiuram mixture	0.7	0.3	0.5	4.6	2.7	3.4
p-Phenylenediamine	0.0	0.0	0.0	1.9	2.7	2.4
Cobalt chloride	0.7	1.4	1.1	2.3	2.7	2.5
Benzocaine	—	—	NT	0.4	0.7	0.6
Caine mix	0.0	0.0	0.0	—	—	NT
Formaldehyde[d]	—	—	NT	1.9	2.2	2.1
Colophony	0.4	1.0	0.7	4.6	5.4	5.1
Quinoline mix	0.4	0.3	0.4	1.9	0.5	1.0
Balsam of Peru	0.7	1.4	1.1	3.4	5.4	4.6
PPD black rubber mix	0.4	0.0	0.2	1.2	0.0	0.5
Wool alcohols	0.4	0.0	0.2	1.2	1.7	1.5
Mercapto mix	0.7	0.0	0.4	1.2	0.2	0.6
Epoxy resin	0.4	0.7	0.5	0.8	0.2	0.5
Paraben mix	0.4	0.3	0.4	0.8	0.2	0.5
p-tert Butylphenol formaldehyde resin	1.1	1.0	1.1	0.4	1.2	0.9
Fragrance mix	1.1	1.0	1.1	6.1	7.1	6.7
Ethylenediamine dihydrochloride[e]	0.4	0.0	0.2	0.8	0.7	0.7
Quaternium 15	0.4	0.0	0.2	0.0	0.0	0.0
Nickel sulfate	2.2	11.1	6.7	4.2	16.1	11.5
MCI/MI (chloro-methyl- and methyl-isothiazolinone)	0.4	1.0	0.7	0.4	0.7	0.6
Mercaptobenzothiazole	0.4	0.0	0.2	1.2	0.2	0.6
Primin[f]	—	—	NT	0.4	1.5	1.0
Thiomersal[g]	3.6	3.1	3.4	—	—	NT
Carba mix[h]	0.7	0.0	0.4	—	—	NT

*   From Nielsen, N. H. and Menné, T., *Acta Dermato-Venereol.*, 72, 456, 1992.
[a]   Menné, unpublished.
[b]   Patch tested with the ready-to-apply TRUE test, Pharmacia (Sweden).
[c]   Test substances from Hermal (Germany).
[d]   Formaldehyde not included in the TRUE test at the time of study.
[e]   Ethylenediamine dihydrochloride excluded from the European Standard series as of August 1992.
[f]   Primin not included in the TRUE test at the time of study.
[g]   Thiomersal not included in European Standard series.
[h]   Carba mix was excluded from the European Standard series January 1989.

TABLE 14-4
Main Allergens Related to Occupational Exposure

Allergens	Sources of Exposure
Acrylates	Adhesives; bone cement; dental products; UV-curing lacquers; etc.
Amines	Hardeners/curing agents for epoxy resin
Chromate	Cement; leather; pigments
Cobalt	Paints/lacquers
Colophony	Adhesives; dental products; paper; tin solder, etc.
Epoxy resin	Adhesives; paints; electric insulation
Formaldehyde	Disinfectants; preservatives; laboratory chemicals; formaldehyde resins; funeral service
Formaldehyde releasers	Metal working fluids; paints; adhesives
Formaldehyde resins	Adhesives; paints/lacquers; impregnated textiles and paper; inks
Isocyanates	Adhesives; paints; fillings; polyurethane foams
Medicaments	Health care; veterinarians
Nickel	Coins; nickel-plated objects; contaminated oils; etc.
Paraphenylenediamine	Hair dyes; rubber additive
Plastics/resins	Adhesives; paints; fillings; containers; etc.
Preservatives	Water based products; metal-working fluids; paints; adhesives; cleaning agents; cosmetics; polishes; skin protection creams; process water; etc.
Rubber additives	Rubber gloves; rubber tubing; washers; etc.

Occupational contact dermatitis is most frequently seen in occupations such as cleaning[24] and health care, the construction and metal industries, and among hairdressers and other service people.[25]

In a study on occurrence of contact allergens in registered chemical products, paints/lacquers, curing agents (hardeners) for two-component products, cleaning agents, binders, adhesives/glues, and toiletries were the most frequently registered product categories for the selected contact allergens included in the study.[26]

Domestic work can cause contact with a variety of different contact allergens, many of which will be identical to exposure from occupationally used products. Table 14-5 lists some common allergens in domestic work and examples of sources of exposure.

Exposure to contact allergens from hobby activities will be equal to occupational exposure if the same sorts of products are used. Allergens in leisure time activities can be sports equipment, artist paints, materials for handicraft, etc. See Table 14-6 for more activities.

Allergic contact dermatitis induced by topical medicaments can arise as complication of preexisting eczema or dermatosis causing exacerbation or spread to other sites, e.g., in leg ulcers and stasis eczema. Topical medicaments causing sensitization varies from one country to another and over time, depending on the use of preparations. Examples of allergens are given in Table 14-7.

TABLE 14-5
Main Allergens Related to Domestic Work

Allergens	Sources of Exposure
Chromium	Leather; foot wear
Colophony	Shoe polish; crayons; plasticine; paper; etc.
Flowers/plants	Gardening; house plants
Nickel	Nickel-plated objects
Plastics/resins	Adhesives; paints; containers; etc.
Preservatives	Cleaning agents; polishes; personal care products
Rubber additives	Gloves and other rubber objects
Woods	Repairs; handicraft

Cosmetics, toiletries, or personal care products defined as preparations applied to the skin, hair, mouth, nails, or mucous membranes for cleansing, perfuming, protection, or changing appearance are widely used and often recognized as causes of contact allergies. Many cases of cosmetic allergy are detected with the European Standard series. Of the 23 substances or mixtures included in the current European Standard series, 9 may be used in cosmetics.[27]

As for other product categories, regional differences in product preference, etc., can cause geographical variations in frequency of sensitization to cosmetic ingredients. Examples of allergens in cosmetics and personal care products which may cause sensitization are given in Table 14-8.

B. INFORMATION ON EXPOSURE TO CONTACT ALLERGENS

Information on occurrence of contact allergens in chemical products can be derived from sources such as textbooks and scientific journals, chemical analysis, inquiries to manufacturers or suppliers, from data bases with information on product composition, or from product labeling. See Table 14-9.

TABLE 14-6
Main Allergens Related to Hobby and Leisure Time Activities

Allergens	Sources of Exposure
Chromium	Leather; foot wear
Colophony	Adhesive tapes; plasticine; paper; violin bow resin; etc.
Dyes/pigments	Crayons; artists paints; textiles
Flowers/plants	Gardening; house plants
Formaldehyde	Textile resins; preservatives in various products
Nickel	Nickel-plated objects
Plastics/resins	Adhesives; paints; containers; etc.
Preservatives	Paints; personal care products
Rubber additives	Gloves; sports equipment
Woods	Handicraft

TABLE 14-7
Main Allergens Related to Topical Medicaments

Antibiotics
Antihistamines
Antimicrobials
Balsams
Benzocaine
Colophony
Ethylenediamine
Formaldehyde releasers
Lanolin
Parabens
Preservatives
Tars

1. Textbooks and Scientific Journals

Several textbooks on dermatology and contact dermatitis present information on occurrence of contact allergens in different product categories, occupations, or types of use based on published papers and the experience of the authors.[13-23] For some substances, information from textbooks and literature will be rapidly outdated due to changes in lifestyle and development of new products and materials. This can especially be expected for some industrially used substances and products, although this can be a problem for household and personal care products, too. On the other hand, textbooks are rapidly utilized.

2. Chemical Analysis

For some contact allergens, the availability of simple analytical methods or spot tests makes it practicable to analyze products or materials for the occurrence of suspected contact allergens. For example, nickel in metal objects can be detected by the dimethylglyoxime spot test, and formaldehyde by the

TABLE 14-8
Main Allergens Related to Cosmetics and Personal Care Products

Allergens	Sources of Exposure
Colophony	Mascara
Dyes	Hair dyes; miscellaneous cosmetics
Fragrances	
Glyceryl thioglycolate	Permanent waving
Lanolin	
Paraphenylenediamine	Hair dyes
Preservatives	Creams; lotions; shampoos; liquid soap; etc.
Formaldehyde releasers	(i.e., most cosmetics and personal care products)
Isothiazolinones	
Parabens	
UV-filters	Sunscreens

TABLE 14-9
Common Sources of Information on Exposure to Contact Allergens

Exposure Categories	Textbooks (Ref.)	Analysis[a]	Data Bases (Ref.)	Labeling[b]
Adhesives	14,18,20,21	(+)	see *	(x)
Cement	14-16,18,22			x
Cleaning agents	21	(+)	PROBAS,26	(x)
			INFODERM[c],44	
Cosmetics and skin care products	13,14,16-19, 21,23	(+)	CODEX,43 DALUK,45 FDA,41,42 INFODERM,44	x in U.S. (x) in Europe
Metal working fluids	14-16,18,20-22	(+)	see *	(x)
Dyes	13-15,18	(+)	see *	—
Medicaments and drugs	13,14,17,18, 21-23	(+)	CODEX,43 DALUK,45 INFODERM,44	(x)
Metals	13-16,18,20-22	+/–	?	—
Paints, varnishes and lacquers	15,16,20,21	(+)	PROBAS,26 see *	(x)
Perfumes	14,15,18,21	?		(x)?
Pesticides	13-15,20,21,23	+	PROBAS,31	(x)
Plastics and resins	13-16,18,20, 21,23	(+)	PROBAS,31 see *	(x)
Preservatives and antimicrobials	13,14,18,21	(+)	INFODERM,44	(x)
Rubber additives	13-15,18,20,21	+		?
Soaps and detergents	20		(PROBAS)	(x)
Solvents	18,20,21	+	see *	(x)
Surface active agents	13			(x)
* Miscellaneous occupational products, unspecified product categories	most textbooks		Clin.Tox.,47 EPA,41,42 OCCALL,46 PROBAS,31	

[a] The indications in this column are preliminary: "+" = analytical methods exist; "(+)" = complex product, may cause problems in the analysis.

[b] The indications in this column are preliminary: "x" = labeling available; "(x)" = labeling possible.

[c] INFODERM has information on preservatives in cleaning agents.

chromotropic acid method or the acetylacetone method.[28] For chromate no standardized analytical method exists, but in the EC cooperation in standardizing measurements for chromium compounds[29] is planned. If reliable analytical methods exist, chemical analysis are an effective way of obtaining information on the occurrence of contact allergens, although many contact allergens cannot be identified by simple spot tests, and in addition, some analytical methods are complicated and require equipment and expertise not available in normal dermatological laboratories. For chemical analysis performed by independent laboratories, the cost and delay in time can be a major disadvantage. Furthermore, verification of occurrence of contact allergens by chemical analysis requires a precise knowledge of which substances to look for, as thorough analysis of the composition of suspected products can be difficult for complex products containing many components.

3. Information from Manufacturers

Extensive resources are often required in order to obtain detailed product information from manufacturers. Time delay and lack of specific chemical information is another obstacle. However, many manufacturers are cooperative and supply detailed information for evaluation of the individual cases and support research within the area.

4. Product Data Bases

Computerized information on ingredients of chemical products collected in data bases can offer a rapid and simple way to obtain information on occurrence of contact allergens, or the composition of products suspected to be the cause of allergic eczema.[26,30] Ideally, access should be facilitated for the relevant users, for example with direct on-line connection or some type of hotline service. The degree of detail, updating of the registered information, etc., can cause problems, and additional difficulties pertaining to confidentiality should be addressed. It seems that the more detailed information on product composition a data base contains, the more restrictions in access are demanded.[31] Data bases useful in the field of contact dermatitis were surveyed by Dooms-Goossens et al.[32]

In Denmark, manufacturers or importers are legally required to notify the Product Register Department of all new and hazardous products (i.e., mostly products with warning labels according to the EC rules) for occupational use, whereupon they receive a Product Registration number (PR No.) before the products can be sold.[33] This means that for all products labeled with a PR No., information on chemical composition, with substances identified by Chemical Abstract Service Registry Numbers (CAS RN), is registered in the Danish Product Register Data Base (PROBAS).[34] In addition, products included in surveys and research projects are also registered. For some product categories, this registration only partly covers the products marketed, although, for product categories such as epoxy- and isocyanate products, asbestos-containing products, pesticides, and cleaning agents, the registration is considered complete for the Danish market.[31]

In Sweden and Norway national data bases on chemical products like the Danish PROBAS have been developed.[35,36] In the U.S. nationwide surveys — National Occupational Hazard Survey (NOHS) and National Occupational Exposure Survey (NOES) have provided data bases containing composition information on occupational products.[37]

Other data bases with registration of composition for chemical products have been described: DERM/INFONET in the U.S. with modules on cosmetics (FDA)[38-40] and occupational products (EPA),[41,42] Codex in Belgium,[43] INFODERM in Finland,[44] DALUK in Sweden,[45] OCCALL in Italy,[46] and Clinical Toxicology of Commercial Products at the University of Rochester[47] (see Table 14-9). Other relevant data bases might exist as our experience is that product data bases either are not described in literature, or seem to be difficult to trace in literature searches.

Computerized information on composition for chemical products offer possibilities to perform surveys of occurrence of allergens in various product categories,[26] trades of use,[48] or to generate individual patch test series based on the patients' exposure to chemical products.

5. Labeling

Labeling would be an obvious way to obtain and provide information on the occurrence of contact allergens in chemical products if all categories of chemical products were labeled with their complete composition.

Labeling cosmetics with lists of components is possible and would be of great benefit for diagnosis, treatment, and advising patients with contact allergy to cosmetics, as it has been shown in the U.S.[27,49,50] The latest proposal for EC Cosmetic Directive[51] includes labeling of cosmetics with a list of ingredients, except for perfumes. Though this Directive is a considerable improvement of the current level of information on cosmetic ingredients, a Working Party of the European Society of Contact Dermatitis has proposed some modifications.[50] Although perfumes consist of numerous substances, labeling of the most frequent sensitizers among these could be seriously considered.

Considering products for occupational and household use these are covered by the 1% limit for labeling of allergens in the EC legislation on labeling chemical products.[52] This concentration limit is not relevant to contact allergens as most cases of induction and elicitation of contact allergy are caused by exposure to chemicals in a concentration below this limit.

C. MISSING INFORMATION ON EXPOSURE

There is a demand for development of standardized analytical methods and for easier access to chemical analysis, e.g., for metal allergens and preservatives.

Ingredient labeling of cosmetics and pharmaceuticals are considered by the European Commission as proposal and draft, respectively. The next goals for dermatologists are ingredient labeling of household and industrial products.[50]

The 1% limit for labeling of ingredients in chemical products[52] is too high, because most allergens will occur in the products in lower concentrations. As many allergens do not appear in lists of dangerous substances[26,53,54] complete listing of ingredients will be the optimal way to provide information on occurrence of contact allergens.

Ubiquitous allergens with numerous sources of exposure makes it necessary to demand general ingredient labeling for all categories of products, as labeling of cosmetics, for example, or occupationally used products alone will not solve the information problems for patients with contact allergy.

Computerized data bases including cosmetic, pharmaceutical, household, and industrial products could provide valuable tools for cross-sectional surveys on occurrence of allergens, and furthermore improve the quality of patient treatment.

III. DOSE–RESPONSE RELATIONSHIPS

In this section, dose–response relationships for induction and elicitation of allergic contact dermatitis in humans is discussed. The available information can be discussed under three main headings.

A.   Experimental sensitization with DNCB (dinitrochlorobenzene).
B.   Experimental sensitization and elicitation with clinical relevant sensitizers.
C.   The clinical experience.

A. EXPERIMENTAL SENSITIZATION IN HUMANS WITH DNCB

Dinitrochlorobenzene (DNCB) is an extremely potent contact sensitizer frequently used in experimental research. This substance is useful in the study of the quantitative aspects of contact sensitization and elicitation as even moderate dosages sensitize nearly 100% of all humans.

Friedman et al.[55] studied the dose response for DNCB both concerning induction and elicitation of contact allergy. The dose used for induction ranged from 62.5 to 1000 μg and those for elicitation from 1.56 to 25 μg DNCB. The proportion of subjects sensitized increased with sensitizing dose; 8% were sensitized of 62.5 μg and 100% were sensitized of 500 μg or more. The authors found a linear relationship between the degree of sensitivity and log sensitizing dose so that, on average, each time the sensitizing dose was halved, the challenge dose required to produce the same response increased one- to five-fold.

White et al.[56] used DNCB to study the effect of concentration and area in extensive human studies. At a concentration of 8.8 to 142 μg/cm² and an area varying between 1.8 to 14.2 cm², the number sensitized depended on concentration and not on the total amount of allergen applied (area). In later studies, using a similar study design, the authors implicitly investigated whether a lower threshold limit for area of exposure existed.[57] By keeping the concentration constant, the authors examined the effect of area by exposing 0.8 and 0.08 cm², respectively. The degree of sensitization was significantly less in the small area group.

Upadhye and Maibach[58] recently reviewed the interrelation between concentration and area for induction of contact sensitization. In most experimental studies the total number sensitized or the degree of sensitization depends first and foremost on concentrations. If the exposed areas are small (or have a low total dose), it can be the limiting factor, whether sensitization occurs or not.

B. EXPERIMENTAL HUMAN SENSITIZATION STUDIES TO COMMON ENVIRONMENTAL SKIN SENSITIZERS

Marzulli and Maibach have thoroughly reviewed the human predictive test methods.[59] The most useful tests are the Draize test and the Maximization test.

The principle in the Draize test is nine to ten repeated occluded exposures to the same site followed by a 2-week rest period and then a 48-hr occluded challenge at a new skin site. The Maximization test, intended for assay of single chemicals, utilizes a high test concentration, and, when not irritating, the addition of sodium lauryl sulfate.

The philosophy behind the Draize test and the modifications of this and the Maximization test differ. The Draize test tries to predict what happens if a larger population is exposed to the chemical in question. The maximization procedure does not attempt to forecast a future incidence of sensitization, but establishes the sensitizing potential of a substance as weak, mild, moderate, strong, and extreme. The included number of volunteers in the Draize test are often 100 to 200, while the maximization procedure only includes 25 persons. Generally, weak sensitizers may be missed with the Draize procedure unless the test concentration is elevated, while the Maximization test may overestimate the sensitizing potential, or miss the potential, because of the small sample size. It is crucial that these tests are only designed, performed, and interpreted by experienced researchers. Such test systems can of course not foresee all types of extreme exposures which may happen in daily life. Therefore, comparison with clinical patch test data are crucial. Kligman[60,61] applied the maximization procedure in human volunteers with different contact allergens. The outcome of a Maximization test for substances included in the diagnostic standard patch test series is shown in Table 14-10. All the moderate to extreme sensitizers are frequent causes of allergic contact dermatitis in humans. The actual number sensitized depends upon the degree of the exposures as detailed in the following sections. The weak sensitizers such as lanolin and neomycin mainly sensitize when applied to diseased skin or when used on intertriginous skin areas with a naturally occurring occlusion (Table 14-11).[62]

Kligman[60,61] in his work with the human Maximization test reached several conclusions, which have not been contradicted by later research. The most important are the following (the selection is ours):

1. Inflammation-producing insults may increase risk of sensitization.
2. Above a certain minimum, sensitization for a fixed surface concentration is not dependent on the size of the area to which the allergen is applied.
3. Sensitization, within limits, is proportional to the surface concentration of allergen and not the total amount of allergen. High concentrations are required for weak allergens.

TABLE 14-10
Experimental Human Contact Sensitization. The Maximization Test.
Allergens Included in the Diagnostic Standard Patch Test

Substance	Induction Concentration %	Challenge Concentration %	Sensitization Rate	Sensitization* Grade
Potassium dichromate	2	0.25	23/23	5
Cobalt sulfate	25	2.5	10/25	4
Nickel sulfate	10	2.5	12/25	3
Formaldehyde	5	1.0	18/25	4
Tetramethylthiuram	25	10	4/25	2
Neomycin	25	10	7/25	2
Vioform®	25	10	0/23	1
Benzocaine	25	10	5/23	2
Lanolin	25	10	0/25	1
p-Phenylenediamine	10	0.5	24/24	5

* 1 = weak, ... , 5 = extreme.
Adapted from Kligman, A. M., *J. Invest Dermatol.*, 47, 375, 1966, and 47, 393, 1966.

4. Within limits, the sensitization rates are roughly proportional to the number of exposures, especially for weaker allergens. With agents not recognized as allergens, even 15 exaggerated exposures were incapable of including sensitization.

Marzulli and Maibach[63] extensively investigated the most commonly used antimicrobials with the Draize procedure on normal human subjects. Based on a comparison with clinical data extrapolation of the Draize test results to a larger population seems generally to be meaningful. Formalin is a commonly recognized sensitizer in patients with eczematous skin diseases. Table 14-12 illustrates the results of the Draize procedure in humans with formalin. A similar fraction seems to be sensitized when induction concentrations of 5 and

TABLE 14-11
Sensitivity to Medicaments According to Site Given as
Percent with Positive Patch Test in 1,029 Patients Tested

	Leg Ulcer		Other Sites	
	Females	Males	Females	Males
Lanolin	25	10	3	3
Neomycin	8	13	0.5	1
Parabens	9	10	1	0.5

Adapted from Wilkinson, J. D., Hambly, E. M., and Wilkinson, D. S., *Acta Dermato-Venereol.*, 60, Suppl. 245, 1980.

TABLE 14-12
Response of Human Subjects Tested with Aqueous Solutions of Formalin (Draize Procedure)

Induction	Challenge	Sensitized	
%	%	Fraction	%
10	1	8/102	7.8
5	1	4/52	7.7

Adapted from Marzulli, F. N. and Maibach, H. I., *J. Soc. Cosmet. Chem.*, 24, 399, 1973.

10% formalin (37% aqueous solution of formaldehyde) are used. The number of sensitized individuals who will react on a challenge dose depends upon the challenge concentration (Table 14-13). Further, to illustrate the clinical relevance of contact sensitization to formalin sensitized individuals were exposed to different formalin-preserved products. Table 14-14 illustrates that a high number of sensitized individuals will react to a formalin-containing product, and further that the actual number of reactions depends upon the product type.

Experimental sensitization (Draize procedure) with sorbic acid (2,4-hexadienoic acid) and chlorocresol sensitize few or none in accordance with the clinical experience that these preservatives are rare sensitizers.[63]

In a later series of studies, Marzulli and Maibach[64] investigated the effect of graded concentrations for the induction of skin sensitization with the Draize procedure. Table 14-15 illustrates the concentration-dependent sensitization risk for benzocaine. A similar dose-response effect could not be illustrated for weaker allergens such as neomycin and dichlorophene.

The findings obtained by both procedures are often (but not always) in agreement, if one seeks a positive or negative answer to skin-sensitization potential. The exceptions must be examined carefully. The Draize procedure,

TABLE 14-13
Elicitation of Skin Reactions in Formalin-Sensitized Subjects (Occluded)

Challenge Concentration (%)	Response	
	Fraction	%
1	4/5	80
0.5	2/5	40
0.2	1/5	20
0.1	1/5	20
0.01	1/5	20

Adapted from Marzulli, F. N. and Maibach, H. I., *J. Soc. Cosmet. Chem.*, 24, 399, 1973.

TABLE 14-14
Reaction of Formalin Sensitized Subjects to
Products Containing Formalin

Product Description	Application Site	Fractional Response
Dry skin lotion (0.5% formaldehyde)	Right face and forearm	5/10
Creme rinse (0.4% formaldehyde)	Left face and forearm	4/10
Bubble bath oil (0.6% formaldehyde)	Right shoulder	2/10

Adapted from Marzulli, F. N. and Maibach, H. I., *J. Soc. Cosmet. Chem.*, 24, 399, 1973.

as currently employed with higher than use concentrations,[59] will identify some allergens not identified with the Maximization test because of the larger number of individuals exposed.[65]

Generally, the results of the mentioned human sensitization procedures are in agreement with the clinical experience. Most important for the induction and elicitation of allergic contact dermatitis are the inherent sensitizing potential of the substance and the exposure concentration.

These points are further discussed in the following sections, where the clinical experience with some selected allergens is discussed in more detail.

C. THE CLINICAL EXPERIENCE
1. Nickel

Nickel is the most common cause of allergic contact sensitization among females. A recent Danish population study found 11.1% of the females and 2.2% of the males to be nickel sensitized.[9]

Primary nickel sensitization is mainly caused by either well-defined domestic or occupational nickel exposures. Trace amounts of nickel in the general environment do not induce clinically overt nickel sensitization.[66] The concen-

TABLE 14-15
Skin Sensitization (Draize Procedure) with Benzocaine
Showing an Increase in the Incidence of Sensitization
with Higher Concentration of Test Material at Induction

Induction Concentration	Challenge Concentration	Response	
%	%	Fraction	%
2	2	0/92	0
10	10	2/173	1.2
20	10	6/99	6.0

Adapted from Marzulli, F. N. and Maibach, H. I., *Food Cosmet. Toxicol.*, 12, 219, 1974.

tration of nickel, cobalt, and chrome in consumer products is so low that primary sensitization is unlikely, although, some sensitized patients may be affected by such products.[67]

While chromate mainly contact sensitizes in its soluble forms, nickel is a contact sensitizer, both as a salt and in the metallic form because many metallic nickel surfaces and alloys corrode easily when exposed to human sweat. This means that nickel-containing or nickel-plated metallic items (such as tools, buttons, costume jewelry, etc.) might release sensitizing amounts of nickel chloride, when exposed to human skin (human sweat) for hours. The nickel ion is both a primary irritant and a moderate skin sensitizer. The irritating effect of the nickel ion is believed to enhance the sensitization risk.[68]

In spite of the high prevalence of nickel allergy in the population, experimental nickel sensitization is not easy. In 1963, Vandenberg and Epstein[69] reported a 9% experimental nickel sensitization rate in 172 subjects exposed to repeated occluded exposures to 25% nickel chloride in 0.1% sodium lauryl sulfate. Further, the exposed skin areas were frozen to increase skin penetration. By repeating the exposure after 4 months, another 5 of 19 re-exposed were sensitized, suggesting that prolonged nickel exposure will raise the sensitization rate. Kligman[61] investigated factors such as concentrations, vehicles, skin area, and race. By optimizing the different variables the highest sensitization rate obtained was 12 of 25 subjects utilizing an induction concentration of 10% $NiSO^4$ and an eliciting concentration of 2.5%.

In the past, when workers were frequently exposed to high concentrations of nickel (2 to 30%) primary occupational nickel sensitization was not uncommon.[70] In 1980, Wall and Calnan[71] described primary nickel sensitization in 7 of 16 operators working in the electroforming industry from repeated exposure to an aqueous solution containing 42 ppm nickel. Factors such as skin irritation and occlusion from gloves might have been contributory elements. Occupational exposure to nickel concentrations ranging from 1 to 20 ppm is not uncommon.[72] Whether such concentrations induce primary nickel sensitization is undocumented.

Nickel release from nickel alloys clinically known to induce nickel sensitization seems to be 10 to 100 times higher compared to alloys known to never or rarely induce primary nickel sensitization[73] (Table 14-16).

Eliciting of nickel dermatitis is unlikely for concentrations below 0.1 to 1 $\mu g/cm^2$ during occluded exposure[73-75] and 15 $\mu g/cm^2$ when nonoccluded.[76] The effect of repeated exposures to normal skin is unexplored. Highly sensitized individuals might react to 0.5 ppm nickel (7.5×10^{-3} g/cm²) when exposed on inflamed skin under occlusion as a single application.[77] We suspect that when larger populations are tested, the threshold level may be less.

The main causes of primary nickel sensitization are domestic nickel exposure by ear piercing, buttons, and costume jewelry. Occupational nickel hand eczema is often secondary to this. Based on the accumulated knowledge, it is now possible to suggest an exposure regulation for the most common nickel

TABLE 14-16
Reactivity to Different Nickel Alloys and
Nickel Release in Synthetic Sweat

Alloy	Nickel Release μg/cm²/week*	% Reactivity in Nickel Sensitized Individuals
"Safe alloys"		
Stainless steel	0.01	3
White gold	0.02	11
Nickel tin	0.1	23
"Sensitizing alloys"		
Nickel silver	20	81
Nickel chemical (2)	32	56
Nickel elec.	40	76
Nickel chemical (1)	45	79
Nickel iron	65	79

* Nickel release after 3 weeks exposure. Different values at week 1 and 6.

Adapted from Menné, T. et al., *Contact Derm.*, 16, 255, 1987.

sensitizing objects. It is believed that such a regulation will prevent most cases of primary sensitization. Since 1990 such a regulation has been in force in Denmark.[78]

2. Chromate

Cement dermatitis is a common occupational skin disease in the construction industry.[79] The dermatitis is caused by the combined effect of skin irritation and sensitization to the hexavalent chromate present in wet cement. The chromate content depends upon the chromate concentration in the raw materials used in the cement production. This explains the great variability of the chromate content in cement from different geographical regions. There seems to be a parallel between the concentration of hexavalent chromate in the local cement and the frequency of allergic cement dermatitis (Table 14-17). In locations with low chromate content the prevalence is approximately 1% and in regions with higher chromate concentrations the prevalence raises to between 9 to 11%.

On the suggestion of Fregert et al.[85] the Scandinavian countries, in the start of the 1980s, added ferrosulphate in low concentrations to cement to reduce the hexavalent chromate to trivalent chromate. The idea with this initiative is that the trivalent chromate is not absorbed, or only to a minor degree, through human skin, and therefore the risk of primary sensitization from this salt is significantly less as compared to hexavalent chromate.[86] Epidemiological studies in Denmark of the same construction sites performed at the beginning of the

TABLE 14-17
Chromate (Cr VI) Concentration and Prevalence* of Allergic Chromate Dermatitis in Different Geographical Regions

Country	Concentration (μg/g) Basic	Concentration (μg/g) Use	Prevalence* (%)	Ref.
U.S. (California)	< 0.1	< 0.02–0.025	1.1	80
Sweden	2–15	0.4–3.8	3.0	81
Norway	6–40	1.2–10	4.6	81,82
Germany	12.5–24	2.5–6	11.1	83
Germany (low)	< 0.4	< 0.08–0.1	0.7	
Denmark 1981	9.6	1.9–2.4	8.9	84
Denmark 1987	< 2	< 0.4–0.5	1.3	

*   Not uniformly defined in all studies.

1980s and at the end of the 1980s, suggest that this measure has reduced the frequency of allergic contact dermatitis significantly (Table 14-17).

3. Isothiazolinone-Based Preservatives

Most preservatives seem to have an inherit contact sensitizing potential. Within the last decade much valuable information has been gathered concerning the preservative system containing 5-chloro-2-methyl-4-isothiazolin-3-one and 2-methyl-4-isothiazolin-3-one (MCI/MI) also named Kathon CG.[87]

Chan et al. (1983)[88] tested MCI/MI in guinea pigs by a modified Buehler's occluded epicutaneous patch test technique. The incidence of delayed contact dermatitis in guinea pigs was dependent on both the induction and the challenge concentrations. When the highest concentrations (2000 ppm) were used for both induction and elicitation, 20 of 20 animals reacted, suggesting a high sensitizing potential. Elicitation was not seen for concentrations below 100 ppm, independent of induction concentration.

In a series of 13 prophetic humans, repeat insult patch testing involving a total of 1450 subjects, no skin sensitization was induced at concentrations of 10, 6, or 5 ppm (1121 subjects) or at 15 ppm (200 subjects) of MCI/MI. Contact sensitization was induced in 1 of 84 subjects at 12.5 ppm and in 2 of 45 subjects at 20 ppm. The authors concluded that use of the preservative system MCI/MI in low concentrations in rinse-off products involves a low risk with respect to contact sensitization.[89]

Schwartz et al. (1987)[90] performed a prospective use test with a lotion containing 15 ppm MCI/MI on 209 healthy individuals with 13 weeks daily exposure. Diagnostic patch testing with MCI/MI was performed before and after the study. None of the individuals were sensitized in this study, and the authors concluded that the risk of sensitization from Kathon CG, when used at 15 ppm in a leave-on product, is minimal.

In spite of this safety assessment, contact allergy to MCI/MI has been seen in eczema patients evaluated at dermatological departments. Particularly high

TABLE 14-18

Contact Allergy to MCI/MI (in percentage) in the General Population and a Contact Dermatitis Referral Department within the Same Geographical Area in 1990

	General Population[a]			Department[b]		
	Male	Female	Total	Male	Female	Total
MCI/MI positive	n = 279 0.4	n = 288 1.0	n = 567 0.7	n = 262 0.4	n = 410 0.7	n = 672 0.6

[a] From Nielsen, N. H. and Menné, T., *Acta Dermato-Venereol.*, 72, 456, 1992.
[b] Menné, unpublished.

frequencies of contact allergy to MCI/MI have been observed in areas where popular moisturizing creams have been preserved with MCI/MI.[91,92] In the Bologna area in Italy, 8.3% of 620 tested eczema patients gave a positive diagnostic patch test to MCI/MI. Most were sensitized from a popular cosmetic product containing 30 ppm of MCI/MI. In a large European multicenter clinical patch test study, the frequency of MCI/MI allergy varied from 0.4 to 11.1%.[93] Large geographical differences were seen. Many variables offer explanation for these differences. Among them patient selections and type of exposures (concentration and leave-on/wash-off products) are the most important.

Studies from the Copenhagen area in Denmark provide information of particular interest because we have data on both exposure and prevalence of contact sensitization in the general population, and in the referral center for eczema patients corresponding to the examined population. Analysis of a random sample of cosmetic and household products in 1988 to 1989 in the Copenhagen area, including a total of 156 products, discloses the presence of MCI/MI in 42% of the examined articles.[94] The concentrations found range from below 1 ppm to 22 ppm. In none of the samples the concentrations exceeded the level of 30 ppm, which during that time was the maximum permitted concentration of MCI/MI in cosmetic products in the EC. From knowledge of consumer patterns, it is realistic to assume that nearly everybody in the examined Copenhagen population have been exposed to one or several MCI/MI-containing products.

Table 14-18 compares the prevalence of contact sensitivity to MCI/MI in the general population and the frequency of MCI/MI sensitivity in the contact dermatitis referral clinic for the same area. Even if the materials are relatively large, the number of MCI/MI sensitive individuals in both samples is so small that the statistical uncertainty is significant. Most of those examined in the general population have healthy skin or only intermittent skin symptoms. From provocative use test studies, it is known that MCI/MI sensitive individuals will react to leave-on products containing MCI/MI in the concentration range from 7 to 15 ppm.[87] The number who react is not only dependent upon the total concentration, but also on the type of product, probably illustrating differences in MCI/MI bioavailability. Even if the percentage sensitized in the population is relatively low, the total number at risk of developing contact dermatitis from MCI/MI is high because millions of consumers are exposed to this preservative system. This large number of sensitized individuals and the widespread use of the preservative system MCI/MI make product labeling an obvious need for secondary prevention of allergic contact dermatitis.

Other preservatives such as formalin and formalin-releasing substances also imply risk of contact sensitization. When MCI/MI has been used to illustrate the dose–response phenomenon, it is because the substance has been extensively investigated recently.

A recent report on the safety assessment of MCI/MI concluded that MCI/MI may be safely used in rinse-off products at a concentration not to exceed 15 ppm and in leave-on cosmetic products at a concentration not to exceed 7.5 ppm.[95]

It seems that the combination of careful toxicological and epidemiological research as illustrated for nickel, chromate, and MCI/MI can define dose–response patterns and suggest exposure levels where the risk of contact sensitization and elicitation of contact dermatitis diminish.

IV. CONCLUSION

International reseachers have agreed classification of contact allergens is needed to provide for operational hazard identification and hazard assessment. The present information systems on exposure assessment are diffuse and insufficient. Legislation is needed to secure public access to systematic and updated information on exposure to contact allergens. Research on dose response is of high priority to achieve the final risk assessment which is the basis for preventive measures.

REFERENCES

1. **De Groot, A. C.,** *Patch Testing. Test Concentrations and Vehicles for 2800 Allergens,* Elsevier, Amsterdam, 1986.
2. **Knudsen, B. B., Wahlberg, J. E., Andersen, I. and Menné, T.,** Classification of contact allergens. Proposal for criteria, *Dermatosen,* 41, 5, 1993.
3. **Bonnevie, P.,** *Aetiologie und pathogenese der eksemkrankheiten. Klinische studien über die ursachen der ekzeme unter besonderer berücksichtigung des diagnostischen wertes der eksemproben,* Barth, Leipzig, 1939.
4. **Fregert, S., Hjorth, N., Magnusson, B., Bandmann, H. J., Calnan, C. D., Cronin, E., Malten, K., Meneghini, C. L., Perilä, V., and Wilkinson, D. S.,** Epidemiology of contact dermatitis, *Trans. St Johns Hosp. Dermatol. Soc.,* 55, 17, 1969.
5. Notice. Revised European standard series from 1 January 1989, *Contact Derm.,* 19, 391, 1988.

6. **Chemotechnique Diagnostics AB.** *Patch Test Allergens. Product Catalogue*, Chemotechnique Diagnostics AB, Malmø, 1992.

7. **Desmond, B.,** Contact dermatitis research groups, in *Textbook of Contact Dermatitis*, Rycroft, R. J. G., Menné, T., Frosch, P. J., and Benezra, C., Eds., Springer-Verlag, Berlin, 1992, 787.

8. **Menné, T., Dooms-Goossens, A., Wahlberg, J. E., White, I. R., and Shaw, S.,** How large a proportion of contact sensitivities are diagnosed with the european standard series, *Contact Dermat.*, 26, 201, 1992.

9. **Nielsen, N. H. and Menné, T.,** Allergic contact sensitization in an unselected Danish population. The Glostrup allergy study, Denmark, *Acta Dermato-Venereol.*, 72, 456, 1992.

10. **Edman, B.,** The usefulness of detailed information to patients with contact allergy, *Contact Derm.*, 19, 43, 1988.

11. **Cronin, E.,** Formaldehyde is a significant allergen in women with hand eczema, *Contact Derm.*, 25, 276, 1991.

12. **Flyvholm, M.-A. and Menné, T.,** Allergic contact dermatitis from formaldehyde. A case study focusing on sources of formaldehyde exposure, *Contact Derm.*, 27, 27, 1992.

13. **Cronin, E.,** *Contact Dermatitis*, Churchill Livingstone, Edinburgh, 1980.

14. **Fregert, S.,** *Manual of Contact Dermatitis*, 2nd ed., Munksgaard, Copenhagen, 1981.

15. **Foussereau, J., Benezra, C., and Maibach, H. I.,** *Occupational Contact Dermatitis. Clinical and Chemical Aspects*, Munksgaard, Copenhagen, 1982.

16. **Maibach, H. I. and Gellin, G. A.,** *Occupational and Industrial Dermatology*, Year Book Medical Publishers, Chicago, 1982.

17. **Nater, J. P. and De Groot, A. C.,** *Unwanted Effects of Cosmetics and Drugs Used in Dermatology*, 2nd ed., Elsevier, Amsterdam, 1985.

18. **Fisher, A. A.,** *Contact Dermatitis*, 3rd ed., Lea & Febiger, Philadelphia, 1986.

19. **De Groot, A. C.,** *Adverse Reactions to Cosmetics*, State University of Groningen, Groningen, 1988.

20. **Adams, R. M.,** *Occupational Skin Disease*, 2nd ed., Saunders, W. B., Philadelphia, 1990.

21. **Fregert, S., Björkner, B., Bruze, M., Dahlquist, I., Gruvberger, B., Persson, K., Trulsson, L., and Zimerson, E.,** *Yrkesdermatologi*, Studentlitteratur, Lund, 1990.

22. **Menné, T. and Maibach, H. I.,** *Exogenous Dermatoses: Environmental Dermatitis*, CRC Press, Boca Raton, FL, 1991.

23. **Rycroft, R. J. G., Menné, T., Frosch, P. J., and Benezra, C.,** *Textbook of Contact Dermatitis*, Springer-Verlag, Berlin, 1992.

24. **Meding, B. and Swanbeck, G.,** Occupational hand eczema in an industrial city, *Contact Derm.*, 22, 13, 1990.

25. **Rycroft, R. J. G.,** Occupational contact dermatitis, in *Textbook of Contact Dermatitis*, Rycroft, R. J. G., Menné, T., Frosch, P. J., and Benezra, C., Eds., Springer-Verlag, Berlin, 1992, 343.

26. **Flyvholm, M.-A.,** Contact allergens in registered chemical products, *Contact Derm.*, 25, 49, 1991.

27. **De Groot, A. C. and White, I. R.,** Cosmetics and skin care products, in *Textbook of Contact Dermatitis*, Rycroft, R. J. G., Menné T., Frosch, P. J., and Benezra, C., Eds., Springer-Verlag, Berlin, 1992, 459.

28. **Fregert, S.,** Physicochemical methods for detection of contact allergens, in *Exogenous Dermatoses: Environmental Dermatitis*, Menné, T. and Maibach, H. I., Eds., CRC Press, Boca Raton, FL, 1991, 74.

29. **Avnstorp, C.,** Cement eczema. An epidemiological intervention study, *Acta Dermato-Venereol.*, Suppl. 179, 1, 1992.

30. **Flyvholm, M.-A.,** Positive patch-test results of unknown relevance explained by a database on chemical products. Proc. 9th Int. Symp. Contact Dermatitis, *Contact Derm.*, 23, 285, 1990.

31. **Flyvholm, M.-A., Andersen, P., Beck, I. D., and Brandorff, N. P.**, PROBAS: The Danish Product Register Data Base — a national register of chemical substances and products, *J. Hazard. Mat.*, 30, 59, 1992.
32. **Dooms-Goossens, A., Dooms, M., and Drieghe, J.**, Computers and patient information systems, in *Textbook of Contact Dermatitis*, Rycroft, R. J. G., Menné, T., Frosch, P. J., and Benezra, C., Eds., Springer-Verlag, Berlin, 1992, 771.
33. **Danish Ministry of Labour.** *Order No. 540 of 2nd September 1982 on Substances and Materials*, Danish Ministry of Labour, Copenhagen, 1984.
34. **Danish Ministry of Labour.** *Order No. 466 of 14th September 1981 on the Register of Substances and Materials*, Danish Ministry of Labour, Copenhagen, 1984.
35. **Statens naturvårdsverk.** *Produktkontroll. Kundgörelse om produktanmälan. SNFS 1979: 2 PK*, Statens naturvårdsverks författningssamling, Sweden, 1979.
36. **Produktregistret.** *Deklarering av kjemiske stoffer og produkter. Veiledning*, Produktregistret, Oslo, 1989.
37. **Seta, J. A., Sundin, D. S., and Pedersen, D. H.**, *National Occupational Exposure Survey. Field Guidelines. DHHS (NIOSH) Publication No. 88-016*, U.S. Department of Health and Human Services, Cincinnati, Ohio, 1988.
38. **Richardson, E. L.**, Update — frequency of preservative use in cosmetic formulas as disclosed to FDA, *Cosmet. Toiletr.*, 96, 91, 1981.
39. **Decker, R. L.**, Frequency of preservative use in cosmetics as disclosed to FDA — 1984, *Cosmet. Toiletr.*, 100, 65, 1985.
40. **Decker, R. L. and Wenninger, J. A.**, Frequency of preservative use in cosmetic formulas as disclosed to FDA — 1987, *Cosmet. Toiletr.*, 102, 21, 1987.
41. **White, R.**, DERM/INFONET — The American Academy of Dermatology's computer based information system for the clinical dermatologist, in *Dermatology in Five Continents. Proceedings of the XVII World Congress of Dermatology, Berlin, May 24–29, 1987*, Springer-Verlag, Berlin, 1987, 831.
42. **Kopf, A. W., Rigel, D. S., White, R., Rosenthal, L., Jordan, W. P., Carter, D. M., Everet, M. A., and Moore, J.**, DERM/INFONET: A concept becomes a reality, *J. Am. Acad. Dermatol.*, 18, 1150, 1988.
43. **Drieghe, J., Dooms-Goossens, A., Dooms, M., and Degreef, H.**, The Codex system, *Semin. Dermatol.*, 8, 96, 1989.
44. **Rantanen, T.**, INFODERM — A microcomputer database system with finnish product files, *Semin. Dermatol.*, 8, 94, 1989.
45. **Edman, B.**, DALUK: The Swedish computer system for contact dermatitis, *Semin. Dermatol.*, 8, 97, 1989.
46. **Nava, C., Venturi, A., Meregalli, E., Bon, E., and Beretta, E.**, Information system in assessment of occupational allergy. The model of the institute of occupational health of Milan, in *Proceedings of the 1st International Workshop on Data Banks in Occupational Health. Villa Ponti, Varese, Italy October 30/31, 1986*, Parmeggiani, L., Roi, R., Aresini, G., and Bino, G., Eds., Eur. Ed., Ispra, 1987, 192.
47. **MacEachran, J. H., Clendenning, W. E., and Gosselin, R. E.**, Computer-derived exposure lists for common contact dermatitis antigens, *Contact Derm.*, 2, 239, 1976.
48. **Brandorff, N. P.**, *Number and Types of Chemicals Used in the Working Environment*, National Institute of Occupational Health, Copenhagen, 1992.
49. **Larsen, W. G.**, Why is the USA the only country with compulsory cosmetic labeling, *Contact Derm.*, 20, 1, 1989.
50. **De Groot, A. C. and White, I. R.**, Cosmetic ingredient labelling in the European Community, *Contact Derm.*, 25, 273, 1991.
51. **Commission of the European Communities.** *Proposal for a Council Directive amending for the 6th time Directive 76/768/EEC on the Approximation of the Laws of the Member States Relating to Cosmetic Products*, 1991.

52. **Danish Environmental Protection Agency.** *Order No. 586 of 8 August 1991 on Classification, Packing, Sale, and Storage of Chemical Substances and Products,* Danish Environmental Protection Agency, Copenhagen, 1991.

53. **Danish Environmental Protection Agency.** *Order No. 589 of 8 August 1991. List of Dangerous Substances,* Danish Environmental Protection Agency, Copenhagen, 1991.

54. **Danish Environmental Protection Agency.** *Order No. 134 of 24 February 1991. Changes to the List of Dangerous Substances,* Danish Environmental Protection Agency, Copenhagen, 1991.

55. **Friedmann, P. S., Moss, C., Shuster, S., and Simpson, J. M.,** Quantitative relationships between sensitizing dose of DNCB and reactivity in normal subjects, *Clin. Exp. Immunol.,* 53, 706, 1983.

56. **White, S. I., Friedmann, P. S., Moss, C., and Simpson, J. M.,** The effect of altering area of application and dose per unit area on sensitization by DNCB, *Brit. J. Dermatol.,* 115, 663, 1986.

57. **Rees, J. L., Friedmann, P. S., and Matthews, J. N. S.,** The influence of area of application on sensitization by dinitrochlorobenzene, *Brit. J. Dermatol.,* 122, 29, 1990.

58. **Upadhye, M. R. and Maibach, H. I.,** Influence of area of application of allergen on sensitization in contact dermatitis, *Contact Derm.,* 27, 281, 1992.

59. **Marzulli, F. N. and Maibach, H. I.,** Contact allergy: predictive testing in humans, in *Dermatotoxicology,* 4th ed., Marzulli, F. N. and Maibach, H. I., Eds., Hemisphere Publishing, New York, 1991, 415.

60. **Kligman, A. M.,** The identification of contact allergies by human assay. II. Factors influencing the induction and measurement of allergic contact dermatitis, *J. Invest. Dermatol.,* 47, 375, 1966.

61. **Kligman, A. M.,** Identification of contact allergies by human assay. III. Maximization test: a procedure for screening and rating of contact sensitizers, *J. Invest. Dermatol.,* 47, 393, 1966.

62. **Wilkinson, J. D., Hambly, E. M., and Wilkinson, D. S.,** Comparison of patch test results in two adjacent areas of England. II. (Medicament), *Acta Dermato-Venereol.,* 60, Suppl. 245, 1980.

63. **Marzulli, F. N. and Maibach, H. I.,** Antimicrobials: Experimental contact sensitization in man, *J. Soc. Cosmet. Chem.,* 24, 399, 1973.

64. **Marzulli, F. N. and Maibach, H. I.,** The use of graded concentrations in studying skin sensitizers: experimental contact sensitization in man, *Food Cosmet. Toxicol.,* 12, 219, 1974.

65. **Marzulli, F. N. and Maibach, H. I.,** Contact allergy: predictive testing of fragrance ingredients in humans by Draize and Maximization test, *J. Environ. Pathol. Toxicol.,* 3, 235, 1980.

66. **Menné, T.,** Nickel Allergy, thesis, University of Copenhagen, 1983.

67. **Basketter, D. A., Briatico-Vangosa, G., Kaestner, W., Lally, C., and Bontinck, W. J.,** Nickel, cobalt and chromium in consumer products: a role in allergic contact dermatitis? *Contact Derm.,* 28, 15, 1993.

68. **Picardo, M., Zompetta, C., de Luka, C., Cristaudo, A., Cannistraci, C., Faggioni, A. and Santucci, B.,** Nickel-keratinocyte interaction: a possible role in sensitization, *Brit. J. Dermatol.,* 122, 729, 1990.

69. **Vandenberg, J. J. and Epstein, W. L.,** Experimental nickel contact sensitization in man, *J. Invest. Dermatol.,* 41, 413, 1963.

70. **Wedroff, N.,** über Ekzeme bei Vernichlern, *Arch. Gewerbe-Pathol. Gewerbe-Hyg.,* 6, 179, 1935.

71. **Wall, L. M. and Calnan, C. D.,** Occupational nickel dermatitis in the electroforming industry, *Contact Derm.,* 6, 414, 1980.

72. **Fischer, T.**, Occupational nickel dermatitis, in *Nickel and the Skin: Immunology and Toxicology*, Maibach, H. I. and Menné, T., Eds., CRC Press, Boca Raton, FL, 1989, 117.
73. **Menné, T., Brandrup, F., Thestrup-Pedersen, K., Veien, N. K., Andersen, J. R., Yding, F., and Valeur, G.**, Patch test reactivity to nickel alloys, *Contact Derm.*, 16, 255, 1987.
74. **Wahlberg, J. E. and Skog, E.**, Nickel allergy and atopy. Threshold of nickel sensitivity and immunoglobulin E determinations, *Brit. J. Dermatol.*, 85, 97, 1971.
75. **Emmett, E. A., Risby, T. H., Jiang, L., Ng, S. K., and Feinmann, S.**, Allergic contact dermatitis to nickel: bioavailability from consumer products and provocation threshold, *J. Am. Acad. Dermatol.*, 19, 314, 1988.
76. **Menné, T. and Calvin, G.**, Concentration threshold of non-occluded nickel exposure in nickel-sensitive individuals and controls with and without surfactant, *Contact Derm.*, 29, 180, 1993.
77. **Allenby, C. F. and Basketter, D. A.**, An arm immersion model of compromised skin (II). Influence on minimal eliciting patch test concentrations of nickel, *Contact Derm.*, 28, 129, 1993.
78. **Menné, T. and Rasmussen, K.**, Regulation of nickel exposure in Denmark, *Contact Derm.*, 23, 57, 1990.
79. **Halbert, A. R., Gebauer, K. A., and Wall, L. M.**, Prognosis of occupational chromate dermatitis, *Contact Derm.*, 27, 214, 1992.
80. **Perone, V. B., Moffitt, A. E., Possick, P. A., Key, M. M., Danzinger, S. J., and Gellin, G. A.**, The chromium, cobalt and nickel contents of American cement and their relationship to cement dermatitis, *Am. Industr. Hyg. Assn. J.*, 35, 301, 1974.
81. **Fregert, S. and Gruvberger, B.**, Chemical properties of cement, *Dermatosen*, 20, 238, 1972.
82. **Høvding, G.**, Cement Eczema and Chromium Allergy, an Epidemiologic Investigation, thesis, University of Bergen, Norway, 1970.
83. **Reifenstein, H., Lück, H., Pätzold, M., and Harms, U.**, Zur Häuftigkeit des Zementeksems bei der Verarbeitung chromatarmer Zemente, *Z. Gesamte Hyg.*, 32, 559, 1986.
84. **Avnstorp, C.**, Prevalence of cement eczema in Denmark before and since addition of ferrous sulfate to Danish cement, *Acta Dermato-Venereol.*, 69, 151, 1989.
85. **Fregert, S., Gruvberger, B., and Sandahl, E.**, Reduction of chromate in cement by iron sulfate, *Contact Derm.*, 5, 39, 1979.
86. **Gammelgaard, B., Fullerton, A., Avnstorp, C., and Menné, T.**, Permeation of chromium salts through human skin in vitro, *Contact Derm.*, 27, 302, 1992.
87. **Bruze, M., Gruvberger, B., Björkner, B., and Kathon C. G.**, An unusual contact sensitizer, in *Exogenous Dermatoses: Environmental Dermatitis*, Menné, T. and Maibach, H. I., Eds., CRC Press, Boca Raton, FL 1991, 283.
88. **Chan, P. K., Baldwin, R. C., Parsons, R. D., Moss, J. N., Stiratelli, R., Smith, J. M., and Hayes, A. W.**, Kathon biocide: manifestation of delayed contact dermatitis in guinea pigs is dependent on the concentration for induction and challenge, *J. Invest. Dermatol.*, 81, 409, 1983.
89. **Cardin, C. W., Weaver, J. E., and Bailey, P.**, Dose-response assessment of Kathon biocide. II. Threshold prophetic patch testing, *Contact Derm.*, 15, 10, 1986.
90. **Schwartz, S. R., Weiss, S., Stern, E., Morici, I. J., Goodman, J. J., and Scarborough, N. L.**, Human safety of body lotion containing Kathon CG, *Contact Derm.*, 16, 203, 1987.
91. **Hannuksela, M.**, Rapid increase in contact allergy to Kathon CG in Finland, *Contact Derm.*, 15, 211, 1986.
92. **Tosti, A.**, Prevalence and sources of Kathon CG sensitization in Italy, *Contact Derm.*, 18, 173, 1988.

93. **Menné, T., Frosch, P. J., Veien, N. K., Hannuksela, M., Björkner, B., Lachapelle, J. M., White, I. R., Vejlsgaard, G., Schubert, H. J., Andersen, K. E., Dooms-Goossens, A., Shaw, S., Wilkinson, J. D., Camarasa, J. G., Wahlberg, J. E., Brandrup, F., Brandao, F. M., Van der Walle, H. B., Angelini, G., Thestrup-Pedersen, K., et al.,** Contact sensitization to 5-chloro-2-methyl-4-isothiazolin-3-one and 2-methyl-4-isothiazolin-3-one (MCI/MI). A European multicenter study, *Contact Derm.*, 24, 334, 1991.

94. **Rastogi, S. C.,** *Indholdet af Kathon CG i kosmetiske produkter og opvaskemidler,* Kemikaliekontrollen, Miljøstyrelsen, Copenhagen, 1989.

95. **Cosmetic Ingredient Review.** Final report on the safety assessment of methylisothiazolinone and methylchlorisothiazolinone, *J. Am. Coll. Toxicol.*, 11, 75, 1992.

Chapter 15

PREVENTION OF ALLERGIC CONTACT SENSITIZATION

Torkil Menné, Mari-Ann Flyvholm, and Howard I. Maibach

CONTENTS

I. INTRODUCTION

Allergic contact sensitization is a delayed type of hypersensitivity reaction (cell mediated, Coomb classification group IV). Contact sensitization can be caused by man-made chemicals, naturally occuring substances, and complex materials (colophony, balsam of Peru) where the exact chemical composition of the hapten is unknown. Most contact sensitizing chemicals have a molecular weight below 600 Da.

Proteins can elicit so-called contact urticaria syndrome and protein contact dermatitis.[1] This can be seen after exposure to rubber gloves and food. These types of reactions will not be dealt with in this chapter.

TABLE 15-1
Contact Sensitivity to Nickel
in the General Population

Age (years)	Men (%)	Women (%)	Total (%)
15–34	2.4	19.6	12.2
35–49	1.1	7.9	4.6
50–69	3.0	2.7	2.9
Total	2.2	11.1	6.7

There are more than 11 million known chemicals (CAS Registry Numbers, 1991). Thousands of new chemicals are synthesized yearly. From a casual consideration, this suggests that preventive strategies might be an overwhelming task. The problems of preventing allergic contact sensitization have been reviewed.[2-11]

II. WHY PREVENT ALLERGIC CONTACT SENSITIZATION?

The frequency of contact allergy to chemicals included in the European Standard Series (Chapters 2, 3, and 5) has been evaluated in a large unselected sample drawn from the general population. 15.2% gave a positive reaction to one or more of the chemicals.[12] Contact allergy to nickel was frequent among females. In the age group of 15 to 34 years, 19.6% of the females had a positive diagnostic patch test to nickel (Table 15-1). Known from earlier studies and confirmed by the present one, nickel sensitized individuals run an increased risk of developing chronic hand eczema.[13,14] As illustrated in Table 15-2 contact allergy was found to correlate with nearly all chemicals included in the Standard Series, mostly in a frequency below 1%. Among patients investigated in dermatological centers, 5 to 23% only react to chemicals not included in the European Standard Series. The true population frequency is therefore higher than the 15.2% figure.[15] Similar figures can be expected internationally, although geographic variation in the frequency of contact allergy to the individual chemical occur.[16]

The majority of sensitized individuals do not have ongoing allergic contact dermatitis.[17] But an allergic sensitization means that the individual runs a risk of developing allergic contact dermatitis if exposed to the chemical in question in a concentration exceeding an individual threshold level as discussed in Chapter 14. A recent population study from Gothenburg showed a prevalence of allergic contact dermatitis on the hands to be 0.4% in males and 1.8% in females.[18] The presence of hand eczema frequently has social and occupational consequences for the individual.[19]

From patch test units it is known that patients with hand eczema represent approximately half of those with allergic contact dermatitis. It can therefore be

TABLE 15-2
The Distribution of Sensitization to Haptens and Mixtures of Haptens
— Percent of Subjects with Sensitivity (95% Confidence Intervals)

Haptens and Mixtures of Haptens	Men (n = 279)	Women (n = 288)	Total (n = 567)
Nickel sulfate	2.2	11.1	6.7
p-Phenylenediamine	0.0	0.0	0.0
Neomycin sulfate	0.0	0.0	0.0
Potassium dichromate	0.7	0.3	0.5
Cainemix	0.0	0.0	0.0
Fragrance mix	1.1	1.0	1.1
Colophony	0.4	1.0	0.7
Epoxy resin	0.4	0.7	0.5
Thiuram mix	0.7	0.3	0.4
Balsam of Peru	0.7	1.4	1.1
Ethylenediamine	0.4	0.0	0.2
Cobalt chloride	0.7	1.4	1.1
p-t-Butylphenol-formaldehyde resin	1.1	1.0	1.1
Parabens	0.4	0.3	0.4
Carba mix	0.7	0.0	0.4
Black-rubber mix	0.4	0.0	0.2
Isothiazoliones*	0.4	1.0	0.7
Quaternium 15	0.4	0.0	0.2
Mercaptobenzothiazole	0.4	0.0	0.2
Wool alcohols	0.4	0.0	0.2
Negative control	0.0	0.0	0.0
Mercapto mix	0.7	0.0	0.4
Thiomerosal	3.6	3.1	3.4
Quinolinemix	0.4	0.3	0.4

* 5-chloro-2-methyl-4-isothiazolin-3-one, 2-methyl-4-isothiazolin-3-one (MCI/MI).

estimated that the prevalence of allergic contact dermatitis in the population is 2 to 4%. Further epidemiological studies are needed to confirm this statement.

Allergic contact dermatitis is a common occupational problem. In most countries occupational skin diseases (irritant and allergic contact dermatitis) amount to 20 to 40% of all occupational diseases. Working days lost because of occupational allergic contact dermatitis are significant. The annual cost of occupational skin diseases in the U.S. was estimated to range between $222 million and $1 billion in 1985.[20]

Contact sensitization is an incurable, often life-long condition.[21,22] The skin disease, allergic contact dermatitis, secondary to contact sensitization, often clears after proper evaluation and instruction of the patient.

It is concluded that prevention of allergic contact sensitization is desirable because of its common occurrence and potential to cause long-lasting skin disease with social and physical consequences for the individual patient and general economic consequences for the society.

III. ENVIRONMENTAL AND GENETIC FACTORS IN ALLERGIC CONTACT SENSITIZATION

When considering prevention of contact sensitization it is of crucial importance to understand the interaction between environmental and genetic factors.

The importance of genetic factors for contact sensitization has been repeatly demonstrated in animal experiments.[23] It even seems that this propensity to sensitization is directed toward specific haptens.[24] The human experience is less convincing. One well-controlled family study indicated that experimental contact sensitization in children was greater when both parents could be sensitized by the same substance compared to children, where only only one parent could be sensitized.[25] A population-based twin study focusing on nickel allergy found a significant genetic effect for the risk of developing this contact sensitivity. Twin studies, using other designs, have failed to show such an association.[22]

As discussed in Chapter 14, particular allergen concentrations seem to be of importance whether contact sensitivity develops or not. Exposure to a potent contact sensitizer such as DNCB (dinitrochlorobenzene) in a concentration exceeding a certain threshold level will sensitize nearly 100% of those exposed after one or few exposures.[26] For a medium-strong sensitizer such as nickel, an experimental sensitization rate of 15 to 50% has been obtained. Also for weaker allergens, the concentration-dependent sensitization risk has been demonstrated (Chapter 14).

When it comes to clinical studies, the primary cause of sensitization can often, but not always, be demonstrated. Most cases of nickel allergy can be attributed to exposure to nickel alloys in close skin contact, which release high concentrations of nickel, when exposed to sweat. Similarly, chromate dermatitis often relates to exposure to hexavalent chromate in wet cement. Sensitization to medicaments is often related to specific exposure situations, as in treatment of leg ulcer and perianal eczema.[27]

Exposure to trace amounts of haptens in the general environment, generally does not produce overt contact sensitization. Investigations of monozygotic female twins, where one or both were nickel sensitive, show that only the twin with a history of contact dermatitis by nickel alloy exposures gives a positive diagnostic patch test to nickel.[28] Further, *in vitro* diagnostic testing failed to demonstrate subclinical nickel sensitization in family members of nickel sensitive individuals.[29] But it cannot be excluded that trace amounts of haptens might induce a subclinical level of contact sensitization. A recent guinea pig study illustrated that percutaneous nonsensitizing concentrations of haptens induced an immune response by preventing subsequent induction of immune tolerance.[30]

For already contact-sensitized individuals, trace amounts of contact allergens (ppm concentrations), particularly when applied on damaged skin, can contribute to the maintenance of the skin disease allergic contact dermatitis.[31]

Even if genetic factors probably are of significance whether the individual develops contact sensitization, exposure-related factors, particularly exposure concentration, seem to be decisive for a number of contact-sensitized individuals in a population. Most of the literature within the field of contact dermatitis supports this view, which establishes the theoretical baseline for the prevention of contact sensitization.

IV. EXAMPLES OF PREVENTION

Distinction is usually made between primary prevention, focusing on the induction of contact sensitization, and secondary prevention, focusing on the eliciting of contact sensitization. In many instances the preventive measures for the two different types are overlapping.

A. PRIMARY PREVENTION

In the 1960s an epidemic of contact dermatitis from dish-washing products occurred in Scandinavia. The epidemic was resolved by a concerted action by dermatologists and manufacturers. Extensive chemical analysis combined with animal predictive testing, identified highly-sensitizing sultones to be present in some products.[32,33] It was clarified that these specific chemicals occurred as an impurity in the manufacturing process, when temperature control was not strictly performed. The evaluation of the problem led to a solution, and similar recurrences have not been seen.

There are several examples where hapten concentrations are legally regulated in attempt to prevent contact sensitization.[7] When it comes to cosmetic products, there is complex regulation, forbidding certain substances and regulating others, i.e., preservatives by a concentration limit.[34] The clinical impact of these regulations have been inadequately documented.

Since the 1950s, chromate in cement has been known to be one of the main causes of allergic chromate dermatitis among construction workers. On the suggestion of Sigfred Fregert, the Scandinavian countries, in the early 1980s, added ferrosulfate in a low concentration to cement, to reduce the hexavalent chromate to trivalent chromate. The idea of this initiative is that the trivalent chromate is not absorbed, or only to a minor degree, through human skin, and therefore the risk of primary sensitization from this salt is significantly less as compared to hexavalent chromate. Epidemiological studies on construction sites performed at the beginning of the 1980s and at the end of the 1980s in Denmark, strongly suggest that this measure has been successful, as the frequency of allergic chromate dermatitis has been dramatically reduced in Denmark.[35]

As outlined in Chapter 14, nickel is a common contact allergen on a global scale. This allergy is caused by intimate skin contact with metal alloys, releasing nickel when exposed to human sweat. Some alloys release high amounts and other alloys low amounts of nickel, under simulated use conditions. Based on such research, some Scandinavian countries have introduced regulations and quality demands when nickel alloys are in prolonged skin contact. It is believed that such a measure might reduce the frequency of nickel allergy in the population significantly. A regulation of nickel exposure among the same outlines is presently being considered in other European countries, and within the EEC.

B. SECONDARY PREVENTION

The cornerstones in the secondary prevention of allergic contact dermatitis (elicitation of contact dermatitis) are based on sufficient diagnostic procedures and patient information systems. The availability of standardized patch test materials is essential. Further, it is crucial that it is possible for the doctor to inform the patient where exposure to the specific allergen can be expected. Of course, it is even more crucial that the patient is able to understand the information and by himself, over the years to come, to identify the allergen in his home and occupational environment. It seems obvious that this type of diagnostic follow-up will work, but has only been evaluated in a limited number of studies. Edman found that the prognosis for patients sensitive to topical medicaments depended upon whether the patients were able to follow the doctor's advice on the occurrence of sensitizers in different products.[36] Later studies have shown that patients with contact allergy to formaldehyde often continued to be exposed to formaldehyde.[37,38] When a careful work-up was done, formaldehyde exposure could be demonstrated in nearly all the patients, which seems to be decisive for the prognosis of their hand eczema.

V. SCIENTIFIC INFORMATION NEEDED TO ESTABLISH PREVENTIVE PROGRAMS

A. THE CHEMICALS

It is necessary to develop labeling systems for all types of products used domestically and occupationally. Presently, information on the occurrence of contact-sensitizing chemicals and the actual use concentration in different products is scanty. Expensive chemical analysis, performed in dermatological centers and university departments, can partly compensate for this lack of knowledge. As these analyses are often complicated, they add an unreasonable burden of costs to the examination of the individual contact dermatitis case. When developing information systems, it is crucial to acknowledge that certain chemicals might cause contact sensitization in the concentration range from a few to 100 ppm. In many countries, chemicals included in products in a concentration below 1% do not require declaration. This type of legislation is obviously insufficient.

B. SENSITIZING POTENTIAL

The sensitizing potential of chemicals can be evaluated by structure activity relationships. This is a computerized comparison of new vs. old chemical structures, where the contact sensitizing potential is known.[39] This might prove to be of value in the future. Also, predictive testing, based on cell culture, needs to be evaluated and further developed. Presently, the only available methods to predict contact sensitization, are by animal and human methods.

C. DIAGNOSTIC TESTING

Standardized clinical diagnostic testing is essential.[27] Committees, particularly the ICDRG (International Contact Dermatitis Research Group) and the EECDRG (European Environmental Contact Dermatitis Reseach Group), The North American Contact Dermatitis Research Group, and The Japanese Contact Dermatitis Research Group have made great efforts to standardize diagnostic patch test materials. This work must be continued and simple and inexpensive screening methods need to be developed. Continuous education of dermatologists to use these diagnostic methods is essential.

D. EPIDEMIOLOGY

Extensive data on the epidemiology of contact sensitization exist. The quality of these data need to be improved. Data concerning patients with skin diseases and studies performed on the general population should be divided. It is important to study data with respect to the most confounding background variables.[40] Computerized systems need to be further developed so that meaningful international comparisons can be made.[41]

E. IDENTIFICATION OF SENSITIZING AND ELICITING ALLERGEN CONCENTRATION

Whether a lower nonsensitizing allergen concentration exists is controversial. Some potent contact sensitizing substances can induce contact sensitization at ppm concentrations. For medium-strong and weak allergens, the concentration range for risk of sensitization is considerably higher. Notwithstanding these theoretical aspects of contact sensitization, the total number sensitized in a population depends upon the exposure concentration from a given hapten. The problem is how to gather information on these concentrations.

The sensitization threshold concentration found in animal studies can only with great caution be extrapolated to the human use situation. Human sensitization assays and human exposure use tests with specific products provide valuable information. If labeling for both domestically and occupationally used products were systematically introduced, this could improve the quality of the published clinical data. One present problem with many clinical patch test studies is that the primary cause of sensitization, either product or allergen concentration, is rarely identified. For most of the common haptens, it is impossible to find the primary sensitizing hapten concentration in the standard textbooks. Extensive research in this area is needed to provide guidelines for industry and make prevention possible.

Similarly, it is important to establish the eliciting threshold concentration in already sensitized individuals to establish guidelines for secondary prevention.

This can be done by dilution patch test series on already sensitized individuals or similar experimental exposure tests (i.e., ROAT).[42]

Perhaps the established lower threshold concentration in sensitized individuals can be suggested as a nonsensitizing concentration. This is a postulate which needs further evaluation.

F. INDUCTION OF TOLERANCE

Immunological tolerance can be induced in animals by noncutaneous hapten exposure. Induction of tolerance in humans to nickel from the exposure to nickel-releasing orthodontic braces in early age has been suggested.[43] In individuals already sensitized, temporary hyposensitization might be achieved by oral intake of the hapten.[44,45]

G. INFORMATION SYSTEMS

Information systems and product labeling are essential both for primary and secondary prevention of allergic contact dermatitis. Different examples are given in Chapter 14. There is a great need to further develop this type of information system. Modern computer technology will be helpful in this area.[41]

H. PROTECTIVE GLOVES

While different glove materials protect against skin irritation and mechanical skin injury, most small sensitizing chemicals rapidly penetrate most rubber and plastic gloves. General recommendations for the use of appropriate gloves have been made by Estlander and Jolanki.[46] A helpful reference database on protective gloves has been established.[47] The 4-H glove protects against epoxy and acrylate monomer.[48]

I. PRE-EMPLOYMENT TESTING

No method exists which can predict an eventual individual propensity to contact sensitization to a given chemical. When patch testing with strong sensitizing chemicals, active sensitization from the test cannot completely be excluded. Pre-employment testing is therefore not a method to prevent contact sensitization.

VI. WAYS OF PREVENTING CONTACT SENSITIZATION

1. Replacement of certain chemicals or particular products.
2. Regulation of exposure (concentration) to sensitizing chemicals, either generally or in specific products, or during particular work processes.
3. Optimal diagnostic and informations systems and education of either groups or individuals.
4. Individually oriented preventive methods, including gloves, barrier creams, and protective clothing.

The problems of contact sensitization have been identified for a number of years, and different types of preventive measures have been tried. Some have been successful, but a number of chemicals still give problems to a significant number of people. Different strategies should be considered, whether it comes to common environmental chemicals, or chemicals with rare specific exposures. Chemicals frequently used in both the domestic and occupational environment need to be regulated by society, either with suggestion of replacement or regulation of the exposure concentration. For rare chemicals, it is often sufficient to focus on specific occupational processes, and educate the individuals exposed in no-touch techniques or introduce personal, individually oriented preventive measures.

REFERENCES

1. **Lahti, A. and Maibach, H. I.,** Immediate contact reactions, in *Exogenous Dermatoses: Environmental Dermatitis*, Menné, T. and Maibach, H. I., Eds., CRC Press, Boca Raton, FL, 1990, 21.
2. **Fregert, S.,** Contact allergens and prevention of contact dermatitis, *J. Allergy Clin. Immunol.*, 78, 1071, 1986.
3. **Wahlberg, J. E.,** Prophylaxis of contact dermatitis, *Semin. Dermatol.*, 5, 255, 1986.
4. **Adams, R. M. and Fisher, A. A.,** Contact allergen alternatives, *J. Am. Acad. Dermatol.*, 4, 951, 1986.
5. **Robinson, M. K., Stotts, J., Danneman, P. J., and Nusair, T. L.,** A risk assessment process for allergic contact sensitization, *Fd. Chem. Toxic.*, 27, 479, 1989.
6. **Adams, R. M.,** *Prevention, Rehabilitation, Treatment in Occupational Skin Diseases*, 2nd ed., Adams, R. M., Ed., W. B. Saunders, Philadelphia, 1990, chap. 16.
7. **Hjorth, N. and Menné, T.,** Prevention of allergic contact sensitization: a historical perspective, in *Exogenous Dermatoses*, Menné, T. and Maibach, H. I., Eds., CRC Press, Boca Raton, FL, 1990, chap. 30.
8. **National Institute for Occupational Safety and Health,** Prevention of occupational skin disorders. Part 1. *Am. J. Contact Derm.*, 1, 1990.
9. **National Institute for Occupational Safety and Health,** Prevention of occupational skin disorders: a proposed national strategy for the prevention of dermatological conditions. Part 2. *Am. J. Contact Derm.*, 1, 116, 1990.
10. **Lachapelle, J-M.,** Principles of prevention and protection in contact dermatitis, in *Textbook of Contact Dermatitis*, Rycroft, R. J. G., Menné, T., Frosch, P. J., and Benezra, C., Eds., Springer-Verlag, Berlin, 1992, chap. 17.
11. **Ayala, F.,** Prevention of occupational dermatitis, *Clin. Dermatol.*, 10, 189, 1992.
12. **Nielsen, N. H. and Menné, T.,** Allergic contact sensitization in an unselected Danish population. The Glostrup allergy study, Denmark, *Acta Dermato-Venereol.*, 72, 456, 1992.
13. **Menné, T., Borgan, J., and Green, A.,** Nickel allergy and hand dermatitis in a stratified sample of the Danish female population: an epidemiological study including a statistic appendix, *Acta Dermato-Venereol.*, 62, 35, 1982.
14. **Wilkinson, D. S. and Wilkinson, J. D.,** Nickel allergy and hand eczema, in *Nickel and the Skin; Immunology and Toxicology*, Maibach, H. I. and Menné, T., Eds., CRC Press, Boca Raton, FL, 1989, chap. 13.

15. **Menné, T., Dooms-Goossens, A., Wahlberg, J. E., White, I. R., and Shaw, S.,** How large a proportion of contact sensitivities are diagnosed with the European standard series? *Contact Derm.,* 26, 201, 1992.

16. **Menné, T., Frosch, P. J., Veien, N. K., et al.,** Contact sensitization to 5-chloro-2-methyl-4-isothiazolin-3-one and 2-methyl-4-isothiazolin-3-one (MCI/MI), *Contact Derm.,* 24, 334, 1991.

17. **Nielsen, N. H. and Menné, T.,** Skin symptoms in contact sensitized individuals in an unselected Danish population. The Glostrup allergy study, unpublished observations.

18. **Meding, B. and Swanbeck, G.,** Epidemiology of different types of hand eczema in an industrial city, *Acta Dermato-Venereol.,* 69, 227, 1989.

19. **Meding, B. and Swanbeck, G.,** Consequences of having hand eczema, *Contact Derm.,* 23, 6 1990.

20. **Mathias, C. G. T.,** The cost of occupational skin disease, *Arch. Dermatol.,* 121, 332, 1985.

21. **Lejman, E., Stoudemayer, T., Grove, G., and Kligman, A. M.,** Age difference in poison ivy dermatitis, *Contact Derm.,* 11, 163, 1984.

22. **Menné, T. and Wilkinson, J. D.,** Individual predisposition to contact dermatitis, in *Textbook of Contact Dermatttis,* Rycroft, R. J. G., Menné, T., Frosch, P. J., and Benezra, C., Eds., Springer-Verlag, Berlin, 1992, Chapter 5.

23. **Andersen, K. E. and Maibach, H. I.,** Guinea pig sensitization assays, in *Contact Allergy Predictive Tests in Guinea Pigs,* Andersen, K. E. and Maibach, H. I., Eds., Karger, Basel, 1985, 263.

24. **Polak, L., Barnes, J. M., and Turk, J. L.,** The genetic control of contact sensitization to inorganic metal compounds in guinea pigs, *Immunology,* 14, 707, 1968.

25. **Walker, F. B., Smith, P. D., and Maibach, H. I.,** Genetic factors in human allergic contact dermatitis, *Int. Arch. Allergy,* 32, 453, 1967.

26. **Friedmann, P. S., Moss, C., Shuster, S., and Simpson, J. M.,** Quantitative relationships between sensitizing dose of DNCB and reactivity in normal subjects, *Clin. Exp. Immunol.,* 53, 706, 1983.

27. **Andersen, K. E., Burrows, D., and White, I. R.,** Allergens from the standard series, in *Textbook of Contact Dermatitis,* Rycroft, R. J. G., Menné, T., Frosch, P., and Benezra, C., Eds., Springer-Verlag, Berlin, 1992, chap. 13.

28. **Menné, T. and Holm, N. V.,** Nickel allergy in a female twin population, *Int. J. Dermatol.,* 22, 22, 1983.

29. **Silvennoinen-Kassinen, S.,** The specificity of nickel sulphate reaction in vitro: a family study and a study of chromium-allergic subjects, *Scand. J. Immunol.,* 13, 231, 1981.

30. **Van Hoogstraten, I. M. W., von Blomberg, B. M. E., Boden, D., Kraal, G., and Scheper, R. J.,** Non-sensitizing percutaneous skin contacts prevent subsequent induction of immune tolerance, *J. Invest. Dermatol.,* 102, 80, 1994.

31. **Allenby, C. F. and Basketter, D. A.,** An arm immersion model of compromised skin. II. Influence on minimal eliciting patch test concentrations of nickel, *Contact Derm.,* 28, 129, 1993.

32. **Magnusson, B. and Gilje, O.,** Allergic contact dermatitis from a dish-washing liquid containing laurylethersulphate, *Acta Dermato-Venereol.,* 53, 136, 1973.

33. **Ritz, H. L., Conner, D. S., and Sauter, E. D.,** Contact sensitization of guinea-pigs with unsaturated and halogenated sultones, *Contact Derm.,* 1, 349, 1975.

34. Council Directive 76/768/EEC, July 27, 1976.

35. **Avnstorp, C.,** Cement eczema. An epidemiological intervention study, *Acta Dermato-Venereol.,* Suppl. 179, 1992.

36. **Edman, B.,** The usefulness of detailed information to patients with contact allergy, *Contact Derm.,* 19, 43, 1988.

37. **Cronin, E.,** Formaldehyde is a significant allergen in women with hand eczema, *Contact Derm.,* 25, 276, 1991.

38. **Flyvholm, M.-A. and Menné, T.,** Allergic contact dermatitis from formaldehyde. A case study focusing on sources of formaldehyde exposure, *Contact Derm.*, 27, 27, 1992.

39. **Benezra, C., Sigman, C. C., Bagheri, D., Tucker Helmes, C., and Maibach, H. I.,** A systematic search for structure activity relationships of skin sensitizers. II. Para-phenylenediamines, *Semin. Dermatol.*, 8, 88, 1989.

40. **Coenraads, P. J. and Smit, J.,** Epidemiology, in *Textbook of Contact Dermatitis*, Rycroft, R. J. G., Menné, T., Frosch, P. J., and Benezra, C., Eds., Springer-Verlag, Berlin, 1992, chap. 6.

41. **Dooms-Goossens, A., Dooms, M., and Drieghe, J.,** Computers and patient information systems, in *Textbook of Contact Dermatitis*, Rycroft, R. J. G., Menné, T., Frosch, P. J., and Benezra, C., Eds., Springer-Verlag, Berlin, 1992, chap. 20.

42. **Hannuksela, M. and Salo, H.,** The repeated open application test (ROAT), *Contact Derm.*, 14, 221, 1986.

43. **Van Hoogstraten, I. M. W., Andersen, K. E., von Blomberg, B. M. E., et al.,** Reduced frequency of nickel allergy upon oral nickel contact at an early age, *Clin. Exp. Immunol.*, 85, 441, 1991.

44. **Kligman, A. M.,** Hyposensitization against Rhus dermatitis, *Arch. Dermatol.*, 78, 47, 1958.

45. **Sjövall, P., Christensen, O., and Möller, H.,** Oral hyposensitization in nickel allergy, *J. Am. Acad. Dermatol.*, 17, 774, 1987.

46. **Estlander, T. and Jolanki, R.,** How to protect the hands, *Dermatol. Clin.*, 6, 105, 1988.

47. **Mellström, G., Lindahl, G., and Wahlberg, J.,** DAISY: reference database on protective gloves, *Semin. Dermatol.*, 8, 75, 1989.

48. **Roed-Petersen, J.,** A new glove material protective against epoxy and acrylate monomer, in *Current Topics in Contact Dermatitis*, Frosch, P. J., Dooms-Goossens, A., Lachapelle, J.-M., Rycroft, R. J. G., and Scheper, R. J., Eds., Springer, Heidelberg, 1989, 603.

Other Tissue Allergies

Chapter 16

GASTROINTESTINAL TRACT ALLERGY AND INTOLERANCE

Charlotte Madsen

CONTENTS

I. INTRODUCTION

The scope of this book is to describe allergic reactions to low molecular weight chemicals that are not drugs. A chapter on gastrointestinal allergy in this context could be very short because low molecular weight chemicals are rarely the cause of a true allergic reaction in the gut, as they rarely, if ever, cause primary sensitization via or in the gut. To put this into perspective, the following is a short review on gastrointestinal immunity, oral tolerance, protein allergy including nontraditional proteins, oral exposure to contact allergens, and food additive intolerance.

II. THE GASTROINTESTINAL IMMUNE SYSTEM

The gastrointestinal tract has a very large surface area of mucous membrane facing the "outside" world. In accordance with this the gut has an extensive and specialized immune system to protect it from foreign organisms and substances.

The gut associated lymphoid tissue (GALT) can be divided into different compartments:

1. The Peyer's patches are organized lymphoid tissue with an architecture comparable to that of a lymph node. They consist of germinal centers (B-cell area), parafollicular area (T-cell area), and mixed T- and B-cell area beneath the dome epithelium. M-cells, which are specialized epithelial cells involved in antigen uptake, overlie the Peyer's patch. The Peyer's patches are mainly located in the ileum.
2. Cells of the immune system in the lamina propria are T- and B-lymphocytes, plasma cells, macrophages, mast cells, eosinophils, and basophils. The lymphocytes are mainly B-cells committed to IgA synthesis. A small percentage of B-cells synthesize other immunoglobulins, mainly IgM or IgG. The majority of the T-cells in the lamina propria are helper T-lymphocytes (CD4 positive).
3. Intraepithelial lymphocytes (IEL) are situated between the epithelial cells forming the epithelial lining of the gut. They occupy approximately 10% of the total epithelial cell mass and may represent, in total lymphoid mass, approximately the size of the spleen.[1] IEL are the first immune cells to come in contact with antigens in the gut lumen. They are predominantly CD8 positive (suppressor subtype). Hoang et al.[2] have shown that human IEL cells suppress proliferation of autologous lamina propria lymphocytes, but not autologous peripheral blood mononuclear cells, when stimulated with phytohemagglutinin.

III. UPTAKE OF MACROMOLECULES

In normal subjects the uptake of macromolecules is limited by digestion, mucus, peristalsis, and intestinal epithelium. In addition to these nonimmunological mechanisms, IgA antibodies are secreted by the majority of plasma cells in the lamina propria. This is part of the local immune response, providing antibodies to the gut lumen. The plasma cells secrete a dimeric IgA that, connected with a secretory component, is resistant to degradation by pancreatic proteases. Both oral and parenteral immunization with specific antigens can reduce their intestinal uptake. A local antibody can probably inhibit the absorption of soluble protein antigens by decreasing their adherence to intestinal epithelial cells.[3] The precise function of IgA in connection with macromolecular uptake is not known.

Despite the above-mentioned protective measures, intact dietary macromolecules enter the circulation and gain contact with the systemic immune system.

In normal adults, antigens can be measured in quantities of ng/ml serum after oral administration. The antigen may occur in its intact form, either free or antibody-bound.[4] In healthy subjects, serum antibodies to dietary antigens is found both in the IgA, IgG, and IgM immunoglobulin classes. Husby et al.[5] found IgG antibodies to ovalbumin and bovine beta-lactoglobulin in 95% of 6 to 20 year-old healthy subjects.

The production of secretory antibodies and serum antibodies to dietary protein antigens is a normal physiological response following the ingestion of food.[6]

IV. ORAL TOLERANCE

Oral tolerance is characterized by the induction of secretory immunity with concomitant production of systemic tolerance.[7] Systemic tolerance is a specific hypo- or unresponsiveness. It can be shown as a lack of response when the antigen subsequently is injected, painted on the skin, or inhaled. Tomasi[7] speculates that it seems likely that different mechanisms of oral tolerance induction may be involved, depending on the chemical structure of the antigen.

A. PROTEINS

Oral tolerance to protein antigens has been described mainly in laboratory animals.[8] It is not known whether oral tolerance occurs in humans. Subjects without prior symptoms from eating hen's eggs can be sensitized via the respiratory tract to household birds and subsequently show IgE-mediated allergic reactions after ingesting hen's eggs.[9] This suggests that if oral tolerance occurs in humans, it can be broken via the respiratory tract.

B. LOW MOLECULAR WEIGHT CHEMICALS

Oral tolerance to low molecular weight chemicals, that are contact sensitizers, has also been studied in experimental animals. For this type of tolerance there is also evidence of the occurrence in humans.

In 1946 Chase[10] reported the successful induction of tolerance to the contact allergen dinitrochlorobenzene (DNCB) after feeding guinea pigs with the allergen. The tolerance induced was specific for the compound and lasted a year. Coe and Salvin[11] showed that the induced tolerance was dose dependent. Induction of oral tolerance is only possible if the antigen is fed prior to cutaneous contact. In a study with nickel in guinea pigs van Hoogstraten et al.[12] found that even subsensitizing skin contact with nickel, that did not lead to detectable hypersensitivity, completely prevented subsequent oral tolerance induction.

In a retrospective study of 2176 patients, attending 9 European patch test clinics, ear piercing strongly favored development of allergic contact hypersensitivity to nickel. Patients having had oral contacts with nickel-releasing appliances (dental braces) at an early age, but only if prior to ear piercing, showed a reduced frequency of nickel hypersensitivity. Frequencies of other hypersensitivities, in particular to fragrances, were not affected.[13] Why nickel in the diet (chocolate contains 200 µg Ni/100g) does not induce oral tolerance is not clear. It may be a dose–response phenomenon and indicate a more readily accessible route to the immune system through free nickel ions from the metal in the mouth than from nickel in the food. The chemical forms in which nickel occurs in food (nickel species) is not well investigated.[14]

V. ALLERGY TO DIETARY COMPONENTS

As has been shown in the previous paragraphs, the gastrointestinal tract is designed to protect the organism from deleterious effects of foreign material, including the regulation of the immune response. The precise mechanisms of sensitization, humoral or cell mediated, via the gastrointestinal tract are not known, but a number of predisposing factors of IgE-mediated allergy are known.

1. *Genetic predisposition.* The risk of developing atopic disease (asthma, allergic rhinitis, atopic eczema, urticaria) is about 25% having a single predisposition for atopy, i.e., if one of the parents or siblings is atopic. If the predisposition is double, the risk of developing atopic disease is about 50%.[15] In Denmark, about 60% of children with gastrointestinal allergy belong to the third of all children with predisposition for developing atopy.[16]

2. *Maturity of the gut and gut-associated lymphoid tissue.* In early infancy there is an increased uptake of antigen and the gut-associated lymphoid tissue is incompletely developed.[3] The possible consequences of this is exemplified in the results from a prospective study of 1749 newborn children in which 39 developed cow's milk allergy during the first year of life. All 39 had ingested cow's milk formula during the first 3 days of life. Nine of the children were exclusively breast fed except for neonatal exposure to cow's milk formula in an amount corresponding to 0.4 to 3.0 g beta-lactoglobulin.[17]

3. *The nature of the allergen and the frequency of exposure.* Some foods are more likely associated with allergy than others. This is probably due to the nature of the allergen, and the intensity of exposure. Peanut allergy is one of the most common food allergies in the U.S., where peanut butter often is a part of the daily diet. Peanuts are a less important part of a European diet and peanut allergy is not common in Europe.

A. ALLERGY TO PROTEINS

Sensitization to protein antigens via the gastrointestinal tract normally occurs in young children.[18] Common allergens are milk, egg, and fish. In a study in children followed from birth to 3 years of age,[19] 8% reacted to foods (excluding fruit juices). All reactions had disappeared by the age of 3.

In older children and young adults, allergy to proteins in nuts, fruits, and vegetables occurs. Such allergies are probably not caused by intestinal sensitization, but by sensitization to pollen via the respiratory tract. The allergy is caused by IgE antibodies cross-reacting with epitopes in the pollen and in the offending nuts, fruits, and/or vegetables.[9] In a survey of Swedish medical students 9% reported allergy to nuts and 7% reported allergy to apples and related fruits. 78% of the students with birch pollen rhinitis and positive prick test to birch pollen reported clinical sensitivity to nuts and/or apples and related fruits.[20]

The incidence of food allergy in the general population is not known. Even the incidence of severe fatal or life-threatening reactions to food is not known. There is no "International Classification of Disease" code for food anaphylaxis and thus no collection of data for health statistics. Bock[21] has, on the basis of severe food reactions in Colorado, estimated that in the U.S., a minimum of 950 fatal or life-threatening reactions to food may occur each year.

1. Nontraditional Proteins

The following examples from the literature describe life-threatening reactions, where a lot of work has been done to elucidate the offending substance, the mechanism, etc. It is likely that only a minority, even of the serious cases, reach the pages of the medical journals.

The dried outer layer of psyllium seeds are fiber-rich and used in bulk laxatives. Recently it has been used in a breakfast cereal to enhance the fiber content. Anaphylactic reactions after ingestion of psyllium-containing cereal is described in nurses being occupationally sensitized to psyllium dust.[22,23]

Serious reactions, including anaphylaxis, have been reported to cottonseed protein used in food supplements,[24] in bread[25] and probably in a candy bar.[26] Cottonseed protein is known as a strong allergen. Atkins et al.[24] warns that genetic engineering removing the gossypol-containing pigment gland from the cottonseed, has led to renewed interest in using cottonseed in products for human consumption.

A severe systemic reaction to the enzyme papain, used as a meat tenderizer, was found in an atopic patient. There was no indication of an occupational sensitization to papain dust.[27]

An extended use of gluten as a food ingredient may occur. One of the proposed uses is as a surface finishing agent of fresh fruit.[28]

The use in processed foods of partly hydrolyzed proteins,[29] the use of vegetable starch with protein residues, or the use of other "unexpected" allergens, including milk, egg, nuts, etc., without proper labeling, poses a hazard for allergic individuals.

B. ALLERGY TO LOW MOLECULAR WEIGHT CHEMICALS

Descriptions of allergic reactions to low molecular weight chemicals that may have sensitized the organism via the gastrointestinal tract rarely occur.

1. Quinine

One case of anaphylactic shock after ingestion of a quinine-containing beverage was described by Jansen et al.[30] Quinine is primarily a drug. IgE-mediated systemic anaphylactic shock after ingestion of an antipyretic combination product containing quinine is reported.[31] Quinine as a contact sensitizer is also described,[32,33] which could suggest that the primary sensitization may have been via the skin.

2. Sulfites

SO_2 and its sulfite salts have numerous technological purposes in the food and pharmaceutical industries. The most common uses are as preservatives and antibrowning agents. Ingestion of sulfites can induce asthma or anaphylactoid reactions in sensitive individuals. The mechanism or mechanisms are not known. Some, but not all, sensitive subjects have positive reactions in skin prick tests with sulfite solutions and it has been possible to transfer skin test reactivity to nonsensitive subjects (Prausnitz-Küstner reaction). This suggests an IgE-mediated reaction. It has not been possible to demonstrate IgE specific to sulfite or sulfite conjugated with human serum albumin. These findings are reviewed by Taylor et al.[34]

As sulfites are used in pharmaceuticals including preparations for inhaling, it is possible that the route of sensitization is not the gastrointestinal but the respiratory tract, i.e., if sulfites behave like nickel and induce tolerance when given orally.

3. Contact Allergens

Some chemicals causing contact allergy are also normal constituents of food. In dermally-sensitized subjects, ingestion of the contact allergen may cause skin flare reactions or other symptoms, e.g., from the gastrointestinal tract. Patients with unexplained eruptions of nickel hand eczema may benefit from a diet with a low nickel content.[35] The chemicals of fragrances and of food flavors, natural or synthetic, are often identical. Veien et al.[36] found an effect of a "balsam"-restricted diet in patients having a flare of dermatitis after oral challenge with balsam of Peru.

Systemic contact-type dermatitis is reviewed by Menné and Maibach.[37] Systemic reactions after ingestion of contact allergens are described for metals (nickel, chromium, cobalt), medicaments (e.g., neomycin, penicillin, sulfonamides), pentadecyl catecols from poison ivy cross-reacting with oil of cashew nuts, flavors from balsam of Peru cross-reacting with food flavors and orange peel, and for the antioxidant butylated hydroxyanisole used both in topical preparations and food.

VI. NON-ALLERGIC REACTIONS TO FOOD

Non-allergic adverse reactions to food can have different etiologies with a known mechanism, i.e., enzyme deficiencies, pharmacologic active ingredients in food such as biogenic amines or caffeine, microorganisms, or toxins. The mechanism of action behind skin rashes and gastrointestinal symptoms in children caused by oranges, tomatoes, strawberries, or other fruits is not known.[18,19] In food additive intolerance the etiologic agent may be identified, but the mechanism of action is not known.

A. INTOLERANCE TO FOOD ADDITIVES
1. Additives
The one thing food additives have in common is that they are added to food. Chemically and functionally they are a very heterogenous group consisting of: enzymes (proteins), flavors (low molecular weight chemicals, lmc), colors (lmc), preservatives (lmc), emulsifiers (lmc), antioxidants (lmc, some of them vitamins), sweeteners (lmc), etc. The substances can either originate from plants or animal tissue, or be identical to naturally occurring substances, but be chemically produced or synthetic. The food additives most often described to cause adverse reactions are colors, both natural and synthetic, preservatives (sulfites, benzoates), and antioxidants (BHA and BHT).

2. Symptoms
The symptoms most consistently described in connection with food additive intolerance are symptoms from the skin[38-41] and asthma.[42] Gastrointestinal symptoms are also described.[41,43,44] The evidence to suggest that food additives play a role in childhood hyperactivity is very limited.[45]

3. Mechanisms
The mechanisms of food additive intolerance is not known. Tartrazine has been the additive of choice when researchers have tried to elucidate the mechanism of food additive intolerance. Aspirin inhibits the cyclo-oxygenase pathway of prostaglandin synthesis. A similar mechanism of action has been proposed for tartrazine. Larsen[46] concludes in his review that the evidence for this is conflicting. A later article[47] describing the inhibition of cyclo-oxygenase using platelet aggregation as a model has not solved the uncertainties. Aspirin, tartrazine, and other food additives can inhibit platelet aggregation, but tartrazine is used in a 100-fold concentration compared to aspirin, to produce a comparable reaction. Attempts to show an immunologic basis is reviewed by Robinson,[48] who concludes that there are data both to support specific (antibody-mediated) and nonspecific activation pathways.

Only minute amounts (<2%) of tartrazine and other water soluble azo-dyes are absorbed intact. The azo-bond is reduced by the gut microflora. This could suggest that an effect is caused by a metabolite[46] and makes *in vitro* studies with the parent compound difficult to interpret. In an *in vitro* model with guinea pig ileum, Hutchinson et al.[49] recently showed that tartrazine enhanced intestinal contraction. The significance of this finding is not clear.

Murdoch et al.[50] have shown that a large cumulative dose of tartrazine (200 mg) in normal subjects resulted in a significant rise in plasma and urinary

TABLE 16-1
Results of a British Study of Prevalence
of Food Additive Intolerance

	No. of Subjects	%
Questionnaire	30,000	
Answers	18,582	
Reporting food additive intolerance		7.4%
Selected for double-blind challenge	132	
Completing double-blind challenge	81	
Positive in double-blind challenge	3	
Positive in population		0.026%

histamine levels. The histamine release was not accompanied by systemic symptoms. This finding suggests a pharmacologic effect of tartrazine. The same group[51] has shown significant rises in plasma histamine paralleling clinical skin symptoms in a patient after oral provocation with tartrazine, carmoisine, sunset yellow, or amaranth. Schaubschlager et al.[52] have shown histamine release, induced by sodium benzoate, from mast cells in the gastric mucosa of patients positive to benzoate after oral provocation. These recent findings support the hypothesis of a pharmacologic effect of tartrazine and benzoic acid of yet unknown mechanism.

4. Prevalence

A theoretical estimate of the prevalence of food additive intolerance was done in an EEC report[53] in 1982. The estimate was based on results from selected patient materials calculating the assumed frequency in the total population. The symptoms considered were chronic urticaria and angioedema in adults and asthma, rhinitis and nasal polyps in adults and children. The calculations were based on assumptions on cross-reactions between acetylsalicylic acid and certain additives which are, for the most part, not considered valid any more.[54] The result was a possible frequency, for the most common manifestations, of 0.03% to 0.15%.

Since then two population studies have emerged; one is British[55] and the other is Danish.[56,57]

In the British study[55] a questionnaire was sent out to 11,388 households comprising an estimated 30,000 people (Table 16-1). Of the 18,582 responses 7.4% reported problems with food additives. Half of these respondants had a personal history of atopy. Nearly half reported symptoms from the musculo-skeletal system, behavioral/mood change, or headache. Selected for double-blind placebo-controlled challenge were subjects over 4 years of age and with a history of reproducible clinical symptoms after ingestion of food additives. Of the 132 entering the trial, 81 persons completed the challenge and 3 subjects showed consistent positive reactions: a 50-year old atopic male whose headache was provoked by annatto, a 31-year old nonatopic female who had upper

TABLE 16-2
Results of a Danish Study of Prevalence
of Food Additive Intolerance

	No. of Subjects	%
Questionnaire (school children)	4,952	
Answers	4,279	
Reporting food additive intolerance		6.6%
Selected/completing open challenge	27 (335)	
Positive in open challenge	17 (23)	
Positive in population based on strata, open challenge		2%
Double-blind challenge		1%

Note: The numbers in brackets are from the out-patient study.

abdominal pain after challenge with annatto, and a 5-year old atopic boy with eczema showing a change in mood after azo-dye challenge. The authors assume that the 81 persons were representative of the 132 entering the trial and the prevalence was calculated to be 4.9 out of 18, 582, i.e., 0.026%. Using the scorings from a diary card, including persons positive to placebo, the prevalence estimate was 0.23%. The food additives used for challenge were synthetic colors: amaranth, sunset yellow, tartrazine, quinoline yellow, carmoisine, and green S; natural colors: indigo carmine and annatto; antioxidants: butylated hydroxyanisole and butylated hydroxytoluene; preservative: sodium benzoate.

The Danish population study is the combined results from a study in four pediatric out-patient allergy clinics[56] and a study in schoolchildren[57] (Table 16-2). A total of 606 children were challenged openly with a mixture of food additives in a red and a green "lemonade." The lemonades contained synthetic colors: tartrazine, quinoline yellow, ponceau 4R, azorubine, patent blue, and sunset yellow; natural colors: turmeric, annatto, beta-carotene, chantaxantine, and beetroot color; preservatives: sulfite, sorbic acid, benzoic acid, and propionic acid; synthetic flavors: anethole and ethylvanilline; natural flavors: orange oil and lemon oil, and sugar and citric acid for adjustment of taste.

The challenged children were 98 (0%) controls, 213 (0.9%) with asthma, 186 (1.6%) with rhinitis, 181 (13%) with atopic eczema, 61 (23%) with urticaria, and 41 (19%) with other symptoms mainly from the gastrointestinal tract. The percentage in parentheses is the percent in the group diagnosed positive after open challenge. Many of the children had more than one symptom. The challenge-positive children were asked to participate in a double-blind, placebo-controlled challenge with the same additives in capsules. Twenty-eight of 40 children who were diagnosed as positive after open challenge participated in the double-blind challenge.

The children in the out-patient study were first-time referred to allergological examination. In the schoolchildren study, children were selected for challenge from the results of a questionnaire forwarded to the parents of all schoolchildren aged 5 to 16 years in the Viborg municipal district. They were asked about reactions of hypersensitivity, symptoms, the suspected allergens including food additives, and contact with the hospital allergy clinic.[58] 4,274 (86.4%) answered the questionnaire. As a result of the questionnaire the distribution of hypersensitivity reactions and suspicion to food additives were known both in the challenged children and in the background population. On the basis of this, the population was divided into different strata. The prevalence of food additive intolerance in each stratum was calculated. A combination of the results of the strata gave the prevalence estimate in the whole population. The prevalence of food additive intolerance in the group of children that had been referred to a hospital allergy clinic was estimated using the results from the outpatient study. Based on the results from the open challenge the estimated prevalence is 2.04% (SD 0.37). The results from the double-blind challenges were too few to allow for a comparable calculation, but as 50% of the children who were positive in the open challenge were positive in the double-blind challenge, the prevalence based on double-blind challenges is estimated to be about 1% with 6.6% suspected food additive intolerance to induce adverse reactions.

The frequencies of food additive intolerance found in the above studies vary by a factor of 100. This may be explained by substantial differences in the study populations.

In the EEC report only chronic urticaria, angioedema, asthma, rhinitis, and nasal polyps in adults and only respiratory symptoms in children were considered relevant. This leaves out more subjective symptoms and skin symptoms in children. Skin symptoms were by far the most frequent symptoms of food additive intolerance in the Danish study. In the British study nearly 50% reported subjective symptoms, i.e., from the musculo-skeletal system, behavioral/mood change, or headache. These types of symptoms were not represented in the challenged children in the Danish study. In this study only 1% stated subjective symptoms such as joint pain, mood change, or headache. This could reflect differences in age groups and/or in questioning in the two studies. Inclusion criteria for challenge in the British study was a history of reproducible clinical symptoms after ingestion of food additives. This criteria of inclusion will probably exclude most children with atopic skin symptoms and other subjects with atopic symptoms where the susceptibility to triggering factors, including food additives, varies with the state of the disease. Neither the EEC report nor the British study include SO_2-induced asthma. The Danish study shows that SO_2-induced asthma probably is not a major problem in children.

A common conclusion from the above is that food intolerance is found in adults with atopic symptoms from the respiratory tract and skin and the prevalence is less than 0.15%. In adults and children with reproducible symptoms, including subjective symptoms such as headache and behavioral/mood change the prevalence is even lower (0.026%). Food additive intolerance is primarily found in atopic children with cutaneous symptoms where the additive is aggravating an existing disease. The prevalence of food additive intolerance in children aged 5 to 16 is 1 to 2%. In the majority of children, the additive intolerance is probably transient.[41]

VII. CONCLUSION

Allergic sensitization to low molecular weight chemicals via the gastrointestinal tract is very unusual, probably because the organism is efficiently organized to deal with foreign substances entering this way, developing what is known as oral tolerance. The oral tolerance is a systemic unresponsiveness, protecting the organism from a harmful immunological response.

Food additive intolerance is caused by certain colors, preservatives, and antioxidants by an unknown, but probably nonimmunologic, mechanism. Most often symptoms from the skin or respiratory tract are provoked in atopic subjects. The prevalence of food additive intolerance is found to be 1 to 2% in a population of children. In a population of children and adults identified in a different way, a much lower prevalence was found. The intolerance may be transient.

The multitude of substances that can cause adverse food reactions, and the lack of good diagnostic tests, makes it a very difficult task to arrive at the right diagnosis. The trend toward increasing consumption of semi-finished or finished foods using new or nontraditional ingredients and new ways of processing, makes it increasingly difficult for the clinician, the dietitian, and the patient to find out what has actually been eaten.

Authorities should be aware of potential hazards of nontraditional food, and the novel use of traditional foods. It should be mandatory to label foodstuffs to show the nature of their ingredients, especially potent allergens.

REFERENCES

1. **Croitoru, K. and Bienenstock, J.,** Mucosal immunity, in *Food Allergy: Adverse Reactions to Food and Food Additives,* Metcalfe, D. D., Sampson, H. A., and Simon, R. A., Eds., Blackwell Scientific Publications, Boston, 1991, chap. 1.
2. **Hoang. P., Dalton, H. R., and Jewell, D. P.,** Human colonic intra-epithelial lymphocytes are supressor cells, *Clin. Exp. Immunol.,* 85, 498, 1991.
3. **Murphy, M. S. and Walker, W. A.,** Antigen absorption, in *Food allergy: Adverse Reactions to Food and Food Additives*, Metcalfe, D. D., Sampson, H. A., and Simon, R. A., Eds., Blackwell Scientific Publications, Boston, 1991, chap. 4.
4. **Husby, S.,** Dietary antigens: uptake and humoral immunity in man, *Acta Pathol. Microbiol. Immunol. Scand.,* 96, (Suppl. 1), 1, 1988.

5. **Husby, S., Schultz Larsen, F., and Hyltoft Petersen, P.,** Genetic influence on the serum levels of naturally occuring human IgG antibodies to dietary antigens. Quantitative assessment from a twin study, *J. Immunogenetics,* 14, 131, 1987.

6. **Rieger, C. H. L.,** The physiological humoral immunologic response to alimentary antigens, in *Food Allergy in Infancy and Childhood,* Harms, H. K. and Wahn, U., Eds., Springer-Verlag Berlin, 1989, chap. 4.

7. **Tomasi, T. B.,** Oral tolerance, *Transplantation,* 29, 353, 1980.

8. **Strobel, S.,** Mechanisms of gastrointestinal immunoregulation and food induced injury to the gut, *Eur. J. Clin. Nutr.,* 45, (Suppl. 1), 1, 1991.

9. **Pauli, G., de Blay, F., Bessot, J. C., and Dietemann, A.,** The association between respiratory allergies and food hypersensitivities, *Allergy Clin. Immunol. News,* 4, 43, 1992.

10. **Chase, M. W.,** Inhibition of experimental drug allergy by prior feeding of the sensitizing agent, *Proc. Soc. Exp. Biol. Med.,* 61, 257, 1946.

11. **Coe, J. E. and Salvin, S. B.,** The specificity of allergic reactions. VI. Unresponsiveness to simple chemicals, *J. Exp. Med.,* 117, 401, 1963.

12. **van Hoogstraten, I. M. W., Andersen, K. E., von Blomberg, B. M. E., Boden, D., Bruynzeel, D. P., Burrows, D., Camarasa, J. G., Dooms-Goossens, A., Kraal, G., Lahti, A., Menné, T., Rycroft, R. J. G., Shaw, S., Todd, D., Vreeburg, K. J. J., and Wilkinson, J. D.,** Reduced frequency of nickel allergy upon oral nickel contact at an early age, *Clin. Exp. Immunol.,* 85, 441, 1991.

13. **van Hoogstraten, I. M. W., Kraal, G., Boden, D., Vreeburg, K. J. J., von Blomberg, B. M. E., Bruynzeel, D. P., and Scheper, R. J.,** Persistent immune tolerance induced by oral feeding of metal allergens, *J. Invest. Dermatol.,* 92, 537, 1989.

14. **Huusfeldt Larsen, E.,** Personal communication.

15. **Kjellman, N.-I. M.,** Atopic disease in seven-year old children in relation to family history, *Acta Paediatr. Scand.,* 66, 465, 1977.

16. **Østerballe, O.,** Personal communication.

17. **Høst, A., Husby, S., and Østerballe, O.,** A prospective study of cow's milk allergy in exclusively breast-fed infants, *Acta Paediatr. Scand.,* 77, 663, 1988.

18. **Kaajosari, M.,** Food allergy in Finnish children aged 1 to 6 years, *Acta Paediatr. Scand.,* 71, 815, 1982.

19. **Bock, S. A.,** Prospective appraisal of complaints of adverse reactions to foods in children during the first 3 years of life, *Pediatrics,* 79, 683, 1987.

20. **Foucard, T.,** Allergy and allergy-like symptoms in 1,050 medical students, *Allergy,* 46, 20, 1991.

21. **Bock, S. A.,** The incidence of severe adverse reactions to food in Colorado, *J. Allergy Clin. Immunol.,* 90, 683, 1992.

22. **Drake, C. L., Moses, E. S., and Tandberg, D.,** Systemic anaphylaxis after ingestion of psyllium-containing breakfast cereal, *Am. J. Emergency Med.,* 9, 449,1991.

23. **James, J., Cooke, S., and Sampson, H.,** Anaphylactic reactions to a psyllium-containing cereal, *J. Allergy Clin. Immunol.,* 88, 402, 1991.

24. **Atkins, F. M., Wilson, M., and Bock, S. A.,** Cottonseed hypersensitivity: new concerns over an old problem, *J. Allergy Clin. Immunol.,* 82, 242, 1988.

25. **Malanin, G. and Kalimo, K.,** Angioedema and urticaria caused by cottonseed protein in whole-grain bread, *J. Allergy Clin. Immunol.,* 82, 261, 1988

26. **O'Neil, C. E. and Lehrer, S. B.,** Anaphylaxis apparently caused by a cottonseed-containing candy ingested on a commercial airliner, *J. Allergy Clin. Immunol.,* 84, 407, 1989.

27. **Mansfield, L. E. and Bowers C. H.,** Systemic reaction to papain in a nonoccupational setting, *J. Allergy Clin. Immunol.,* 71, 371, 1983.

28. **Koch, B.,** Personal communication, 1992.

29. **Taylor, S. L.,** Food additives — chemistry, uses, and potential for sensitivity reactions, presented at the Annual Meeting of the American Academy of Allergy and Immunology, Florida, March 6 to 11, 1992.

30. **Jansen, A. P. H., Oei, H. D., van Uffelen, R., and van Toorenberger, A. W.**, Anaphylactic shock due to allergy to quinine-containing beverage, presented at VII International Food Allergy Symposium, Toronto, July 13 to 15, 1990.

31. **Pin, I., Dor, P. J., Vervloet, D., Senft, M., and Charpin, J.**, Hypersensibilite immediate a la quinine, *Presse Medicine*, 14, 967, 1985.

32. **Pevny, I. and Hutzler, D.**, Gruppenallergien bzw. mehrfachreactionen bei patienten mit allergie gegen hydroxychinolin-derivate, *Derm. Beruf. Umwelt.*, 36, 91, 1988.

33. **Nater, J. P. and de Groot, A. C.**, *Unwanted Effects of Cosmetics and Drugs in Dermatology*. Excerpta Medica, Amsterdam, 1983, chap. 4.

34. **Taylor, S., Bush, R. K., and Nordlee, J. A.**, Sulfites, in *Food Allergy: Adverse Reactions to Food and Food Additives*, Metcalfe, D. D., Sampson, H. A., and Simon, R. A., Eds., Blackwell Scientific, Boston, 1991, chap. 17.

35. **Veien, N. K.**, Restriction of nickel intake in the treatment of nickel-sensitive patients, in *Metabolic Disorders and Nutrition Correlated with Skin, Current Problems in Dermatology vol 20*, Vermeer, B. J., Wuepper, K. D., van Vloten, W. A., Baart de la Faille, H., and van der Schroeff, J. G., Eds., Karger, Basel, 1991.

36. **Veien, N. K., Hattel, T., Justesen, O., and Nørholm, A.**, Oral challenge with balsam of Peru, *Contact Dermatitis*, 12, 104, 1985.

37. **Menné, T. and Maibach, H. I.**, Systemic contact-type dermatitis, in *Dermatotoxicology*, Marzulli, F. N. and Maibach, H. I., Eds., Hemisphere Publishing, Washington, 1991.

38. **Juhlin, L.**, Recurrent urticaria: clinical investigation of 330 patients, *Brit. J. Dermatol.*, 104, 369, 1981.

39. **van Bever, H. P., Docx, M., and Stevens, W. J.**, Food and food additive in severe atopic dermatitis, *Allergy*, 44, 588, 1989.

40. **Goodman, D. L., McDonnell, J. T., Nelson, H. S., Vaughan, T. R., and Weber, R. W.**, Chronic urticaria exacerbated by the antioxidant food preservatives, butylated hydroxyanisole (BHA) and butylated hydroxytoluene (BHT), *J. Allergy Clin. Immunol.*, 86, 570, 1990.

41. **Pollock, I. and Warner, J. O.**, A follow-up study of childhood food additive intolerance, *J. Royal Coll. Physicians London*, 21, 248, 1987.

42. **Schwartz, H. J.**, Asthma and food additives, in *Food allergy: Adverse Reactions to Food and Food Additives*, Metcalfe, D. D., Sampson, H. A., and Simon, R. A., Eds., Blackwell Scientific, Boston, 1991, chap. 22.

43. **Gross, P. A., Lance, K., Whitlock, R. J., and Blume, R. S.**, Additive allergy: allergic gastroenteritis due to yellow dye #6 (sunset yellow), *Ann. Intern. Med.*, 111, 87, 1989.

44. **Wilson, N. and Scott, A.**, A double-blind assessement of additive intolerance in children using a 12 day challenge period at home, *Clin. Exper. Allergy*, 19, 267, 1989.

45. **Pollock, I.**, Hyperactivity and food additives, in *Food Allergy and Food Intolerance. Nutritional Aspects and Developments. Bibl. Nutr. Dieta No. 48*, Somogyi, J. C., Müller, H. R., and Ockhuizen, T., Eds. Karger, Basel, 1991, chap. 8.

46. **Larsen, J. C.**, Absorption and transformation of xenobiotics, in *Allergic Responses and Hypersensitivities Induced by Chemicals*. Proceedings of a Joint WHO/CEC Workshop, Frankfurt am Main 12 to 15 October, 1982, Interim document 12, WHO, 1983.

47. **Williams, W. R., Pawlowicz, A., and Davies, B. H.**, Aspirin-like effects of selected food additives and industrial sensitizing agents, *Clin. Exp. Allergy*, 19, 533, 1989.

48. **Robinson, G.**, Tartrazine — the story so far, *Fd. Chem. Toxicol.*, 26, 73, 1988.

49. **Hutchinson, A. P., Carrick, B., and Nicklin, S.**, Adverse reactions to synthetic food colours: interactions between tartrazine and muscarinic acethylcholine receptors in isolated guinea-pig ileum, *Toxicol. Lett.*, 60, 165, 1992.

50. **Murdoch, R. D., Pollock, I., and Naeem, S.**, Tartrazine induced histamine release in vivo in normal subjects, *J. Royal Coll. Physicians London*, 21, 257, 1987.

51. **Murdoch, R. D., Pollock, I., Young, E., and Lessof, M. H.**, Food additive induced urticaria: studies of mediator release during provocation tests, *J. Royal Coll. Physicians London*, 21, 262, 1987.

52. **Schaubschlager, W. W., Becker, W. M., Schade, U. Zabel, P., and Schlaak, M.**, Release of mediators from human gastric mucosa and blood in adverse reactions to benzoate, *Int. Arch. Allergy Appl. Immunol.*, 96, 97, 1991.

53. **Commision of the European Communities.** Report of a working group on adverse reactions to ingested additives. Reports of the Scientific Committee for Food. Twelfth series: EUR 7823, Brussels, 1982.

54. **Madsen, C.,** Prevalence of food additive intolerance, *Human Exp. Toxicol.*, 13, 393, 1994.

55. **Young, E., Patel, S., Stoneham, M., Rona, R., and Wilkinson, J. D.,** The prevalence of reactions to food additives in a survey population, *J. Royal Coll. Physicians London*, 21, 241, 1987.

56. **Fuglsang, G., Madsen, C., Halken, S., Jørgensen, M., Østergaard, P. A., and Østerballe, O.,** Adverse reactions to food additives in children with atopic symptoms, *Allergy*, 49, 31, 1994.

57. **Fuglsang, G., Madsen, C., Saval, P., and Østerballe, O.,** Prevalence of intolerance to food additives among Danish school children, *J. Pediatr. Allergy Immunol.*, 4, 123, 1993.

58. **Saval, P., Fuglsang, G., Madsen, C., and Østerballe, O.,** Prevalence of atopic disease among Danish school children, *J. Pediatr. Allergy Immunol.*, 4, 117, 1993.

Chapter 17

KIDNEY AND ALLERGY

Lucette Pelletier, Blanche Bellon, and Philippe Druet

CONTENTS

I. INTRODUCTION

Many toxins or pollutants are able to induce immunologically mediated injury in humans. Kidney diseases will be chosen as examples to illustrate the way by which chemicals may act.[1-4] However, on some occasions, we will refer to data obtained from immunologically mediated diseases that affect other organs, such as the liver. An immune origin of chemically induced nephropathies is evoked when a small percentage of individuals is affected due to both genetic factors (genes encoding for enzymes involved in the metabolism of the chemical or genes encoding for MHC class II molecules have been implicated). Due to the differences of toxicological effects of a chemical, immunologically mediated disorders cannot be predicted. As a rule, manifestations do not appear after the first contact; priming is necessary. Extrarenal manifestations are often associated with the renal disease. Withdrawal of the toxin leads to remission of the disease. Renal biopsies are of major interest to assist diagnosis. Indeed, mononuclear infiltrates or immunoglobulin deposits detected under histological examination support an immune mechanism of induction. Although these nephropathies have been described for a long time, very little is known about their induction and effector mechanisms. Drugs or chemicals may induce hypersensitivity reactions in which the immune response is directed against the

Processing and presentation of a foreign antigen by a macrophage

Processing and presentation of a foreign antigen by a B lymphocyte

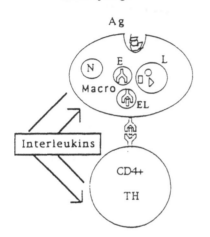

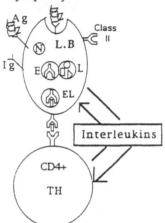

FIGURE 17-1. Processing and presentation of an antigen by macrophages or B-lymphocytes. E = endosomes; L = lysosomes; N = nucleus. Antigen (Ag) or antigen bound to its receptor on B-cells, i.e., the surface immunoglobulin (Ig) on B-cells, is internalized. Class II molecules of the MHC (E) fix peptides derived from the antigen after fusion of E and L (EL). The complex class II peptide is reexpressed at the membrane of the antigen presenting cell. It is recognized by the T-cell receptor of the specific CD4+ T-helper cell which is activated, secretes interleukins, proliferates, and differentiates.

toxin, or autoimmune manifestations in which the immune response is directed against an autoantigen. However, it is difficult to differentiate between these two mechanisms. In any case, whether effector cells or antibodies are deleterious, the pivotal role is played by T-cells. On one hand, T-cells act directly and recruit other cell types; on the other hand, they provide help to B-cells to produce antibodies.

In this chapter, we shall try to determine the pathogenesis of the different chemically induced nephropathies, and to do this we shall describe some experimental situations and the underlying mechanisms that can be relevant. Finally, we shall discuss some mechanisms of chemically induced B-cell polyclonal activation in which a renal structure is only one of the targets of the immune response.

II. THE IMMUNE RESPONSE BASIS OF HYPERSENSITIVITY OR AUTOIMMUNITY

The first step of an immune response is the processing of the exogenous antigen by antigen presenting cells (APCs) (Figure 17-1). These cells are essentially macrophages-monocytes expressing MHC class II molecules on their surface. Dendritic cells, Langerhans cells, and some epithelial cells such

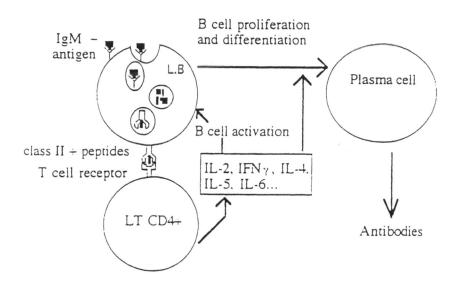

FIGURE 17-2. Cooperation between T- and B-cells for antibody production. IgM at the B-cell surface may fix its specific antigen. The complex is then internalized and antigen gives rise to peptides. Peptides bind to class II MHC molecules and the complex is expressed at the cell surface where it will be recognized by a CD4+ helper T-cell having the specific T-cell receptor. The activated T-cell produced cytokines at play in the activation, proliferation, and differentiation of B-lymphocytes into plasma cells.

as those of the renal proximal tubule may function as APCs.[5] These cells internalize the antigen nonspecifically.[6] B-cells that also act as APCs process antigens they specifically recognize through surface immunoglobulin (B-cell antigen receptor).[7] By themselves, chemicals are not immunogenic; they behave as haptens and need to bind a carrier to induce an immune reaction. Therefore, only reactive chemicals or those which are metabolized into reactive products will be immunogenic. Exogenous antigens are processed by APCs and peptides derived from these antigens are reexpressed on the cell membrane associated with MHC class II molecules.[7-9] The peptide is then recognized by a specific CD4+ helper T-lymphocyte which is then activated.[10,11] This activated T-cell produces interleukins, proliferates, and allows the differentiation of B- or cytotoxic CD8+ cells (Figures 17-1, 17-2).[11] B-cells activated by a cognate interaction with T-cells proliferate and differentiate under the control of various interleukins in plasma cells secreting antibodies. Depending upon the cytokine production and their functions, CD4+ T-cells are divided into Th1 and Th2 cells.[12,13] Th1 cells produce IL-2, gamma interferon, and TNF-β, are responsible mainly for delayed-type hypersensitivity reactions, and can cooperate with B-cells for antibodies of the IgG2a isotype in mice. Th2 cells produce IL-4, IL-5, and IL-6 and are mainly involved in B-cell assistance for the production of IgA, IgG1, and IgE antibodies. These two

TABLE 17-1
Role of Some Cytokines in the Immune Response

Cytokine	Produced By	Effects
IL-1	Activated macrophages	Promotion of coagulation Inflammatory reactions
TNF-α,	Activated macrophages	Inflammatory reactions
IL-6	Activated macrophages	Inflammatory reactions
IL-10	Macrophages/Th2	Increased ability of antigen presentation by B-cells
IL-2	Th1	Proliferation and differentiation of T-cells
IFN-γ	Th1	Macrophage activation, increased ability of antigen presentation by APC; allows some nonprofessional APC to present antigens, cytotoxic effects
TNF-β	Th1	Inflammatory reactions, cytotoxicity against some epithelial cells
IL-4	Th2	Proliferation and differentiation of Th2 and B-cells (IgE). Differentiation of eosinophils and basophils
IL-5	Th2	Differentiation of eosinophils

subsets also exist in humans.[14] Several functions of cytokines produced by macrophages and/or T-cells are summarized in Table 17-1.[15] There are four main types of hypersensitivity reactions, and drugs and chemicals may induce at least one of these.[16] Type I, II, and III hypersensitivity reactions are associated with the production of antibodies, while the fourth one is cell-mediated. These reactions are described elsewhere. Thus, we shall just mention immediate hypersensitivity (type I), in which IgE elicited by the antigen binds to the Fcε receptor of mast cells or eosinophils, leading to their activation. Finding IgE-containing plasma cells or eosinophils in the interstitium of some cases of acute interstitial nephritis also supports a role for immediate hypersensitivity in immunologically mediated chemical-induced nephritis. Type II hypersensitivity is characterized by cytotoxic antibodies, and type III by the deposition of immune complexes or by the formation of these immune complexes *in situ*. Glomerulopathies associated with drug-induced antitubular basement membrane (TBM) or antiglomerular basement membrane (GBM) antibodies probably represent examples of the latter situation.[17,18] Type IV hypersensitivity could be involved in several forms of drug-induced acute interstitial nephritis.

Finally, a B-cell polyclonal activation may be responsible for production of pathogenic autoantibodies (Figure 17-3).[19] In a normal situation, autoantigens do not trigger an autoimmune response, mainly because autoaggressive T-cells are either eliminated or inactivated.[20-23] In pathological conditions, a toxin might act by activating B-cells directly or by giving rise to autoreactive T-cells able to cooperate with all or numerous B-cells to produce antibodies, including those with a renal specificity.

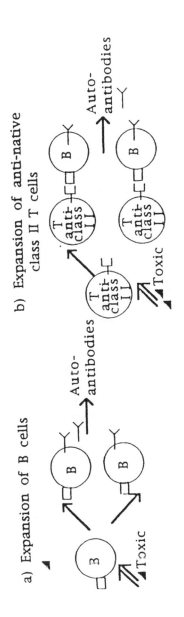

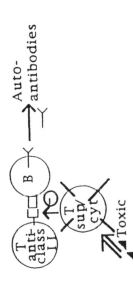

FIGURE 17-3. Theoretical mechanisms of toxic-induced B-cell polyclonal activation. (a) Direct expansion of B-cells including autoreactive B cells. (b) Induction of anti-self class II T-cells (T-anti-class II). The cognate interaction of the T-cell receptor of this autoreactive cell with self-class II at the surface of B-cells leads to a polyclonal activation of B-cells. (c) A defect at the T-suppressor (Tsup/cyt) level can also lead to the expansion of anti-class II T-cells and thus to B-cell polyclonal activation.

TABLE 17-2
Drugs Responsible for Immunologically Mediated Nephropathies

Renal Disease	Drugs
Glomerulopathies	
Minimal changes	Lithium, NSAID, Hg, gold salts, D-penicillamine
Membranous GN	Gold salts, D-penicillamine, and other drugs with a thiol group (captopril,...)
TIN	Antibiotics: Penicillin (methicillin,...), cephalosporins, rifamycin, sulfamides, NSAID, analgesics, anticoagulants, diuretics, cimetidine, allopurinol, phenytoin

Note: NSAID = nonsteroidal anti-inflammatory drugs; TIN = Tubulointerstitial nephritis.

III. IMMUNOLOGICALLY MEDIATED RENAL DISEASES INDUCED BY CHEMICALS

Chemicals may induce glomerulopathies, tubular interstitial nephritides, or both. These immunologically mediated renal diseases are genetically controlled, and susceptibility is dependent upon genes encoding for enzymes involved in the toxin's metabolism or upon genes encoding for MHC class II molecules. Extrarenal symptoms, such as fever, rash, or biological abnormalities (hypereosinophilia and hyperIgE), are frequently associated with these nephropathies.

A. MINIMAL CHANGE DISEASE
The nephrotic syndrome with minimal changes (MC) is characterized by a heavy proteinuria associated with glomeruli optically normal and fusion of foot processes of epithelial glomerular cells. There are no Ig deposits. Lithium[1,16,24-27] and nonsteroidal anti-inflammatory drugs (NSAID) may induce such glomerulopathies (Table 17-2).[15,28,29] A role for cytokines was proposed[30-32] and it was recently supported by the finding of a rise of IL-4 in patients with childhood MC disease (MCD);[33] however, the way cytokines interfere in the nephrotic syndrome is still unclear. Recently, T-cell hybridomas have been obtained from patients with MCD, and culture supernatants have been injected in rats. The rats developed transient proteinuria and histological changes analogous to those found in MCD.[32] The factor at play is partially characterized and presents a tumor necrosis-like activity for tumoral cells from epithelial origin. It is therefore tempting to speculate that this factor is toxic for glomerular epithelial cells. Whether or not drugs responsible for MCD also act through the release of cytokines is unknown. In that respect, it has been recently suggested that adriamycin-induced MCD in the rat would require the presence of T-cells.[34,35] If this observation is confirmed, it would be interesting to study the role of cytokines.

B. IMMUNE COMPLEX TYPE GLOMERULONEPHRITIS

Membranous glomerulopathies which are characterized in immunofluorescence by subepithelial glomerular granular IgG deposits are often due to immune responses triggered by a toxin (Table 17-2). Heavy metals like gold or mercury may induce such immunologically mediated glomerular damages. Mercury compounds exist either as inorganic mercury or organomercurials in which Hg is covalently bound to carbon such as in methylmercury. Hg is mainly inhaled and is oxidized from HgO to Hg^{2+} in erythrocytes, a process that requires catalase and hydrogen peroxides and is antagonized by glutathione peroxidase.[36-38] Individual variations in the ability to oxidize HgO, depending upon genetic background and environmental factors, would condition susceptibility to toxic effects. Methylmercury, present in fish, is mainly absorbed through the gastrointestinal tract and undergoes partial biotransformation to inorganic Hg by demethylation.[36] Besides the toxic effects on tubular epithelial cells, mercury-containing drugs and even mercury-containing skin lightening creams may induce immunologically mediated glomerulopathy.[1,39] Clinical presentation is initially that of a heavy proteinuria or a nephrotic syndrome with normal renal function and normal blood pressure. Histologically, this glomerulopathy can be a membranous glomerulopathy with granular IgG deposits by immunofluorescence or an MCD thought to be mediated by T-lymphocytes. In some cases, immunoglobulins were found deposited linearly along the glomerular basement membrane and were or were not associated with granular IgG deposits.[1] This suggests that, in some patients, antiglomerular basement membrane antibodies may be produced. An increased prevalence of antilaminin antibodies (laminin is a component of glomerular basement membrane) was reported in workers exposed to mercury vapor,[40] but it was not confirmed in a later study.[41]

Gold salts are used to treat patients with rheumatoid arthritis, but they cause proteinuria and the nephrotic syndrome in 6 to 17.2% and 2.6 to 5.3%, respectively, of the patients. Among the patients who developed proteinuria, 89.5% had a membranous glomerulonephritis, and 9.6% had minimal glomerular changes.[16,42,43] There is a causal relationship between the occurrence of membranous glomerulonephritis since such glomerular lesions have been rarely reported among rheumatoid patients who did not receive gold salts or related drugs. The occurrence of membranous glomerulonephritis is not correlated with the cumulative dose of gold, with the duration of treatment, or with the gold salts used. Proteinuria is only observed in some patients and is not dose-related, which suggests susceptibility may be genetically determined. This is supported by the fact that patients with HLA-B8 or DRW3 antigens are at higher risk.[44]

Several drugs containing a sulphydryl group are able to induce immune type glomerulonephritis.[1,16] Among them, D-penicillamine is responsible for the greatest number of cases. Rheumatoid patients account for the largest number of patients treated with this drug, and 7 to 20% of them demonstrated proteinuria, whereas the nephrotic syndrome was present in 49.6% of them. The characteristics of proteinuria or of the nephrotic syndrome are similar to those mentioned for gold salts. Renal biopsies often demonstrated a membranous glomerulonephritis quite similar to that induced by gold salts (85.2%) and rarely a mesangioproliferative glomerulonephritis (10.4%); MCD and focal necrotizing glomerulonephritis and a rapidly progressive glomerulonephritis were observed occasionally.[45] The sulphoxidation status and the role of HLA DR3 have been emphasized[44,46]; indeed, D-penicillamine-induced membranous nephropathy is 32 times more frequent in HLA DR3 individuals than in non-HLA DR3 ones. Other drugs with a sulphydril group have been associated with an immune type glomerulonephritis. Proteinuria was detected in patients treated with captopril, thiopronine, or pyrithioxine, associated with a membranous glomerulonephritis or MCD.[1,16]

The presence of immunoglobulins in the glomeruli of patients with a membranous glomerulopathy can result from deposition of circulating immune complexes, but there is no experimental proof at this time. It is now clear that most experimental immune-type complex glomerulopathies are due to circulating antibodies that fix on glomerular structures or on planted antigens. The specificity of the glomerular-bound immunoglobulins is unknown in the human situation, but it has been shown in experimental ones that, for example, antibodies against angiotensin converting enzyme[47] or against GBM[48] may trigger a membranous glomerulopathy. The former might first bind to endothelial structures and then relocalize at the subepithelial site.[47] The fixation of anti-GBM antibodies on its antigen might also result in a redistribution of anti-GBM antibodies as suggested for the membranous glomerulopathy induced by $HgCl_2$ in Brown-Norway (BN) rats.[48] This phenomenon might be due to the fixation of anti-idiotypic antibodies on the antilaminin antibodies fixed in the glomeruli. Alternatively, fixation of antilaminin antibodies on laminin might trigger an increase in the production of this compound that would lead to a preferential localization of antibodies on the external aspect of the GBM. Numerous antigens might function as "planted" antigens, such as cationic molecules, because they bind to anionic charges of the GBM. Histones are an example of "planted" antigens. They may bind DNA and allow anti-DNA antibodies to fix histone-bound DNA.[49] Such mechanisms might explain the presence of antinuclear antibodies in the immune deposits eluted from the kidneys of Balb/c mice injected with $HgCl_2$.[50]

Whatever the specificity of glomerular-bound antibodies, it is tempting to speculate that they are responsible for the disease. They can act in several ways. They could activate the cascade of complement[51] or interfere with the function of the structure they bind[52] (Table 17-3). For example, it is possible that fixation of an antibody on a renal cell leads to cell death and an increase of glomerular permeability, as has been proposed in Heymann's nephritis.[51] In this respect, it is noteworthy that glomerular cells may undergo apoptosis (programmed death).[53] Anti-GBM antibodies that can be induced by chemicals

TABLE 17-3
Mechanisms Responsible for Membranous Glomerulopathy

Deposition of circulating immune complexes
Immune response against a glomerular structure
   Antigen expressed on epithelial cells: gp 330
   Laminin
   Endothelial structure
Immune response against a "planted" antigen
Cationic proteins: histones (which can bind DNA), lectins...

in experimental animals (see below) might be responsible for a membranous glomerulopathy, because the fixation of the antibody elicits a redistribution of the antigen. It might lead to an alteration of the internal geometry of the GBM and to proteinuria.[48] It is also possible that, upon fixation, some antibodies directed against epithelial antigens containing sialic residues elicit a retraction of foot processes of glomerular epithelial cells and abnormalities of permeability.[54] Though in some models transfer of antibodies is sufficient to induce proteinuria, it has been proposed in other ones that cytokines such as IL-1, TNF-β, and/or interferon γ are also required for proteinuria to occur.[54,55]

C. CELL-MEDIATED GLOMERULOPATHY

A recent study suggested that T-cells responsible for delayed-type hypersensitivity reactions can be involved in glomerular injury. The authors induced an immune response against a hapten, namely TNP, in such a way that only Th1 cells were stimulated and that anti-TNP antibodies were not produced.[56] Thereafter, they injected, in the renal artery, TNP coupled to cationic BSA that localized in the kidney. A glomerulopathy marked by a cellular influx without antibody deposits and a transient proteinuria was observed. The cellular influx, consisting mainly of mononuclear phagocytes expressing MHC class II molecules, might mediate inflammatory reactions and be responsible for proteinuria by this bias.[56]

D. DRUG-MEDIATED ALLERGIC
TUBULAR INTERSTITIAL NEPHRITIS

Drug-mediated allergic tubular interstitial nephritis (TIN) is a well-known side effect of drug therapy.[29,57-66] This etiology represents 0.8 to 8% of all cases of acute renal failure. Several of the many drugs that can induce an allergic TIN are examplified in Table 17-2. Methicillin and cephem antibiotics are frequently implicated,[57] and a cross-sensitization with penicillin derivatives and cephalosporins may exist.[66,67] Most of the patients exhibit a renal failure, enhanced serum IgE levels, hypereosinophiluria, and extrarenal signs such as fever, skin eruptions, and diarrhea. Features found in one series of patients with drug-induced allergic nephritis are given in Table 17-4.[29] In some cases of methicillin-induced allergic TIN, IgG deposits are found along the tubular

TABLE 17-4
Features of Some Drug-Induced Allergic Nephritides

Agents	Penicillin, cephem, NSAID, minocyclin	
Clinics	"Allergic symptoms"	7/10
	Renal failure	10/10
	Proteinuria	9/10
	Hematuria	4/10
Biology	Increase in serum IgE level	4/5
	Hypereosiniphilia	6/9
Histology	TIN	10/10
	Glomeruli	
	Normal	6/10
	MG	2/10
	Proliferation	1/10
	Thrombotic m.a.	1/10
Proliferative Assay with Toxin		10/10

Note: m.a. = microangiopathy; NSAID = nonsteroidal anti-inflammatory drugs; TIN = tubulointerstitial nephritis.

basement membrane. They probably recognized either DPO (dimethoxyphenyl penicilloyl hapten, a metabolite of methicillin) or a self-TBM component modified by DPO.[62,66] The clinical and biological features of these drug-induced allergic nephritis have led Churg et al.[68] to propose that immediate (IgE) hypersensitivity reactions were involved, but there is no proof to date. In NSAID-induced-TIN, association with the nephrotic syndrome is common.[16,69] Renal failure is always present and may be very severe. The glomerulopathy responsible for the proteinuria is mostly a MCD. Histologically, glomeruli exhibited minimal changes and diffuse or focal interstitial infiltrations with lymphocytes, plasma cells, and frequently eosinophils. Studies of the lymphocyte subpopulations of the interstitial infiltrates showed 80% of T-cells, mostly of the cytotoxic/suppressor subset, and 20% of B-cells which were mainly IgE-bearing cells.[70]

In these situations, the role of antibody deposition is always discussed. However, a model in which anti-hapten T-cells alone might be responsible for an interstitial nephritis has recently been reported. In this study, BN rats were immunized with an hapten (ABA) in such a way that they developed only hypersensitivity reactions in the absence of antibody production. Perfusion of ABA in the renal artery of these rats led to a granulomatous diffuse interstitial nephritis associated with a proliferative glomerulonephritis.[71] Anti-ABA Th1 cells might act by releasing cytokines such as gamma interferon responsible for expression of MHC class II molecules on renal cells able to present ABA to T-cells. Such a mechanism would amplify the inflammatory reaction. The disease was focally reproduced by passive transfer of T-lymphocytes without any production of anti-ABA antibodies.

Another model, in contrast, supports the major importance of antibody deposition in the appearance of immunoallergic TIN. Thus, Joh et al. showed that, in normal mice, injection of anti-cephalotin IgG antibodies followed by intrarenal challenge of cephalotin coupled to a protein, led to TIN.[72] By using a similar procedure (injection of anti-hapten antibodies and thereafter infusion of the hapten in the kidney), they demonstrated that the isotype of the antibodies is critical: IgG2a but not IgE anti-hapten antibodies induced the disease. A role of complement activation following the complex hapten (or drug)-anti-hapten (drug) antibody formation was suggested by authors to explain the inflammatory lesions.

It will be important to isolate T-cells from the interstitial infiltrates to characterize their specificity and the profile of cytokines produced in drug-interstitial nephritides. Indeed, cytokines could be at play in the appearance of MCG that is often associated with this. This approach was fruitful in the study of T-cells infitrating livers of drug-induced allergic hepatitis; T-cell hybridomas were obtained that recognized not the drug itself but a self protein–drug complex; they produced mainly interferon γ and IL-5.[73] This would explain the inflammatory reaction since IFN-γ activates macrophages, and eosinophilia since IL-5 is the cytokine promoting differentiation of eosinophils. Similar results might be obtained with T-cells from drug-induced immunoallergic TIN.

IV. B-CELL POLYCLONAL ACTIVATION

Some chemicals may induce kidney diseases in predisposed individuals by inducing a dysfunction of the immune system and a B-cell polyclonal activation responsible for the production of antibodies. Thus, D-penicillamine may induce autoimmune manifestations such as myasthenia gravis or systemic lupus erythematosus. It is of note that these manifestations occur in patients different from those who develop a membranous glomerulopathy. Development of experimental models of mercury-, gold salt-, or D-penicillamine-induced autoimmunity argue that these compounds can be deleterious by acting on the immune system. Nontoxic amounts of mercuric chloride (1 mg/kg, s.c., 3 times a week) induce in BN rats an enlargement of spleen and lymph nodes due to proliferation of both CD4$^+$ T-helper/inducer and B-cells.[74-76] A hyperimmunoglobulinemia which affects mainly IgE[77] appears after 1 week, and several autoantibodies of the IgG isotype are found.[78-80] Autoantibodies directed against GBM, DNA, IgG, type II collagen, and thyroglobulin as well as antibodies specific for exogenous antigens (trinitrophenol hapten, sheep red blood cells ...) are detected from the end of the first week.[74,80,81] Anti-GBM antibodies are responsible for an autoimmune glomerulopathy. All the autoimmune abnormalities peak on weeks 2 to 3 and then spontaneously decrease in surviving rats.[74,78-80] Moreover, these rats become resistant to further injections of HgCl$_2$. HgCl$_2$ given intravenously, orally, or intratracheally induces the disease in a similar way.[82] Exposure to mercury vapor also induces the autoimmune disease in BN rats.[83] Methylmercury, pharmaceutical ointments, and solutions containing organic mercury are effective when these products are applied on wounds or even on normal skin.[84]

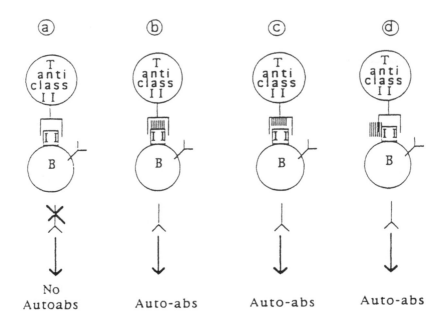

FIGURE 17-4. A toxin could increase the affinity between the T-cell receptor (TCR) of an anti-self class II T-cell and class II molecules, by binding class II (a), the TCR (b), or both (c). In all the cases, it would lead to the polyclonal activation of B-cells.

In the susceptible strain, the effects of $HgCl_2$ depend upon dosage.[79] Thus, the increase in serum IgE concentration is positively correlated to the dosage. Susceptibility to mercury-induced autoimmunity is genetically controlled, inherited as an autosomal dominant trait, and governed by three to four genes, one of which is localized with MHC.[85] Mice bearing the H-2^S haplotype are also susceptible to mercury-induced autoimmunity, and the role of MHC class II I-A and I-E regions in susceptibility or resistance, respectively, to mercury-induced autoimmunity in mice has been underlined.[86] BN rats inoculated with D-penicillamine[87,88] or gold salts[89] display quite similar autoimmune disorders. Moreover, one idiotype was characterized on several antiglomerular basement antibodies. This public idiotype is also found on anti-GBM antibodies during the course of D-penicillamine and gold salt-induced autoimmune disease in BN rats. This strongly suggests that the same B-cell clones are stimulated during the polyclonal B-cell activation occurring during these autoimmune diseases and that the mechanisms leading to autoimmunity are at least partly common.[90] In $HgCl_2$-injected BN rats autoimmune disorders are more intense than in gold salt- and D-penicillamine-treated BN rats. A proteinuria with or without a nephrotic syndrome is found in all the rats from the first group vs. 50% in the second and none in the third. T-cells play a major role in the appearance of autoimmunity in BN rats.[91] Indeed, transfer of T-cells from BN rats injected with $HgCl_2$ is sufficient to induce the disease in normal syngeneic recipients. T-cells responsible for the induction of immune abnormalities are probably

T-cells specific for self class II molecules.[92,93] The way these cells are induced is unclear. In the normal situation, the affinity of the TCR specific for native MHC class II molecules expressed on B-cells is too low to allow activation of B-cells. Chemicals such as $HgCl_2$, D-penicillamine, or gold salt might increase affinity of the autoreactive anti-class II T-cells for B-cells by binding to MHC class II molecules, or to the TCR, or both (Figure 17-4), thus inducing activation and differentiation of B-cells. A chemical might also induce DNA demethylation and therefore gene expression. Such a mechanism might explain the increase in the expression of class II on B-cells induced by $HgCl_2$.[94,95] It is also possible that $HgCl_2$ initially triggers the gene for IL-4, since mRNA for IL-4 is found in BN lymphocytes only after 2 h of culture in the presence of $HgCl_2$.[96] The increase in class II expression could allow the activation of B-cells by anti MHC class II T-cells. The major production of IgE during drug-induced autoimmunity in BN rats suggests that autoreactive T-cells produce IL-4 and therefore belong to the Th2 subset. In susceptible mice, $HgCl_2$-induced IgE increase was prevented by a treatment with an anti-IL-4 monoclonal antibody, therefore demonstrating the role of Th2 cells.[97]

V. CONCLUSION

Prediction of drugs able to mediate allergic reactions is an important problem of public health. Until now prediction was not possible before administration of the drug to the patients. Before elaboration of efficient tests, understanding of the underlying mechanisms is critical. Amazingly, though immunoallergic drug-induced nephropathies were known for a long time, still little is known about their causal mechanisms. The recent development of experimental models of drug-induced glomerulonephropathies and tubulointerstitial nephritides will provide useful tools. Progress gained in the understanding of drug-induced hepatitides will also be of interest to approach mechanisms of drug-induced nephropathies. Indeed, in the former, the specificity of drug-induced antibodies was defined and T-cell clones are now available. This might be accurate for nephropathies since several toxins such as tienilic acid can induce both nephropathies and hepatitides.

REFERENCES

1. **Fillastre, J. P., Druet, P., and Mery, J. P.,** Proteinuria associated with drugs and substances of abuse, in *The Nephrotic Syndrome*, Cameron, J. S. and Glassock, R. J., Eds., Marcel Dekker, New York, 1988, 697.
2. **Bigazzi, P. E.,** Autoimmunity induced by chemicals, *Clin. Toxicol.*, 26, 125, 1988.
3. **Kammüller, M. E., Bloksma, N., and Seinen, W.,** Autoimmunity and toxicology. Immune dysregulation induced by drugs and chemicals, in *Autoimmunity and Toxicology*, Kammüller, M. E., Bloksma, N., and Seinen, W., Eds., Elsevier, New York, 1989, 3.

4. **Tournade, H., Pelletier, L., Glotz, D., and Druet, P.,** Toxiques et autoimmunité, *Med. Sci.*, 5, 303, 1989.
5. **Hagerty, D. T. and Allen, P. M.,** Processing and presentation of self and foreign antigens by the renal proximal tubule, *J. Immunol.*, 148, 2324, 1992.
6. **Ziegler, K. and Unanue, E. R.,** Identification of a macrophage antigen-processing event required for I-region restricted antigen presentation to T-lymphocytes, *J. Immunol.*, 127, 1861, 1981.
7. **Davidson, H. W., Reid, P. A., Lanzavecchia, A., and Watts, C.,** Processed antigen binds to newly synthetized MHC class II molecules in antigen-specific B lymphocytes, *Cell*, 67, 105, 1991.
8. **Buus, S., Sette, A., Colon, S. M., Jenis, D. M., and Grey, H. M.,** Isolation and characterization of antigen-Ia complexes involved in T cell recognition, *Cell*, 47, 1071, 1987.
9. **Kourilsky, P. and Claverie, J. M.,** MHC-antigen interactions: what does the T cell receptor see? *Adv. Immunol.*, 45, 107, 1989.
10. **Schwartz, R. H.,** T-lymphocyte recognition of antigen-Ia complexes in association with gene products of the major histocompatibility complex, *Ann. Rev. Immunol.*, 3, 237, 1985.
11. **Paul, W. E.,** *Fundamental Immunology*, Raven Press, New York, 1989.
12. **Mosmann, T. R., Cherwinski, H., Bond, M. W., Giedlin, M. A., and Coffman, R. L.,** Two types of murine helper T cell clones I. Definition according to profiles of lymphokines activities and secreted proteins, *J. Immunol.*, 136, 2348, 1986.
13. **Mosmann, T. R. and Coffman, R. L.,** Heterogeneity of cytokine suppression patterns and functions of helper T cells, *Adv. Immunol.*, 46, 111, 1989.
14. **Romagnani, S.,** Human TH1 and TH2 subsets: doubt no more, *Immunol. Today*, 12, 256, 1991.
15. **Banchereau, J.,** Interleukines et autres cytokines impliquées dans la réponse immunitaire et la réponse allergique, in *Allergologie*, Charpin, J. and Vervloet, D., Eds., Médecine Sciences-Flammarion, Paris, 1992, 213.
16. **Druet, P., Jacquot, C., Baran, D., Kleinknecht, D., Fillastre, J. P., and Mery, J. P.,** Immunologically mediated nephritis induced by toxin and drugs, in *Nephrotoxicity in the Experimental and Clinical Situation*, Bach, P. H. and Lock, E. A., Eds., Martinus Nijhoff, Amsterdam, 1987, 727.
17. **Dixon, F. J., Feldman, J. D., and Vazquez, J. J.,** Experimental glomerulonephritis: the pathogenesis of a laboratory model resembling the spectrum of human glomerulonephritis, *J. Exp. Med.*, 113, 899, 1961.
18. **Hoedemaeker, P. J., Fleuren, G. J., and Weening, J. J.,** Experimental models of the nephrotic syndrome, in *The Nephrotic Syndrome*, Cameron, J. S. and Glassock, R. J., Eds., Marcel Dekker, New York, 1988, 89.
19. **Druet, P.,** Induction and regulation of autoimmune experimental glomerulonephritis, in *Immunology & Medicine*, Pusey C. D., Ed., Kluwer Academic Publisher, Dordrecht, 1991, 29.
20. **Kappler, J. W., Roehm, N., and Marrack, P.,** T cell tolerance by clonal elimination in the thymus, *Cell*, 49, 273, 1987.
21. **Jenkins, M. K.,** The role of division in the induction of clonal anergy, *Immunol. Today*, 13, 69, 1992.
22. **Jones, L. A., Chin, L. T., Longo, D. L., and Kruisbeek, A. M.,** Peripheral clonal elimination of functional T cells, *Science*, 250, 1726, 1990.
23. **Lo, D., Freedman, J., Hesse, S., Brinster, R. L., and Sherman, L.,** Peripheral tolerance in transgenic mice: tolerance to class II MHC and non-MHC transgene antigens, *Immunol. Rev.*, 122, 87, 1991.
24. **Richman, A. V., Masco, H. L., Rifkin, S. I., and Acharya, M. K.,** Minimal-change disease and the nephrotic syndrome associated with lithium therapy, *Ann. Intern. Med.*, 92, 70, 1980.
25. **Alexander, F. and Martin, J.,** Nephrotic syndrome associated with lithium therapy, *Clin. Nephrol.*, 15, 267, 1981.

26. **Moskovitz, R., Springer, P., and Urquhart, M.,** Lithium-induced nephrotic syndrome. *Am. J. Psychiatry*, 138, 382, 1981.
27. **Depner, T. A.,** Nephrotic syndrome secondary to lithium therapy, *Nephron*, 30, 286, 1982.
28. **Abraham, P. A. and Keane, W. F.,** Glomerular and interstitial disease induced by non-steroidal anti-inflammatory drugs, *Am. J. Nephrol.*, 4, 1, 1984.
29. **Joh, K., Aizawa, S., Yamaguchi, Y., Inomata, I., Shibazaki, T., Sakai, O., and Hamaguchi, K.,** Drug-induced hypersensitivity nephritis: lymphocyte stimulation testing and renal biopsy in 10 cases, *Am. J. Nephrol.*, 10, 220, 1990.
30. **Shaloub, R. J.,** Pathogenesis of lipoid nephrosis: a disorder of T-cell function, *Lancet*, 2, 556, 1974.
31. **Lagrue, G., Xhemermont, S., Branellec, A., Hirbec, G., and Weil, B.,** A vascular permeability factor elaborated from lymphocytes. 1. Demonstration in patients with nephrotic syndrome, *Biomedicine*, 23, 37, 1975.
32. **Koyama, A., Fujizaki, M., Igarashi, M., and Narita, M.,** A glomerular permeability factor produced by human T cell hybridomas, *Kidney Int.*, 40, 453, 1991.
33. **Cho, B.-S., Lee, C.-E., and Pyun, K.-H.,** Elevation of interleukin (IL-4) activities and mRNA expression in childhood minimal change nephrotic syndrome, *J.A.S.N.*, 2, 591, 1991.
34. **Amore, A., Mazzuco, G., Cavallo, F., Forni, G., Motta, M., Gianoglio, B., Peruzzi, L., Porcellini, M. G., Roccatello, D., Piccoli, G., and Coppo, R.,** Adriamycin model in nude mice: evidence of an immunological involvement, XXIXth Congress of the European Dialysis and Transplant Association, 1992.
35. **Desassis, J. F., van den Born, J., and Berden, J. H. M.,** Anti-proteinuric effect of cyclosporin A (CsA) in adriamycin (ADR) nephropathy in the rat, XXIXth Congress of the European Dialysis and Transplant Association, 1992.
36. **Berlin, M.,** Mercury, in *Handbook on the Toxicology of Metals*, Friberg, L., Nordberg, G. F., and Vouk, V. B., Eds., Elsevier, Amsterdam, 1986, 387.
37. **Lind, B., Friberg, L., and Nylander, M.,** Preliminary studies on methylmercury biotrans-formation and clearance in the brain of primates: II Demethylation of mercury in brain, *J. Trace Elem. Exp. Med.*, 1, 49, 1988.
38. **Halbach, S., Ballatori, N., Clarkson, T. W.,** Mercury vapor uptake and hydrogen peroxyde detoxification in human and mouse red blood cells, *Toxicol. Appl. Pharmacol.*, 96, 517, 1988.
39. **Kibukamusoke, J. W., Davies, D. E., and Hutt, M. S. R.,** Membranous nephropathy due to skin lightening cream, *Br. Med. J.*, 2, 646, 1974.
40. **Lauwerys, R., Bernard, A., Roels, H., Buchet, J. P., Gennart, J. P., Mahieu, P., and Foidart, J.-M.,** Anti-laminin antibodies in workers exposed to mercury vapour, *Toxicol. Lett.*, 17, 113, 1983.
41. **Bernard, A. M., Roels, H. R., Foidart, J-M., and Lauwerys, R. L.,** Search for anti-laminin antibodies in workers exposed to cadmium, mercury vapour or lead, *Int. Arch. Occup. Environ. Health*, 59, 303, 1987.
42. **Hall, C.,** Gold and D-penicillamine induced renal disease, in *The Kidney and Rheumatic Disease*, Bacon, P. A. and Hadler, N. M., Eds., Butterworth, London, 1982, 246.
43. **Katz, W. A., Blodgett, R. C., and Pietrusko, R. G.,** Proteinuria in gold-treated rheuma-toid arthritis, *Ann. Intern. Med.*, 101, 176, 1984.
44. **Wooley, P. H., Griffin, J., Panayi, G. S., Batchelor, J. R., Welsh, K. I., and Gibson, T. J.,** HLA-DR antigens and toxic reaction to sodium aurothiomalate and D-penicillamine in patients with rheumatoid arthritis, *N. Engl. J. Med.*, 303, 300, 1980.
45. **Sadjadi, S. A., Seelig, M. S., Berger, A. R., and Milstoc, M.,** Rapidly progressive glomerulonephritis in a patient with rheumatoid arthritis during treatment with high-dosage D-penicillamine, *Am. J. Nephrol.*, 5, 212, 1985.
46. **Emery, P., Panayi, G. S., Huston, G., Welsh, K. L., Mitschell, S. C., Shah, R. R., Idle, J. R., Smith, R. L., and Waring, R. H.,** D-penicillamine induced toxicity in rheumatoid arthritis: the role of sulphoxidation status and HLA-DR3, *J. Rheum.*, 11, 626, 1984.

47. **Brentjens, J. R. and Andres, G.,** Interaction of antibodies with renal cell surface antigens, *Kidney Int.,* 35, 954, 1989.
48. **Aten, J., Veninga, A., Bruijin, J. A., Prins, F. A., De Heer, E., and Weening, J. J.,** Antigenic specificities of glomerular-bound antibodies in membranous glomerulopathy induced by mercuric chloride, *Clin. Immunol. Immunopathol.,* 63, 89, 1992.
49. **Schmiedke, T. M. J., Stöckl, F. W., Weber, R., Sugisaki, Y., Batsford, S. R., and Vogt, A.,** Histones have high affinity for the glomerular basement membrane. Relevance for immune complex formation in lupus nephritis, *J. Exp. Med.,* 169, 1879, 1989.
50. **Hultman, P. and Eneström, S.,** Mercury-induced antinuclear antibodies in mice: characterization and correlation with renal immune complex deposits, *Clin. Exp. Immunol.,* 71, 269, 1988.
51. **Quigg, R. J., Cybulsky, A. V., and Salant, D. J.,** Effect of nephritogenic antibody on complement regulation in cultured rat glomerular epithelial cells, *J. Immunol.,* 147, 838, 1991.
52. **Kawachi, H., Orisaka, M., Matsui, K., Iwanaga, T., Toyabe, S., Oite, T., and Shimuzu, F.,** Epitope-specific induction of mesangial lesions with proteinuria by a MoAb against mesangial cell surface antigen, *Clin. Exp. Immunol.,* 83, 399, 1992.
53. **Savill, J.,** Apoptosis: a mechanism for regulation of the cell complement of inflamed glomeruli, *Kidney Int.,* 41, 607, 1992.
54. **Mendrick, D. L. and Rennke, H. G.,** I. Induction of proteinuria in the rat by a monoclonal antibody against SGP-115/107, *Kidney Int.,* 33, 818, 1988.
55. **Montinaro, V., Hevey, K., Aventaggio, L., Fadden, K., Esparza, A., Chen, A., Finbloom, D. S., and Rifai, A.,** Extrarenal cytokines modulate the glomerular response to IgA immune complexes, *Kidney Int.,* 42, 341, 1992.
56. **Oite, T., Shimizu, S., Kagami, S., and Morioka, T.,** Hapten specific cellular immune response producing glomerular injury, *Clin. Exp. Immunol.,* 76, 463, 1989.
57. **Kleinknecht, D., Vanhille, P., Morel-Maroger, L., Kanfer, A., Lemaitre, V., Mery, J. P., Laederich, L., and Callard, P.,** Acute interstitial nephritis due to drug hypersensitivity. An up-to-date review with a report of 19 cases, in *Advances in Nephrology,* Grünfeld, J. P. and Maxwell, M. H., Eds., Year Book Medical Publishers, Chicago, 1983, 277.
58. **Ooi, B. S., Jao, W., First, M. R., Mancilla, R., and Pollak, V. E.,** Acute interstitial nephritis. A clinical and pathologic study based on renal biopsies, *Am. J. Med.,* 59, 614, 1975.
59. **Kleinknecht, D., Landais, P., and Goldfarb, B.,** Analgesic and non-steroidal anti-inflammatory drug-associated acute renal failure: a prospective collaborative study, *Clin. Nephrol.,* 25, 275, 1986.
60. **McCluskey, R. T. and Bhan, A. K.,** Immunologic mechanisms in drug-induced acute interstitial nephritis, in *Nephrotoxic Mechanisms of Drugs and Environmental Toxins,* Porter, G. A., Ed., Plenum Medical, New York, 1982, chap. 34.
61. **Linton, A. L., Clark, W. L., Driedger, A. A., Turnbull, D. I., and Lindsay, R. M.,** Acute interstitial nephritis due to drugs. Review of literature with a report of 9 cases, *Ann. Intern. Med.,* 93, 735, 1980.
62. **Border, W. A., Lehman, D. H., Egan, J. D., Sass, H. J., Glade, J. E., and Wilson, C. B.,** Anti-tubular basement membrane antibodies in methicillin associated interstitial nephritis, *N. Engl. J. Med.,* 291, 381, 1974.
63. **Lehman, D. H., Wilson, C. B., and Dixon, F. J.,** Extraglomerular immunoglobulin deposits in human nephritis, *Am. J. Med.,* 58, 765, 1975.
64. **Mayaud, C., Kourilsky, O., Kanfer, A., and Sraer, J. D.,** Interstitial nephritis after methicillin, *N. Engl. J. Med.,* 292, 1132, 1975.
65. **Ditlove, J., Weichmann, P., and Bernstein, M.,** Methicillin nephritis, *Medicine,* 56, 483, 1977.
66. **Colvin, R. B., Burton, N. E., and Hyslop, N. E.,** Penicillin associated interstitial nephritis, *Ann. Intern. Med.,* 81, 404, 1974.

67. **Girard, J. P.,** Common antigenic determinants of penicillin G, ampicillin and the cephalotins demonstrated in man, *Int. Arch. Allergy. Appl. Immunol.,* 33, 428, 1968.
68. **Churg, J., Cotran, R. S., and Sinniah, R. S.,** Tubulointerstitial nephritis induced by immediate (IgE type) hypersensitivity, in *Renal Disease: Classification and Atlas of Tubulointerstitial Disease,* Churg, J., Cotran, R. S., and Sinniah, R. S., Eds., Igakushoin, Tokyo, 1985, 82.
69. **Abt, A. B. and Gordon, J. A.,** Drug-induced interstitial nephritis: coexistence with glomerular disease, *Arch. Intern. Med.,* 145, 1063, 1985.
70. **Stachura, I., Jayakumar, S., and Bourke, E.,** T and B lymphocyte subsets in fenoprofen nephropathy, *Am. J. Med.,* 75, 9, 1983.
71. **Rennke, H. G., Klein, P. S., and Mendrick, D. C.,** Cell-mediated immunity (CMI) in hapten-induced interstitial nephritis and glomerular crescent formation in the rat, *Kidney Int.,* 37, 428, 1990.
72. **Joh, K., Shibasaki, T., Azuma, T., Kobayashi, A., Miyahara, T., Aizawa, S., and Watanabe, N.,** Experimental drug-induced allergic nephritis mediated by antihapten antibody, *Int. Arch. Allergy Appl. Immunol.,* 88, 337, 1989.
73. **Tsutsui, H., Terano, Y., Sakagami, C., Hasegawa, I., Mizoguchi, Y., and Morisawa, S.,** Drug-specific T cells derived from patients with drug-induced allergic hepatitis, *J. Immunol.,* 149, 706, 1992.
74. **Hirsch, F., Couderc, J., Sapin, C., Fournié, G., and Druet, P.,** Polyclonal effect of $HgCl_2$ in the rat, its possible role in an experimental autoimmune disease, *Eur. J. Immunol.,* 12, 620, 1982.
75. **Pelletier, L., Pasquier, R., Guettier, C., Vial, M. C., Mandet, C., Nochy, D., Bazin, H., and Druet, P.,** $HgCl_2$ induces T and B cells to proliferate and differentiate in BN rats, *Clin. Exp. Immunol.,* 71, 336, 1988.
76. **Aten, J., Bosman, C. B., Rozing, J., Stjnen, T., Hoedemaeker, P. J., and Weening, J. J.,** Mercuric chloride-induced autoimmunity in the Brown Norway rat. Cellular kinetics and major histocompatibility complex antigen expression, *Am. J. Pathol.,* 133, 127, 1988.
77. **Prouvost-Danon, A., Abadie, A., Sapin, C., Bazin, H., and Druet, P.,** Induction of IgE synthesis and potentiation of anti-ovalbumin IgE response by $HgCl_2$ in the rat, *J. Immunol.,* 126, 699, 1981.
78. **Sapin, C., Druet, E., and Druet, P.,** Induction of anti-glomerular basement membrane antibodies in the Brown-Norway rat by mercuric chloride, *Clin. Exp. Immunol.,* 28, 173, 1977.
79. **Druet, P., Druet, E., Potdevin, F., and Sapin, C.,** Immune type glomerulonephritis induced by $HgCl_2$ in the Brown Norway rat, *Ann. Immunol. (Inst. Pasteur),* 129C, 777, 1978.
80. **Bellon, B., Capron, M., Druet, E., Verroust, P., Vial, M. C., Sapin, C., Girard, J. F., Foidart, J. M., Mahieu, P., and Druet, P.,** Mercuric chloride-induced autoimmunity in Brown Norway rats: sequential search for anti-basement membrane antibodies, *Eur. J. Clin. Invest.,* 12, 127, 1982.
81. **Pusey, C. D., Bowman, C., Morgan, A., Weetman, A. P., Hartley, B., and Lockwood, C. M.,** Kinetics and pathogenicity of autoantibodies induced by mercuric chloride in the Brown Norway rat, *Clin. Exp. Med.,* 81, 76, 1990
82. **Bernaudin, J.-F., Druet, E., Druet, P., and Masse, P.,** Inhalation or ingestion of organic or inorganic mercurials produces autoimmune disease in rats, *Clin. Immunol. Immunopathol.,* 20, 129, 1981.
83. **Hua, J., Pelletier, L., Berlin, M., and Druet, P.,** Autoimmune glomerulonephritis induced by mercury vapour exposure in the Brown-Norway rat, *Toxicology,* 79, 119, 1993.
84. **Druet, P., Teychenne, P., Mandet, C., Bascou, C., and Druet, P.,** Immune type glomerulonephritis induced in the Brown-Norway rat with mercury containing pharmaceutical products, *Nephron,* 28, 145, 1981.

85. **Druet, E., Sapin, C., Günther, E., Feingold, N., and Druet, P.,** Mercuric chloride induced anti-glomerular basement membrane antibodies in the rat. Genetic control, *Eur. J. Immunol.*, 7, 348, 1977.
86. **Mirtcheva, J., Pfeiffer, C., Bruijn, J. A., Jaquesmart, F., and Gleichmann, E.,** Immunological alterations induced by mercury compounds. III. H-2A acts as an immune response and H-2E as an immune "suppression" locus for HgCl$_2$-induced anti-nucleolar antibodies, *Eur. J. Immunol.*, 19, 2257, 1989.
87. **Donker, A. J., Rocco, C. V., Vladutiu, A. O., Brentjens, J. R., and Andres, G. A.,** Effects of prolonged administration of D-penicillamine or captopril in various strains of rats. Brown-Norway rats treated with D-penicillamine develop autoantibodies, circulating immune complexes and disseminated intravascular coagulation, *Clin. Immunol. Immunopathol.*, 30, 142, 1984.
88. **Tournade, H., Pelletier, L., Pasquier, R., Vial, M. C., Mandet, C., and Druet, P.,** D-penicillamine-induced autoimmunity in Brown-Norway rats: similarities with HgCl$_2$-induced autoimmunity, *J. Immunol.*, 144, 2985, 1990.
89. **Tournade, H., Guéry, J. C., Pasquier, R., Nochy, D., Hinglais, N., Guilbert, B., Druet, P., and Pelletier, L.,** Experimental gold-induced autoimmunity, *Nephrol. Dial. Transplant.*, 6, 621, 1991.
90. **Guéry, J. C., Tournade, H., Pelletier, L., Druet, E., and Druet, P.,** Rat anti-glomerular basement membrane antibodies in toxin-induced autoimmunity and in chronic graft-versus-host reaction share recurrent idiotypes, *Eur. J. Immunol.*, 20, 101, 1990.
91. **Pelletier, L., Pasquier, R., Rossert, J., Vial, M. C., Mandet, C., and Druet, P.,** Autoreactive T cells in mercury disease. Ability to induce the autoimmune disease, *J. Immunol.*, 140, 750, 1988.
92. **Rossert, J., Pelletier, L., Pasquier, R., and Druet, P.,** Autoreactive T cells in mercury-induced autoimmune disease. Demonstration by limiting dilution analysis, *Eur. J. Immunol.*, 18, 1761, 1988.
93. **Castedo, M., Pelletier, L., and Druet, P.,** A role for autoreactive anti-class II T cells in gold salt-triggered B cell polyclonal activation in the Brown-Norway (BN) rat, *J.A.S.N.*, 3, 578, 1992.
94. **McCabe, M. and Lawrence, D. A.,** The heavy metal lead exhibits B cell-stimulatory factor by enhancing B cell Ia expression and differentiation, *J. Immunol.*, 145, 671, 1990.
95. **Dubey, C., Bellon, B., Hirsch, F., Kuhn, J., Vial, M. C., Goldman, M., and Druet, P.,** Increased expression of class II major histocompatibility complex molecules on B cells in rats susceptible or resistant to HgCl$_2$-induced autoimmunity, *Clin. Exp. Immunol.*, 86, 118, 1991.
96. **Prigent, P. et al.,** Mercuric chloride, a chemical responsible for TH$_2$-mediated autoimmunity in Brown-Norway rats directly triggers T cells to produce IL-4, *J. Clin. Invest.*, in press.
97. **Ochel, M., Vohr, H. W., Pfeiffer, C., and Gleichmann, E.,** IL-4 is required for the IgE and IgG1 increase and IgG1 autoantibody formation in mice treated with mercuric chloride, *J. Immunol.*, 146, 3006, 1991.

GLOSSARY OF TERMS IN ALLERGOLOGY

This glossary was developed by an international group of experts within the framework of the project and it does not necessarily represent the decisions of the World Health Organization or the Commission of the European Community. It was developed for the purpose of clarification of different terms used in the publication with an attempt to reflect the current status of scientific knowledge in the area of immunology, and allergology in particular.

GLOSSARY

ADHESION MOLECULES: Molecules, belonging mainly to the immunoglobulin-, selectin-, or integrin-families of molecules (e.g., LFA-1, ICAM-1), expressed on the membrane of various cells of the immune system. Interactions with each other as receptors and corresponding ligands facilitate cooperation (cross-talk) of cells, signal transduction and information transfer between cells.

ALLERGEN: An antigen that provokes allergy.

ALLERGIC CONTACT DERMATITIS: Dermatitis in an individual with a specific contact allergy when exposed to the specific allergen (hapten) in a concentration exceeding that individual's threshold level.

ALLERGY: A specific immune response resulting in adverse health effects.

ANERGY: Loss of immune responsiveness, e.g., delayed type hypersensitivity (DTH), responses to universally encountered antigens (e.g., candidal antigens).

ANTIBODY-DEPENDENT CELL-MEDIATED CYTOTOXICITY (ADCC): Lysis of various target cells coated with antibody by Fc receptor-bearing killer cells, including large granular lymphocytes (NK cells), neutrophils, eosinophils, and mononuclear phagocytes.

ANTIBODY: An immunoglobulin produced by B-cells and plasma cells after exposure to an antigen, with the capacity to interact specifically with the epitopes that elicited its production.

ANTIGEN PRESENTING CELLS: Any cell expressing MHC gene products, with the capacity to process antigen. Macrophages, dendritic cells, B-lymphocytes, and Langerhans cells are termed as professional or constitutive antigen presenting cells. However, other cells (such as endothelial cells) can acquire the ability to present antigen in certain pathological conditions.

ANTIGEN PROCESSING AND PRESENTATION: The antigens are processed (cleaved by enzymes) in various compartments of antigen presenting cells. The immunogenic peptides interact with the binding sites of MHC class II products (exogenous antigens) or with those in MHC class I products (endogenous antigens, including viruses). The processed antigen-MHC complex is recognized by the antigen receptor complex of T-helper cells.

ANTIGEN: Any compound recognized by antigen-receptor bearing lymphocytes. Antigens induce immune responses or tolerance. Antigens inducing immune responses only with the help of T-cells are T-dependent antigens while those which do not need T-help are T-independent antigens.

ANTIGENIC DETERMINANT: A single antigenic site (epitope) usually exposed on the surface of a complex antigen. Epitopes are recognized by antigen-receptors on T- or B-cells (T-cell epitopes or B-cell epitopes).

ASTHMA: A disorder characterized by variable air flow limitation. Most cases are associated with hyperresponsiveness and inflammatory changes in the airways.

ATOPIC ECZEMA: Chronic skin disease in individuals with propensity to develop IgE-mediated allergy.

ATOPY: A genetic predisposition toward development of IgE-mediated immediate hypersensitivity reactions against common environmental antigens.

AUTOIMMUNE DISEASE: A disease involving immune responses against self antigens, resulting in pathological change.

AUTOIMMUNITY: Immune responses against self (autologous) antigens.

B-LYMPHOCYTES: Bone marrow derived lymphocytes, expressing an antigen-receptor complex composed of membrane-bound immunoglobulin (mIg) and associated molecular chains. B-cell receptors interact with epitopes directly (no MHC-restriction). B-lymphocytes produce antibody and are efficient antigen presenting cells. They are the precursors of plasma cells.

BASOPHIL: A circulating granular leukocyte having prominent cytoplasmic granules when stained with dyes that indicate a basic pH. The granules contain histamine and chondroitine sulphates. After binding of antigen to membrane-bound IgE via FceRI receptors they release histamine, PAF and leukotrienes, and other inflammatory mediators.

BRONCHOALVEOLAR LAVAGE: Harvesting of cells and fluid from the lung, commonly by gentle lavage.

BRONCHOPROVOCATION: Use of inhaled triggers (cold air, histamine methacholine allergen, etc.) to assess the responsivity of the airways.

CARRIER: An immunogenic macromolecule (usually protein) to which a hapten is attached, allowing the hapten to be immunogenic.

CD16: Low affinity Fcg receptor (FcgRIII) expressed mainly on NK cells, granulocytes, and macrophages, mediating ADCC.

CD23: Low affinity Fce receptor induced by IL-4 and expressed on activated B-cells and macrophages.

CD3: A molecule composed of five polypeptide chains associated with the heterodimer T-cell receptor (TCR), forming the T-cell receptor complex (TCR/CD3); CD3 transduces the activating signals when antigen binds to the TCR.

CD4: A cell surface antigen belonging to the immunoglobulin superfamily of molecules. Marker of T helper cells. As adhesion molecule interacts with the nonpolymorphic part of MHC class II gene product.

CD8: A cell surface molecule belonging to the immunoglobulin superfamily of molecules. Marker of suppressor and cytotoxic T-cells. As adhesion molecule interacts with the MHC class I gene product.

CELL-MEDIATED OR CELLULAR IMMUNE RESPONSE: That form of immune response in which T-lymphocytes mediate the effects.

CLASS I MHC GENE PRODUCTS: Antigens encoded by the MHC class I genes are expressed on all nucleated cells. They present antigen-derived peptides of endogenous origin.

CLASS II MHC GENE PRODUCTS: Antigens encoded by the MHC class II genes are expressed on antigen presenting cells. They present antigen in a form that can stimulate lymphocytes.

CLONAL ANERGY: A form of self tolerance developing as a consequence of negative selection during the thymic selection processes. Clones of thymocytes whose antigen receptors (TCR) bind with high affinity to self antigens in association with MHC molecules are inactivated.

COMPLEMENT SYSTEM: A group of serum proteins with the capacity to interact with each other when activated. The chain reaction of the activated complement components results in formation of a lytic complex and several biologically active low molecular weight peptides (anaphylatoxins). The system can be activated by antigen-antibody complexes (classical pathway) and by other components, e.g., bacteria (alternative pathway). As an effector mechanism of the humoral immune response, the activated complement system facilitates opsonization, phagocytosis and lysis of cellular antigens.

CONTACT SENSITIVITY: An eczematous epidermal reaction which may develop when a hapten is applied to the skin of a previously sensitized individual. Also called hypersensitivity.

CROSS-REACTIVITY: The ability of an antibody, specific for one antigen, to react with a second antigen; a measure of relatedness between two antigenic substances, and/or polyspecificity of the antibody molecule (e.g., some rheumatoid factors).

CYTOKINE: see INTERLEUKIN.

CYTOTOXIC T-CELL (CYTOLYTIC T-CELL): CTL. A subpopulation of T-cells with the capacity to lyse target cells displaying a determinant in association with MHC gene products, recognized by its antigen receptor complex (TCR/CD3).

DELAYED TYPE HYPERSENSITIVITY (DTH): A form of T-cell-mediated immunity in which the ultimate effector cell is the activated mononuclear phagocyte (macrophage); the response of DTH appears fully over 24 to 48 hr. Previous exposure is required. Examples include response to Mycobacterium tuberculosis (tuberculin test) and contact dermatitis.

DENDRITIC CELL: An ubiquitous cell type characterized by extended cytoplasmic protrusions and a high expression of adhesion molecules and class II MHC gene products facilitating antigen presentation to specific lymphocytes.

DERMATITIS: Inflammatory skin disease showing redness, infiltration, scaling and sometimes vesicles and blisters.

DESENSITIZATION: Generally transient state of specific non-reactivity in previously sensitized individual, resulting from repeated antigen exposures.

ECZEMA: Same as dermatitis. Particularly applied to dermatitis on the hands (hand eczema).

ELICITATION: Production of a cell mediated or antibody mediated allergic response by exposure of a sensitized individual to an allergen.

ENDOCYTOSIS: The uptake by a cell of a substance from the environment by invagination of its plasma membrane; it includes both phagocytosis mediated by receptors and pinocytosis.

ENZYME-LINKED IMMUNOSORBENT ASSAY (ELISA): An assay in which an enzyme is linked to an antibody and a labeled substance is used to measure the activity of bound enzyme and, hence, the amount of bound antibody. With a fixed amount of immobilized antigen, the amount of labeled antibody bound decreases as the concentration of unlabeled antigen is increased, allowing quantification of unlabeled antigen (COMPETITIVE ELISA); with a fixed amount of one immobilized antibody, the binding of a second, labeled antibody increases as the concentration of antigen increases, allowing quantification of antigen (SANDWICH ELISA).

EOSINOPHIL: A circulating granular leukocyte having prominent granules which stain specifically by eosin and containing numerous lysosomes. Expressing IgE Fc receptors, eosinophils are important effector cells in immune reactions to antigens that induce high levels of IgE antibodies (e.g., parasites). Eosinophils are also abundant at sites of immediate hypersensitivity reactions.

EPIDEMIOLOGY: The study of the distribution and determinants of health-related states or events in specified populations, and the application of this study to control of health problems.

EPITOPE: Antigenic determinant.

Fc RECEPTORS: Receptors expressed on wide range of cells, interacting with the Fc portion of immunoglobulins belonging to various isotypes. Membrane bound Fc receptors mediate different effector functions (endocytosis, antibody-dependent cellular cytotoxicity [ADCC]), induce mediator release. Both the membrane bound and soluble form of Fc receptors regulate antibody production of B-cells.

FEV I: Forced expiratory volume in 1 second. Physiological measurement of the volume of air expired in one second with a maximal respiratory effort.

FVC: Forced ventillatory capacity. The physiological measurement of lung volume associated with a complete respiratory effort.

GLOMERULOPATHY: ("Glomerulonephritis"). Disease of the glomeruli which may show either thickening of the basement membrane — membranous glomerulopathy associated with IgG deposits — due to the accretion of proteins, or "minimal change glomerulopathy," in which there is functional damage but little structural change by light microscopy.

HAPTEN: A nonimmunogenic low molecular weight compound which becomes immunogenic after conjugation with a carrier protein or cell and in this form induces immune responses. Antibodies bind the hapten alone in the absence of carrier.

HELPER T-CELLS: A functional subpopulation of T-cells (expressing CD4 antigen) that help to generate cytotoxic T-cells and cooperate with B-cells in the production of an antibody response. Helper T-cells recognize antigen in association with MHC class II gene products. Depending on their capacity to produce various cytokines one can functionally differentiate Th1 (IL-2, IL-3, and IFN-g producing) and Th2 (IL-3, IL-4, and IL-6 producing) cells.

HLA (Human Leukocyte Antigen): The major human histocompatibility complex situated on chromosome 6. Human HLA-A, -B, and -C (resembling mouse H-2K, D, and L) are class I MHC molecules, whereas HLA DP, -DQ and -DR (resembling mouse I-A and I-E) are class II MHC molecules.

HUMORAL IMMUNE RESPONSE: That form of immune response in which specific antibodies induce the effector functions (such as phagocytosis and activation of the complement system).

HYPERSENSITIVITY PNEUMONITIS (HPS): Also known as extrinsic allergic alveolitis. This is an immune-mediated inflammatory disease of the lung parenchyma caused by exposure to an inhaled chemical allergen or organic dust.

HYPERSENSITIVITY: Abnormally increased response to a stimulus.

IAR: Immediate-onset allergic response.

IgE BINDING Fc RECEPTORS: The high affinity IgE binding FceR type I is expressed on mast cells and basophils. Interacts with IgE antibodies with high affinity. The cross-linking of these receptors results in release of mediators (such as histamine). The receptor is composed of alpha, beta, and gamma chains; the alpha chain contains the IgE binding site, while the gamma chain is responsible for signal transfer. The low affinity IgE binding Fc receptor (CD23) is expressed on B-cells, its soluble (truncated) form is generated by proteolytic cleavage and regulates IgE production of B-cells.

IMMEDIATE TYPE HYPERSENSITIVITY: A form of antibody mediated immunity that takes place in minutes to hours after the administration of antigen. Previous exposure is required. An example is allergic rhinitis to pollen antigen.

IMMUNODEFICIENCY: Defects in one or more components of the immune system leading to inability to eliminate or neutralize non-self compounds. Congenital or primary immunodeficiencies are genetic or due to developmental disorders (such as congenital thymic aplasia). Acquired or secondary immunodeficiencies develop as a consequence of malnutrition, malignancies, immunosuppressive compounds, radiation, or infection of immunocompetent cells with human immunodeficiency virus (HIV). Defects of the nonspecific defense system may also result in immunodeficiency.

IMMUNOGLOBULIN (Ig): Immunity-conferring portion of the plasma- or serum-gammaglobulins. Various isotypes (classes and subclasses) of immunoglobulins have a common core structure of two identical light (L) and two identical heavy (H) polypeptide chains which contain repeating homologous units folded in common globular motifs (Ig domains). The amino acid sequences of the N-terminal domains are variable (V-domains) in contrast to the more conserved constant regions (C-domains). The V-domains contain the complementarity-determining regions (CDRs) forming the antigen-binding sites, whereas the C-domains trigger several effector functions of the immune system.

IMMUNOGLOBULIN GENE SUPERFAMILY: Genes encoding proteins containing one or more Ig domains (homology units) which are homologous to either Ig V or C domains. Cell surface and soluble molecules mediating recognition, adhesion or binding functions in and outside the immune system, derived from the same precursor, belong to this family of molecules (e.g., Ig, TCR, MHC class I and II, CD4, CD8, FcgR, NCAM, PDGFR).

IMMUNOSURVEILLANCE: The mechanism by which the immune system recognizes and destroys malignant cells (mutant clones) before the formation of an overt tumor.

INTERLEUKIN: A nonantibody protein secreted by cells of the immune system, with the capacity to induce responses in other cells which express the corresponding cytokine receptor. Some nonimmune cells can also secrete cytokines (keratinocytes, testicular cells, etc.). The pleiotropic effect of cytokines depends on the cellular distribution of their receptors.

LANGERHANS CELLS: Bone marrow derived epidermal cells with a dendritic morphology, expressing CD1 marker and containing the cytoplasmic organelle, called the Birbeck granule. They are strongly positive for class II MHC antigen and capable for antigen presentation.

LAR: Late-onset allergic response.

LARGE GRANULAR LYMPHOCYTES: (LGL) Large lymphocytes with numerous cytoplasmic granules, having the capacity to lyse a variety of virus infected and tumor-cells cells without obvious antigenic stimulation. A subset consists of natural killer (NK) cells. LGL also mediate ADCC.

LYMPHOCYTE: Small, bone marrow derived cell with virtually no cytoplasm, with the ability to migrate and exchange between the circulation and tissues, to home to sites of antigen exposure, and to be held back at these sites. The only cells that specifically recognize and respond to antigens (mainly with the help of accessory cells). Lymphocytes consist of various subsets differing in their function and products (e.g., B-lymphocytes, helper T-cells, cytolytic T-cells).

MACROPHAGES: Mononuclear cells, derived from monocytes settled in tissues. Activated by different stimuli they may appear in various forms such as epitheloid cells and multinucleate giant cells. Macrophages found in different organs and connective tissues have been named according the specific locations e.g., as microglia, alveolar macrophages, Kupffer cells. Macrophages may function as antigen presenting cells, effector cells of cell-mediated immunity, and phagocytes eliminating opsonized antigens.

MAST CELL: Tissue bound mononuclear granular cells with staining affinity for basic dyes at low pH. The specific granules contain mediators of allergic inflammation, e.g., histamine. Upon stimulation with antigen via membrane bound IgE antibodies they release pre-formed and newly generated mediators. Two types of mast cells exist. Tryptase-containing T-mast cells are mainly associated with mucosal epithelial cells. Chymase-containing TC mast cells are long living tissue cells.

MONOCYTE: Bone marrow derived mononuclear phagocytic leukocyte, with bean-shaped nucleus and fine granular cytoplasm containing lysosomes, phagocytic vacuoles, and cytoskeletal filaments. Once transported to tissues they develop into macrophages.

NATURAL KILLER CELLS (NK CELLS): A subset of lymphocytes found in blood and lymphoid tissues, derived from the bone marrow and appearing as LGL. NK cells possess the capacity to kill certain tumor cells or virus-infected normal cells. The killing is not induced by specific antigen and is not restricted by MHC molecules.

NEPHROPATHY: ("Nephritis"). Disease of the kidney that may involve either or both the glomeruli (specialized structures where blood is filtered) and the renal tubules (connected structures where the composition of the filtrate is greatly modified in accordance with the physiological needs of the body). Thus, the term includes "glomerulopathy" and "tubular interstitial nephritis."

NEPHROTIC SYNDROME: A clininal disease in which damage to glo-meruli has caused leaky filtration, resulting in major loss of protein from the body.

NEUTROPHIL (Polymorphonuclear Leukocyte): Granular leukocytes having a nucleus with three to five lobes and fine cytoplasmic granules stainable by neutral dyes. The cells have properties of chemotaxis, adherence to immune complexes, and phagocytosis. The cells are involved in a variety of infla-matory processes including late-phase allergic reactions.

OCCUPATIONAL ASTHMA: Asthma caused by a sensitizing agent present in the workplace, after a period of asymptomatic exposure.

OPSONIZATION: Coating of antigens with antibody and/or complement components. The interaction of opsonized complexes with Fc- or complement-receptors facilitates their uptake by the receptor-bearing phagocytic cells.

PEFR: Peak expiratory flow rate. A physiological measure of the maximum air flow.

PLASMA CELL: A terminally differentiated B-lymphocyte with little or no capacity for mitotic division, that can synthesize and secrete antibody. Plasma cells have eccentric nuclei, abundant cytoplasm, and distinct perinuclear haloes. The cytoplasm contains dense rough endoplasmic reticulum and a large Golgi complex.

PREVENTION FOR ALLERGY: Primary prevention: control of the exposures inducing allergy. Secondary prevention: detection of allergy at an early stage with the aim of preventing progression to clinical disease. Tertiary prevention: care of persons with clinical allergic disease so as to prevent complications.

PSEUDO-ALLERGY: Nonimmunologic hypersensitivity with clinical symptoms mimicking those of allergic diseases.

RADIOALLERGOSORBENT TEST (RAST): A solid-phase radioimmunoassay for detecting IgE antibody specific for a particular antigen.

RADS: Reactive Airways Dysfunction Syndrome is a syndrome characterized by reversible airflow limitation and complicating bronchial hyper-responsiveness induced by acute exposure to high concentrations of nonsensitizing irritant gases at work.

RENAL FAILURE: ("Uraemia"). General description of a late stage in any kidney disease in which there are disorders due to the retention of substances normally excreted in the urine, and loss of essential materials normally retained by the kidney.

SENSITIZATION: Induction of specialized immunological memory in an individual by exposure to an allergen.

STEM CELL: Pluripotent cells, representing 0.01% of bone marrow cells, having the capacity for self renewal, and committed to differentiate along particular lineages, e.g. erythroid, megacaryocytic, granulocytic, monocytic, and lymphocytic. Cytokines stimulate the proliferation and maturation of distinct precursors.

SUPPRESSOR T-LYMPHOCTYE: A subpopulation of T-lymphocytes that inhibits the activation phase of immune responses. They are CD8$^+$, their growth and differentiation may be dependent on CD4$^+$ cells.

SURVEILLANCE PROGRAMS: Programs to identify the frequency of chemical sensitivity and occupational asthma by studying the incidence of these conditions and relating them to exposure to allergens. Programs such as SWORD (U.K.) and SENSOR (U.S.) are already in operation.

SURVEILLANCE: Ongoing scrutiny, generally using methods distinguished by their practicability, uniformity, and frequently their rapidity, rather than by precision. Its main purpose is to detect changes in trend or distribution in order to initiate investigative or control measures.

TOLERANCE: Persistent condition of specific immunological unresponsiveness, resulting from previous nonsensitizing exposure to the antigen.

TUBULOPATHY: ("Tubular interstitial nephropathy"). Disease of the kidney in which there is predominant damage to renal tubules associated with mononuclear cell infiltration in the adjacent interstitial tissue, resulting eventually in renal failure.

INDEX

A